Technological Breakthroughs and Future Business Opportunities in Education, Health, and Outer Space

Angus Hooke
Australian Institute of Higher Education, Australia

IGI Global
PUBLISHER of TIMELY KNOWLEDGE

A volume in the Advances in Human and Social Aspects of Technology (AHSAT) Book Series

Published in the United States of America by
IGI Global
Business Science Reference (an imprint of IGI Global)
701 E. Chocolate Avenue
Hershey PA, USA 17033
Tel: 717-533-8845
Fax: 717-533-8661
E-mail: cust@igi-global.com
Web site: http://www.igi-global.com

Library of Congress Cataloging-in-Publication Data

Names: Hooke, Angus, editor.
Title: Technological breakthroughs and future business opportunities in education, health, and outer space / Angus Hooke, editor.
Description: Hershey, PA : Engineering Science Reference, [2021] | Includes bibliographical references and index. | Summary: "Based on four areas of focus (technology; health, education, and outer space; the contributions of technologies; future business opportunities in the upcoming generations) this book provides both an historical overview of changes in technologies and their impact on business in our lives, as well as suggests plausible forecasts of developments in technologies and associated business opportunities in the near and medium term future"-- Provided by publisher.
Identifiers: LCCN 2020054413 (print) | LCCN 2020054414 (ebook) | ISBN 9781799867722 (hardcover) | ISBN 9781799867739 (paperback) | ISBN 9781799867746 (ebook)
Subjects: LCSH: Technological innovations--History. | Technological innovations--Forecasting. | New business enterprises.
Classification: LCC HC79.T4 T4278 2021 (print) | LCC HC79.T4 (ebook) | DDC 338/.064--dc23
LC record available at https://lccn.loc.gov/2020054413
LC ebook record available at https://lccn.loc.gov/2020054414

This book is published in the IGI Global book series Advances in Human and Social Aspects of Technology (AHSAT) (ISSN: 2328-1316; eISSN: 2328-1324)

British Cataloguing in Publication Data
A Cataloguing in Publication record for this book is available from the British Library.

All work contributed to this book is new, previously-unpublished material. The views expressed in this book are those of the authors, but not necessarily of the publisher.

For electronic access to this publication, please contact: eresources@igi-global.com.

Advances in Human and Social Aspects of Technology (AHSAT) Book Series

Mehdi Khosrow-Pour, D.B.A.
Information Resources Management Association, USA

ISSN:2328-1316
EISSN:2328-1324

Mission

In recent years, the societal impact of technology has been noted as we become increasingly more connected and are presented with more digital tools and devices. With the popularity of digital devices such as cell phones and tablets, it is crucial to consider the implications of our digital dependence and the presence of technology in our everyday lives.

The **Advances in Human and Social Aspects of Technology (AHSAT) Book Series** seeks to explore the ways in which society and human beings have been affected by technology and how the technological revolution has changed the way we conduct our lives as well as our behavior. The AHSAT book series aims to publish the most cutting-edge research on human behavior and interaction with technology and the ways in which the digital age is changing society.

Coverage

- Cyber Behavior
- Technology and Freedom of Speech
- Technoself
- ICTs and social change
- Public Access to ICTs
- Gender and Technology
- End-User Computing
- Computer-Mediated Communication
- Technology and Social Change
- Technology Adoption

IGI Global is currently accepting manuscripts for publication within this series. To submit a proposal for a volume in this series, please contact our Acquisition Editors at Acquisitions@igi-global.com or visit: http://www.igi-global.com/publish/.

The Advances in Human and Social Aspects of Technology (AHSAT) Book Series (ISSN 2328-1316) is published by IGI Global, 701 E. Chocolate Avenue, Hershey, PA 17033-1240, USA, www.igi-global.com. This series is composed of titles available for purchase individually; each title is edited to be contextually exclusive from any other title within the series. For pricing and ordering information please visit http://www.igi-global.com/book-series/advances-human-social-aspects-technology/37145. Postmaster: Send all address changes to above address.

Titles in this Series

For a list of additional titles in this series, please visit:
http://www.igi-global.com/book-series/advances-human-social-aspects-technology/37145

Technological Influences on Creativity and User Experience
Joshua Fairchild (Creighton University, USA)
Information Science Reference • © 2021 • 305pp • H/C (ISBN: 9781799843542) • US $195.00

Machine Law, Ethics, and Morality in the Age of Artificial Intelligence
Steven John Thompson (University of California, Davis, USA & University of Maryland Global Campus, USA)
Engineering Science Reference • © 2021 • 266pp • H/C (ISBN: 9781799848943) • US $295.00

Examining the Socio-Technical Impact of Smart Cities
Fenio Annansingh (York College, City University of New York, USA)
Information Science Reference • © 2021 • 231pp • H/C (ISBN: 9781799853268) • US $190.00

Human-Computer Interaction and Technology Integration in Modern Society
Hakikur Rahman (Institute of Computer Management and Science, Bangladesh)
Engineering Science Reference • © 2021 • 347pp • H/C (ISBN: 9781799858492) • US $195.00

Information Technology Applications for Crisis Response and Management
Jon W. Beard (Iowa State University, USA)
Information Science Reference • © 2021 • 310pp • H/C (ISBN: 9781799872108) • US $195.00

ICT Solutions for Improving Smart Communities in Asia
Noor Zaman (Taylor's University, Malaysia) Khalid Rafique (AJK Information Technology Board, Pakistan) and Vasaki Ponnusamy (Universiti Tunku Abdul Rahman, Malaysia)
Engineering Science Reference • © 2021 • 377pp • H/C (ISBN: 9781799871149) • US $235.00

Information Technology Trends for a Global and Interdisciplinary Research Community
Francisco J. García-Peñalvo (University of Salamanca, Spain)
Information Science Reference • © 2021 • 374pp • H/C (ISBN: 9781799841562) • US $195.00

Latin American Women and Research Contributions to the IT Field
Adriana Peña Pérez Negrón (Universidad de Guadalajara, Mexico) and Mirna Muñoz (CIMAT - Unidad Zacatecas, Mexico)
Engineering Science Reference • © 2021 • 459pp • H/C (ISBN: 9781799875529) • US $195.00

IGI Global
PUBLISHER of TIMELY KNOWLEDGE

701 East Chocolate Avenue, Hershey, PA 17033, USA
Tel: 717-533-8845 x100 • Fax: 717-533-8661
E-Mail: cust@igi-global.com • www.igi-global.com

Editorial Advisory Board

Table of Contents

Detailed Table of Contents

Primitive autonomous robotic vehicles were first developed during World War I and deployed during wars throughout the 20th century. More recently, autonomous vehicles with cognition were enabled by new technologies, especially in sensors, processors, and software, along with advances in humanoid and other legged robots. The autonomous car, a transformative and disruptive technology, will lead over the coming decades to the development of ubiquitous autonomous robots with increasing levels of cognition for many different applications. These robots will fill nearly every economic sector and occupational niche. The potential impact of autonomous intelligent robots on society, by the end of the century, will lead to ethical and moral dilemmas, as well as impacting jobs across the globe. There is a need to provide alternative sources of income, or alternative employment, for unemployed humans. There is also a need to consider the consequence of the possible emergence of robot self-awareness, consciousness, and free will.

Chapter 10
Toward Benevolent AGI by Integrating Knowledge Graphs for Classical Economics, Education, and Health: AI Governed by Ethics and Trust-Based Social Capital 163
Don MacRae, HIAlba, Australia, Australia

This chapter considers the contributions that may be made to the evolution of an ethical and compassionate form of artificial general intelligence (AGI) by integrating knowledge graphs for classical economics, education, health, and space science. Classical economics offers pathways to the evolution of a form of AGI characterised as moral, ethical, and providing a capacity for building trust-based social capital of societies. Pathways that can be elucidated by applications of machine and deep learning to knowledge graphs of the fields of classical economics, ethics, and social capital. A network platform based on the application of distributed ledger technology is proposed to provide the basis for eliciting insights from interdependences between an ever expanding digital-quantum cloud hosting similarly empowered knowledge graphs for an ever increasing myriad of advancing fields. Knowledge graph sketches for education, health, space science, and other fields germane to building social capital serve to illustrate this proposed process and attendant business opportunities.

Section 3
Output-Oriented Developments

Chapter 11
The Transition to Online Teaching and Learning: 'Like Ducks to Water' 188
Greg Whateley, Universal Business School Sydney, Australia
Alan Manly, Universal Business School Sydney, Australia

The onset of the COVID-19 pandemic in early 2020 provided both a need and an opportunity for educational institutions in Sydney to explore new ways of providing teaching and learning for their students. UBSS was able to respond quickly because it had already prepared to introduce online learning. In mid-2019, the institution had decided to offer an online option and a senior staff member and support staff had been working since then to convert existing product into suitable online format. By the time the demand for online teaching and learning arrived, the institution was already well placed for the conversion.

In 2001, Professors W. Steve Albrecht and Robert J. Sack wrote in their 2001 CPA Journal article that the accounting practice community perceived accounting education to be completely outdated. Fewer students majored in accounting, many accounting firms perceived accounting students higher educational training as outdated and hired fewer accounting majors. In 2021, these issues are still prevalent. This chapter discusses what has changed over the last 20 years and what issues are faced with current accounting graduates. A more agile accounting profession, a new CPA exam approach in 2024, and the use of more advanced analytical and technological tools in the profession only make the challenge to higher education even harder. The discussion then concludes with mitigating strategies to not only improve requisite skills that the profession demands, but also discusses strategies to curtail declining enrollment trends.

One billion people globally live with disabilities that are physical, sensory, psychiatric, neurological, cognitive, or intellectual. Their disabilities are dynamic and can be temporary or permanent, singular or plural, from birth or developed, and can change over time. People with disabilities face barriers to economic, social, political, and cultural participation. Assistive technology, artificial intelligence, and broader technology can amplify their inclusion, participation, and independence. This chapter will highlight emerging and evolving technologies, rooted in machine learning and neural networks, which assist across different disabilities and seek to improve the user's sense of ability and independence. These include Seeing AI app, OXSIGHT, OrCam, Envision smart glasses, and Dot Watch for vision impairment; Ava app and cognitive hearing aid for hearing impairment; Liftware self-stabilising utensils for limited hand mobility; Eyegaze and Tobii – assistive technologies that allow users to control computer and smartphone screens with their eyes; and 3D printed prosthetics.

Progress in biomedical sciences and engineering will take mankind closer to a world without disease. The objective of this chapter is to review and forecast research and technologies that contribute to eliminating all diseases. In addition to improved traditional therapeutics such as medicinal chemicals, antibiotics, vaccines, and antibodies, DNA and RNA therapeutics will play key roles in addressing genetic, infectious, and chronic diseases by decreasing disease facilitating proteins or increasing disease suppressing proteins. Nanorobots will find and treat thrombosis and vascular stenosis caused by lipid deposits in the intima of arteries, as well as remove diseased tissues and repair injured tissues. Nanorobots

Foreword

Technological change is accelerating, it is doing so at an increasing rate. This acceleration in the development and application of new ideas is generating more numerous and more difficult challenges for entrepreneurs, managers of established businesses, policy makers, academics, and the general public. Businesspersons want to know what new products are likely to emerge, what new processes are being developed, and where the new markets are lying. Public policy makers want to know where the new jobs will be, what new demands will be placed on governments, and what types of infrastructure should be developed to maximise benefits in these areas. Academics want to know what new skills future workers will need and what technologies are becoming available to help them equip studecan nts with these skills. And the general public want to know what major changes are going to impact on their lifestyles and the welfare of their families and friends. Will the acceleration of change continue, and will technological changes continue to transform society at the same speed?

This book does not answer all these questions. But it has been written by global experts including the President of TechCast (one of the world's leading forecasting organisations), the Chief Scientist at NASA's Langley Research Center, the Director of the Millennium Project, renowned professors from the Chinese Academy of Sciences and leading universities in Australia, China, New Zealand and the United States, and entrepreneurs with a strong track record of identifying and exploiting future business trends. It provides illuminating insights from those best qualified to make these calls.

The book discusses recent and prospective developments for a range of technologies including biotechnology, information technology, artificial general intelligence, and robotics. It also considers the new business opportunities that may be generated by these technologies, especially in the areas of education, health, and outer space. It considers whether and when we can eradicate all disease, the potential for new technologies to remove the physical and mental constraints currently imposed on the disabled, the scope for extending the healthy life span, and the expanding movement of humanity into outer space.

This book is essential reading for all those with a keen interest in the ways in which technology is changing our immediate and longer-term futures. If we know more about what the future may bring, we will be in a better position to cope with the challenges and to benefit from the opportunities. The book provides the reader with a wealth of information to help us do just that.

Oliver Obst
Centre for Research in Mathematics and Data Science, School of Computer, Data, and Mathematical Sciences, Western Sydney University, Australia

Preface

TECHNOLOGY AND DEVELOPMENT

Since the emergence of Homo habilis about 3 million years ago the economic footprint of humans, approximated by the size of gross world product (GWP), has been expanding at an exponential rate. This has been due to the generally rising growth rates of both the work force (determined overwhelming by population) and labor productivity (determined mainly by technology). However, while both growth rates have been rising for most of human history, the growth of population was the higher of the two until about 50 years ago. This resulted in the share of GWP growth contributed by population growth being more than 90% during the one-million-year scavenging era, 80% in the two-million-year hunting era, and about 60% during the 12,000-year farming era. Annual population growth continued to rise during the early and middle parts of the 200-year-industrial era, peaking at above 2% in the 1960s. However, it has fallen since then, to about 1% in 2020, and is expected to be around zero by the middle of this century. On the other hand, the growth of technology has continued to rise, has moved above the growth of population, and has accounted for about 75% of the growth of GWP during the current information era (commencing AD 2000). For better or worse, the human footprint is continuing to increase at an exponential rate, and technology has replaced population as its driving force.

This book looks at the historical, current, and expected future development of technology generally. It also explores some breakthrough technologies that are expected to have major effects on the lives of humans in the coming decades as well as at the possible business opportunities that must be exploited if the full potential of these technologies is to be realized.

The book is directed at readers who are interested in the future of humans and in the trends leading to this future. It should be of particular value to (1) executives and entrepreneurs (especially in the fields of education, health, and outer space) who want to know where the most attractive business opportunities might lie; (2) government officials seeking to identify where growth and employment opportunities will be so they can plan their programs and activities to take best advantage of these; and (3) academics and students in the disciplines of business, education, engineering, information technology, medicine, and science.

This book focuses on breakthrough technologies in the fields of education, health, and outer space. It is organized into three sections: an overview of historical and prospective technologies and their realized and potential contribution to human welfare; technology-oriented developments focusing on information technology, quantum technology, biotechnology, cognition technology, robots, and artificial general intelligence; and output-oriented developments in the fields of education, health, and outer space. A brief description of the chapters in each section follows.

SECTION 1: THE BIG PICTURE

Chapter 1 focuses on the global production of goods and services and the major technologies that fueled this production from the advent of the first humans to the end of the 21st century. Its major contributions are numerical estimates of GWP from the beginning of human history to the end of 2020 and projections of GWP as well as of GRP (gross regional product) for major regions and GDP (gross domestic product) for major countries to the end of the 21st century. For the historical analysis, the authors identify five technological eras, namely the scavenging (consuming the remains of organisms that have died from causes other than predation), hunting (chasing and killing wild animals), farming (systematic raising of plants and animals), industrial (manufacturing, mining, and chemical processing), and information (processing, storing, retrieving, and communicating data) eras. Using a population times per capita income approach, they estimate historical GWP at just under US $4,400 trillion. They note that approximately half of historical GWP has been produced in just the first two decades of the 21st century. The country analyses include the forecast that the GDP of India will move above that of China in about 2050 and will be 50% larger by 2010.

The second chapter discusses the technological convergence that is allowing humanity to move from slow and erratic biological evolution to fast and precise technological evolution. Technological convergence is based on the merger of different scientific disciplines as a result of the acceleration in the fields of nanotechnology, biotechnology, information technology and cognitive technology (NBIC). It offers the prospect of developments that have only been dreamt of, like biological rejuvenation. The author suggests that humans will transcend their biological limitations to become trans humans and eventual post humans. These radical changes will have profound philosophical implications about what it is to be human and about interactions between humans in an increasingly modified environment. The author suggests that, as we continue to evolve technologically, we will move beyond our home planet of Earth to guarantee the survival of humanity. He also provides a brief chronology of possible developments 2020-99, including the availability of neural implants that allow connection of human brains to the Internet in the 2030's, the capability of mental transfers among brains by 2039, and eradication ageing and changing death from an inevitability to an option by 2045.

Chapter 3 points out that technological convergence is leading to dramatic advances in such areas as plant-based food production, education, and healthcare. It describes the use of robots in weed and pest control and in collecting and analyzing data on soil conditions, the terrain, and equipment performance. Advances in existing and emerging food production technology extend to vertical farming, plant-based meat substitutes, cell-based meat, 3D printing of food, and precision fermentation. In education, there is increasing use of augmented reality and virtual reality. These are being coupled with cost-effective advances in high resolution communication making it possible to conduct business meetings, conference gatherings, and group interactions with holographic images of participants. The growing use of networks, such as social X(R), that interact with digital media feeds in virtual space allow engagement with real people. With X(R) technologies, participants gain a new perspective on connecting learning and real-world engagement. Developments in health care and medicine include telehealth, robotized surgery, the use of drones in the delivery of medicines, and distributed ledger technology (DLT) and block chain to provide treatment for conditions such as depression.

Chapter 4 discusses technology's role in social evolution. The author suggests that humans have always had a love-and-hate relationship with technology. Primitive humans probably were scared the first time they saw fire, the 19th century Luddites feared that technology would destroy jobs, and recent conspiracy theories link 5G to COVID-19. He argues that machines (structures of any kind created by humans) are an integral part of human history and have played and will continue to play a fundamental role in the social evolution of humans. The chapter concludes with the reminder us that the discourse of the merging of man and machine is an old story. However, the existence of unprecedented high multiplicity of merging potentially interconnected drivers with exponential increased complexity making it harder to clearly envision plausible future scenarios. Nevertheless, three scenarios are presented: Scenario 1 (2030): the cloud/global total network hyper connectivity; Scenario 2 (2050): collective consciousness/ brain planetary smart skin; and Scenario 3 (+2050): transcendence/singularity/A new human species.

SECTION 2: TECHNOLOGY-ORIENTED DEVELOPMENTS

The first chapter in this Section observes that Information technology (IT) has developed rapidly since 2000, especially in the areas of digitization, networking, and intelligence. These developments have made, and will continue to make, important contributions to education, health, and other industries. Digitization refers to information representation and processing methods in which the data are encoded. All data is digitized. The essence of digitization is the deep understanding and use of big data generated by the integration of IT with economic and social disciplines. Networking is a physical carrier for the dissemination of information, and its development trend is toward the Internet of Everything (IoT). Intelligence reflects the level of information application, and its development trend is a new generation of intelligence, a so-called super-intelligence, that has the capacity to perform tasks that are well beyond present human capabilities. The authors use the two sectors of education and health to depict the implications of historical and future IT breakthroughs. AI can help adapt education to each student's characteristics and aptitudes by supporting online teaching resources, intelligent diagnosis and recommendations, and a more independent and attractive learning environment. In the area of health, the combination of AI and clinical medical information can lead to intelligent diagnosis and treatment of diseases including cancer and cardiovascular problems.

Chapter 6 reviews the relatively short history of computers and outlines the new possibilities represented by emerging quantum computing technologies. The latter include better security encryption, faster search, and access to greater computational power. The chapter explores potential impacts of quantum technologies for a range of disciplines and areas of life including neuroscience, finance, and personal information storage. In the field of neuroscience, Brain-Computer Interfaces (BCIs) might be able to control brain function by targeting specific mechanisms at a molecular level and taking advantage of possible quantum influences on highly synchronous neuronal firing, using nano-robots to avoid possible contraindications from drugs. The author explores the future for quantum money, discussing new currencies and financial products in relation to blockchain technology as well as hybrid-classical systems and calculations. The inability of quantum states to resist cloning could have important implications for our understanding of consciousness and free will as well as the possibilities for quantum currencies.

Chapter 7 explores the field of biotechnology, which the author disaggregates into five areas: medicine, agriculture, industry, bioinformatics, and environment. Starting with the earliest use of biotechnology by humans in the development and creation of vaccines, he provides a comprehensive history of the use of biotechnology in medicine and points to emerging trends in and applications for biotechnology, such as the growth of gene therapy. He argues that the development of genetic engineering in the 1970s was pivotal in the advancement of modern biotechnology, which has now made significant contributions across a wide range of scientific disciplines and industries. The 1980 United States Supreme Court decision to allow the patenting of genetically modified organisms (GMOs) significantly expanded the commercial viability of biotechnology and acted as a catalyst for a more rapid growth of the GMO industry. The author discusses the emergence of genetically modified crops in the 1990s and the benefits and concerns associated with their use as well as the potential for biotechnology to help solve environmental problems. He also examines the threat of bioterrorism and other potentially negative effects of biotechnology. He concludes by discussing the biotechnological possibilities that may exist in 2030, 2035 and 2040, such as increased efficiency in food production, better human health, and longer lifespans.

Chapter 8 discusses the relatively new discipline of cognitive science. Scientific research into the practice of meditation and the state of mindfulness is providing new insights, highlighting an important and emerging role for the discipline in the 21st century. The author argues that a more thorough understanding of how cognition works could support a fundamental revolution in human development. He examines theoretical and experimental research findings and points out that the roles of representation and computation are fundamental to cognitive science. Ideas explored include using cognitive technologies to identify the full content of cognition and predict behaviors, understanding the lessons taken from virtual reality and cloud computing, and the applications of mindfulness to 'resetting the cognitive algorithm'. Impacts of cognitive technologies are explored in relation to education, health, and space exploration. The author discusses the view of the world as an algorithm that can be dramatically altered to shift our life experience and the possibility of meditation training to transform the possibilities of lengthy space travel in astronauts. He concludes with a description of 'cloud-mind based evolution', a situation in which mindful awareness supports the focused and open cognitive capacities that provide the possibility for an augmented and sustainable future.

The technology of focus in Chapter 9 is robotics. Autonomous robots are explored as a disruptive and transformative technology that could have as significant an impact on society's behavior, values, culture, and economy as steam engines, electric lights and the internet had in the19th and early 20th centuries. The chapter explores how the idea of a 'robot' emerged and was popularized as well as the huge range of devices that fall into this category. Advances in robotic technologies and deep learning architectures and algorithms can improve robotic sensory perception with contributions from AI fields such as natural language, expert systems, genetic algorithms, and mechanical engineering. In healthcare, robots can undertake logistical tasks to support nursing staff and hospitals. Technologies such as unmanned air vehicles (UAVs) have spawned entirely new industries and future developments will likely eliminate the role of drivers in the transportation of goods and services. The chapter concludes with a discussion of robot ethics and the robot mind, arguing that the issue of consciousness is fundamental to any consideration of the future uses of robots.

Chapter 10 focuses on artificial general intelligence (AGI). It identifies three potential outcomes for the evolution of the discipline: "corporate techno-fascism", "decentralized digital democracy (DDD)", and "human free". The author argues that to achieve DDD, which would result in "a future of diverse flourishing creativity" and minimize the threats of the other two potential scenarios, benevolent AGI must be self-sufficient and also understanding of and compassionate towards human desires. He recommends the integration of Knowledge Graphs (KGs) as a basis for representation of interconnected information through the use of Distributed Ledger Technology (DLT). KGs are sketched across a variety of areas including social capital, classical economics, thought leaders, reducing inequality, food production and nutrition, education, health, housing, employment, and belonging. The chapter provides insight into public and private organizations and projects at the forefront of efforts to guide benevolent AGI that can support human wellbeing on a global scale. It also explores concepts such as trust, truth, negotiation, ethics, causality and common sense and their applications in the development of benevolent AGI.

SECTION 3: OUTPUT-ORIENTED DEVELOPMENTS

Education

Chapter 11 describes how one of Australia's leading business schools, Universal Business School Sydney (UBSS), responded to the sudden need to switch from onsite to online learning following the onset of the COVID-19 pandemic in early 2020. The School was able to respond quickly and confidently because, fortuitously, it was already preparing to introduce online learning. In mid-2019, it had decided to offer an online option, and a senior staff member plus support staff had been working since then to convert existing product into suitable online format. When online teaching and learning become a necessity, the institution was already well placed for the conversion. Nevertheless, it faced several challenges in switching to online delivery. The chapter describes these challenges and how the School met them. In a survey conducted in late 2020, more than 90% of MBA students responded that they preferred online to onsite learning.

The second chapter on education focusses on the teaching of Accounting. The authors point out that the profession has changed dramatically in recent decades, from offering basic bookkeeping and tax filing services to providing sophisticated and complex client consultation services. However, there is a growing gap between accounting educational curricula at universities and colleges and the expectations of the profession. Bureaucracy in accounting departments and the business colleges and universities within which these departments operate do not allow timely changes to meet the profession's needs. In addition, regulatory and CPA exam eligibility regulations require time-consuming reviews of most changes. Accounting firms have been big adopters of technology over the last twenty years. They are using tools such as data analytics, artificial intelligence, and robots to improve performance. Failure to incorporate emerging technology tools into accounting curricula is resulting in graduating students not possessing the skills demanded by the profession. The authors argue that higher education needs to adapt more quickly to changing markets to ensure that graduating students are properly prepared to work in the modern accounting profession.

Health

The first chapter in the sub-section on health outputs looks at how technology is improving the lives of people with disabilities. About 15% of the global population live with a disability, a rate that is projected to increase over the next few decades. Assistive technology (AT) aims to maintain and improve the functioning, independence, and overall well-being of those with a disability. The chapter highlights emerging technologies that use machine learning, neural networks, and 3D printing to assist across a range of disabilities. For visually impaired individuals, OrCam MyEye is a voice active device that can be attached to any pair of glasses. It uses digital sensors, speakers, and machine learning algorithms to capture the user's surrounding and convert this information to audio. Dot Watch pairs to the user's smartphone, allowing them to receive phone notifications in braille on the screen. For those with hearing loss, Ava is a captioning program for individual or group conversations that uses a smartphone or computer microphone to listen what is being said and translate the information into written text. BrightSign is a smart glove that translates wearer's sign language into speech and written text. Eye Control and Eye Drive technologies enable users to control other devices using only their eyes.

Chapter 14 focuses on the eradication of disease. The author notes that life expectancy has been increasing linearly, with no tendency to plateau, due to progress made in the biomedical sciences in understanding the human aging process. Mankind can anticipate a largely disease-free world as medicine prevents and treats diseases. Vaccines, aseptic and sterilization, and antibiotics as well as dramatically improved living standards are already making bacterial infection a much less serious threat to life and health. Although drug-resistance is a severe clinical problem, it is surmountable with new antibiotics and other new and emerging therapies as well as better management of their applications. DNA and RNA therapeutics are playing key roles in addressing infectious diseases by decreasing disease-facilitating proteins and increasing disease-suppressing proteins. Advances in nanotechnology and robotics, nanomachines and nanorobots will permit operations to clear diseased tissues inside the body, deliver antibiotics to inflamed tissues directly, and apply aseptic to the focal infection inside the body. Diseases caused by biological pathogens will not completely disappear, but the threat will be much lower than in historical pandemics, because of not only advances in medicine, but also progress in information technology and automation.

Chapter 15 is the first of two chapters that address the issue of aging. This Chapter explores ageing in the context of technology and the emerging 'longevity economy'. It begins by detailing the "historical reversal", a global demographic megatrend described by the United Nations as the shift towards the population of people aged 65 years and older outnumbering those under the age of 15 years. As of 2020, 34 developed countries have experienced the reversal. This number is expected to grow to 90 by 2050 and to include all countries by 2075. The shift is due mainly to increasing longevity, a testament to advances in public health and technology. Other shifting paradigms are the level of digital literacy, technological access, and social license for using technology to support healthy ageing. The chapter details smart technologies that detect bedsores, monitor vital signs, sense patterns of change, detect wandering behaviors and manage fire risk in cooking. As well, apps and robotics assist in the management of health, provision of care, and coping with social isolation. These include AI-enabled hearing aids, robotic exoskeletons to lift and transfer elders, 'pet' robotic plush dogs and cats that provide comfort and companionship, and natural-language conversation robots. The potential for technology as an enabler for life-affirming ageing is infinite, and exploitation of this potential is limited only by cultural and individual imagination.

It is often said that the only certainties in life are death and taxes. The author of Chapter 16, however, describes death as a curable disease and susceptible to advances in science and technology. Barriers include the second law of thermodynamics, of entropy increasing in any natural process, and the Hayflick limit, which holds that a normal human cell can replicate a finite number of times - estimated at between forty and sixty cell divisions. With the development of synthetic biology, belief is growing that aging and death can be overcome. The field of anti-aging and life extension is vast and includes the prospects of microscopic nano-bots traveling through the blood stream and sweeping away disease, brain-machine uploading, and replicating and replacing human organs on a routine basis via 3-D printing. The chapter concludes with a note on the promissory nature of longevity research and investment, as well as the fragile credibility of many of the scientific claims used to bolster radical age extension, key reasons why many ethicists find the prospect of immortality highly controversial.

Chapter 17 addresses mental health, a major and increasing concern in the 21st century. Technology is providing more resources to help improve mental health. Numerous apps have been developed and online platforms have been built to assist in assessing mental disorders, providing feedback and suggestions for management, and subsequent monitoring of patients with a view to enhancing the quality of mental health care provided by traditional face-to-face services. In the current situation, where the world is facing the challenge of Covid-19, technology is making an important contribution to the promotion of mental health.

Outer Space

Chapter 18 draws on forecasts from TechCast, a highly-regarded organisation that uses collective intelligence to forecast emerging technologies, social trends and wild cards for planners and decision makers. The author, who is President of TechCast, suggests that the development of a stable civilization on Earth during the coming decades will lay the required foundation for the Space Age. By 2050, the commercialization of space should be well underway, colonies will established on the Moon and possibly Mars, solar satellites will likely be functioning, and space tourism will have become normal. Attention will then turn to the final frontier of outer space. He suggests that the intellectual resources of roughly 10 billion educated people will produce the breakthroughs in physics needed to travel, not just within our own solar system, but to other star systems in our galaxy.

Chapter 19 has been written by the Chief Scientist at NASA's Langley Research Center. It therefore provides many unique and highly informed insights into space travel. The author notes that, after about 60 years of highly trained astronauts going into space, the related technologies have progressed considerably, and costs have fallen to the point where space tourism will soon be an option for private citizens - initially suborbital for minimal times and earth orbital for up to two weeks. He points out that safety problems associated with being in space for a few weeks have largely been solved, but those related to space tourism over many years as tourism expands beyond Earth orbit to the Moon, Mars, asteroids, and other planets the safety issues still need serious work. He also notes that the imminence of physical tourism is being accompanied by rapidly improving digital reality / immersive virtual presence technology, making feasible space tourism experiences available to most people.

Chapter 20 evaluates the emerging industry of asteroid mining and the pivotal role that the United States is playing in shaping the rules for an extra-terrestrial economy. The authors note that while there are large amounts of valuable mineral resources in space, many legal problems must be solved before these resources can be optimally exploited. They point out that the use of outer space is governed by the Outer Space Treaty-1967 (OST), which has over one hundred signatories. With space exploration expanding to encompass both public and private stakeholders, there is a growing concern that the Treaty's provisions prohibit the assertion of sovereignty and, hence, property rights, in outer space. The United States has developed a legal framework that facilitates commercial asteroid mining and a political strategy that focuses on bilateral space exploration agreements with individual countries. The authors suggest that, due to its dominant position in the space sector, the United States will strongly influence the regulatory roadmap for the future exploitation of resources in space.

Section 1
The Big Picture

Chapter 1
Technology and Gross World Product:
From Homo habilis to Homo techne

Angus Hooke
https://orcid.org/0000-0003-4626-9075
Universal Business School Sydney, Australia

Lauren Alati
Argon & Co, Australia

ABSTRACT

This chapter divides the economic history of humans into technological eras and uses a population multiplied by per capita income approach to estimate gross world product in each era and, therefore, for the history of humankind. It also provides an overview of the major technologies that introduced each era and supported growth during the era. The chapter uses a model developed by the authors to predict gross world product during the remainder of the 21st century (2021-2100). It also considers which economies might have been the largest in the world since the dawn of civilisation about 6,000 years ago. The chapter concludes with the prediction that China, India, and the United States will remain the dominant economic powers during the remainder of the 21st century, that the gross domestic product (GDP) of India will pass that of the United States in the late 2030s and the GDP of China in the late 2040s, and will be more than 50% larger than that of second-placed China by 2100.

INTRODUCTION

This chapter focuses on the global production of goods and services, and on the major technologies that fueled this production, from the advent of the first humans to the end of the 21st century. It identifies five technological eras up to 2020 and discusses the breakthrough technologies that introduced each era as well as some of the smaller but still radical technologies that supported economic activity during the eras. It also sets start-and-end dates for the eras and makes estimates of average population and per

DOI: 10.4018/978-1-7998-6772-2.ch001

capita GWP (referred to hereafter as per capita income) in each era. These allow estimates to be made of GWP in each era and thus in all human history up to 2020. The authors then use their World Income Model (WIM) to make predictions of GWP during 2021-2100. They conclude by considering which economies might have dominated historically and which will be the largest during the remainder of the 21st century. They suggest that China, which topped the global gross domestic product (GDP) ladder for about 2,000 of the last 4,000 years, and which regained the top ranking in 2014, will maintain its lead until the middle of the century. Its GDP will then be overtaken by that of India, and by the end of the century India's economy will be more than 50 percent larger than that of China.

The Framework

GDP is the value, usually expressed in monetary terms, of the goods and services produced in an economy during a specified period. GWP is the sum of the GDPs of all the world's economies. In this chapter, GWP is measured in purchasing power parity (PPP) terms to remove the effects of differences in the cost-of-living across countries and in constant 2020 dollars to eliminate the effects of inflation over time. This PPP dollar is referred to as an international dollar (I$).

The main objective of the chapter is to provide an estimate of the historical GWP of humankind up to and including 2020. The analytical framework comprises technological eras, their duration, and average annual population and per capita income in each era. The authors define a technological era as a period in which a specific set of technologies has a dominating effect on the growth and composition of GWP. Many other definitions have been put forward by economic historians; some, for example, include effects on the ways in which economic activity is structured, work is carried out, and society is organised. Further, even when there is agreement on criteria, there are differing views on whether the effects of new technologies have been sufficient in scale and nature to be credited with ushering in a new era. Fortunately, the analysis and measure of historical GWP in this chapter do not depend on the specific categorisation of eras. Rather, the classification adopted provides a convenient and reasonably familiar framework that allows a large and complex problem to be broken into smaller, more manageable units.

The authors identify five technological eras: the scavenging (consuming the remains of organisms that have died from causes other than predation), hunting (chasing and killing wild animals), farming (systematic raising of plants and animals), industrial (manufacturing, mining, and chemical processing), and information (processing, storing, retrieving, and communicating data) eras. The technologies whose name is given to the era may not have been the technologies that made the greatest contributions to GWP in those eras. For example, in the scavenging era, the technology of foraging (searching for food and other provisions) probably contributed more than scavenging to the 'income' of humans, and in the early part of the industrial era farming accounted for more than half of economic activity. However, it is suggested that they were the technologies that accounted for most of the growth of income in the relevant eras.

The Scavenging Era

The first era begins, appropriately, with the advent of the first humans. Setting a date for this event raises a myriad of conceptual and empirical issues. There are many views on the unique characteristics of humans, on the identity of the first species to possess specified characteristics, and on the time when the first species with these characteristics emerged (Dominguez-Rodrigo 2002). A popular view is that the necessary and sufficient characteristics of humans are an upright bipedal stance, an opposable thumb, an enlarged brain case, a capacity for complex speech, and an ability to make tools.

The authors accept this list of characteristics and the view that Homo habilis (Handy Man) was the first species to possess them all. Other primates that preceded Homo habilis possessed one or more of the characteristics. For example, Australopithecus africanus, perhaps the immediate ancestor of Homo habilis, had a somewhat upright bipedal stance which they could probably sustain for travel over short distances, and had an opposable thumb although not one long enough to meet the ends of their fingers. Some anthropologists also argue that Homo habilis resembles the ancestral Australopithecines more than they do modern humans, and that Homo erectus (Upright man), being the first hominid to have the general shape of modern humans, with shorter arms and longer legs than the Australopithecines, should be recognised as the first human. However, Homo habilis may have been the first primate to place emphasis on tool making and on using tools systematically to improve their material welfare. The authors do not delve further into these issues, partly because they are not qualified to do so, and partly because the results of their analyses would not be greatly affected if they were to adopt other proposed resolutions, at least for the early eras. They therefore adopt the premise that Homo habilis was the first human, that technology is a defining characteristic of humans, and that the arrival of Homo habilis marked the beginning of the first technological era.

Homo habilis lived in Africa's sub-Saharan Rift Valley, which extends from the junction of the Red Sea and the Gulf of Aden through Ethiopia and Kenya to Tanzania. Fossil remains suggest that males were about 157 cm (5' 2") in height and 52 kg (115 lb), while females were 125 cm (4' 1") in height and 32 kg (70 lb) in weight. The brain size of homo habilis was about 650 cubic centimetres (cc), or half that of modern humans, and the bulge of the Broca area suggests that they were capable of at least rudimentary speech. Compared with Australopithecus africanus, they were 5-10 percent taller and heavier and had brain size about 40 percent larger. They also had an opposable thumb which, in contrast with that of africanus, was easily able to reach the tips of all the fingers.

The dominant technology of the Australopithecines was foraging. Homo habilis was also a forager - their diet included berries, figs, shoots, nuts, and eggs. However, they added another technology, scavenging, which enabled them to move from a largely above-ground plant-based diet to one supplemented with below-ground tubers as well as meat and marrow from recently deceased animals. They ate rhinos and elephants, possibly left by carnivores such as sabre-toothed tigers, which lacked the teeth needed to strip all the meat from carcasses. Meat and bone marrow provided more calories and protein than root vegetables which, in turn, provided more than plant vegetables. This allowed the gut to reduce in size, leaving more space and energy for the development of other organs including the brain. Snooks (1989) suggests that carnivores derive a huge competitive edge by "consuming herbivores to gain the fatty acids required for the development of their central nervous systems, rather than using scarce brain capacity to produce those fatty acids in their own bodies."

Homo habilis was able to add scavenging to the earlier dominant technology of foraging because they developed new tools (Facchini 2003). Australopithecus africanus probably also used tools (sticks and unmodified stones) to increase the effectiveness of their foraging technology. However, there is no evidence that africanus made tools. Homo habilis, with their larger brain size, were able to imagine new tools. With their longer thumb they were able to exert considerable pressure while holding objects, an ability that allowed them to turn their creative ideas into reality. Homo habilis not only made tools, they also used intermediate tools to make final-product tools.

Homo habilis exploited what is referred to as Mode 1 technology, which includes the making and use of core tools, choppers, and scrapers. These are often called Oldowan tools, since the first discoveries were made at Oldoway (now Olduvai) Gorge, in Tanzania. Homo habilis made new stone tools by using roundish rocks (hammerstones) to strike other more angular rocks (cores) so as to remove one or more rock fragments (flakes). They used their stone tools to dig for tubers. They used the choppers and scrapers to slice and dice the flesh from animals as well as to break open animal bones and extract the bone marrow. They may also have used the bones of animals as clubs and daggers to attack small animals and defend themselves against larger predators.

To estimate the GWP of the scavenging era, it is necessary to specify the duration of the era. The earliest fossil of Homo habilis, a jawbone found in Ethiopia, is dated 2.8 million years ago (MYA). However, it is unlikely that this fossil is from an original member of the species. The authors therefore date the appearance of Homo habilis, and the beginning of the scavenging era, at a slightly earlier, round number date – 3 MYA. The scavenging period came to an end when hunting became the main source of economic growth. This is normally associated with the arrival of Homo erectus (Upright man), who may have been the immediate descendant of Homo habilis. Homo erectus is believed to have emerged in Africa about 2 MYA, and the authors set this as the end date of the scavenging era. They therefore set the duration of the scavenging era at 1 million years.

Based on the geographical region in which Homo habilis lived and the number of hectares needed to support one person surviving on a foraging and scavenging technology, it has been estimated that the population of Homo habilis might have been about 2,000. The authors adopt this as their estimate of average population during the era. Since climate fluctuated during this period, population would have been smaller in the cooler and drier periods when food would have been scarce and larger in the warmer and wetter periods when food was more abundant.

The per capita income of Homo habilis would have been extremely low by current standards (though, looking forward from the standpoint of the Australopithecines, it would have seemed quite high). To find a benchmark, the authors commence with data on one of the world's poorest countries, the Central African Republic (CAR) which, by coincidence, is only a few hundred kilometres from the Rift Valley in Kenya. In 2020, the CAR had a population of just under 6 million, GWP (PPP basis) of I$ 3.4 billion and thus per capita income of about $570. However, the poorest 10 percent of the population shared in only 2 percent of GWP. Their per capita income was therefore about I$ 120. But again, that number is the average of those in the lowest-income decile, and the poorest of these would have had per capita income of I$ 100 or less.

While very low compared to that of the country with the highest per capita income in 2020 (Qatar with I$ 130,000) and to global per capita income in that year (I$ 19,100), the poorest people in the CAR enjoy a higher per capita income today than did Homo habilis several million years ago. Even those in the lowest income decile consume food that is varied, can be cooked, and can be preserved for at least a few days. They wear clothes that keep them warm in cold weather, live in structures that provide security and defence against extreme weather, and have access to at least rudimentary education, medical care, and entertainment. They also consume public goods such as roads, parks, police, and defence. The annual income of Homo habilis, on the other hand, comprised mainly the plants they foraged, the meat and bone marrow they scavenged, and the new tools they made. They lacked the techniques and implements needed to make clothing and permanent housing. Analysis of tooth decay among 26 Homo habilis fossils shows a mean age of 13 years and maximum age of 25 years. Even those in the poorest decile in the CAR have much longer life expectancy (LE) and health-adjusted life expectancy (HALE) than had Homo habilis. Per capita income during the scavenging era would have been much less than I$ 100.

Technically, the 2020 PPP value of per capita income during the scavenging era is the value of the goods and services that income could be exchanged for in New York in July 2020. The uncooked plant and animal food plus the new tools of Homo habilis would have little exchange value in today's Manhattan. The authors, still somewhat arbitrarily, set the per capita income of Homo habilis at I$ 40, of which $30 might have been from the foraging technology inherited from the Australopithecines and $10 from scavenging and tool making.

The average annual GWP of Homo habilis is therefore estimated at I$ 80,000, or about half the annual income of the mean American household in 2020. GWP for the era is estimated at I$ 80 billion (Table 1). This is equivalent to the GDP in 2020 of Cambodia, which is classified as a least developed country and accounts for 0.0004 (0.04%) of GWP.

Table 1. GWP During the Scavenging Era

Av. population	2,000
Av. PCY (I$)	40
Annual GWP (I$)	80,000
Duration (years)	1,000,000
Era GWP (I$ b)	80

Source. Various, including estimates by the authors.

The Hunting Era

About 20 different species/sub-species of human have been identified by anthropologists, with all but Homo habilis having their origins in the hunting era. The authors focus on measuring the GWP of three of these hunting humans – Homo erectus, Homo neanderthalis, and Homo sapiens. They then make an allowance for the probably small contribution to GWP of the remaining Homo species/sub-species.

Homo Erectus

Fossil evidence suggests that Homo erectus lived from about 2 MYA (2 megaannum ago) to 50 TYA. It was the dominant human species until 200 TYA. Homo erectus emerged in Africa but moved quickly to west Asia (1.9 MYA), then to east Asia, south eastern Europe, and Indonesia (it is in Indonesia where the 50,000-year-old fossils were found).

Reflecting their wider geographical distribution and longer existence, Homo erectus showed greater variation in physical characteristics than Homo habilis (Coolidge 2011). Their height ranged from 1.4 to 1.8 m (4'7" to 5'11"), their weight from 51 to 65 kg (112 to 143 lb), and their brain size from 550 to 1,150 cm^3. Recordings in the higher ranges are from later-dated fossils, indicating continuing evolution of the species. Early Homo erectus had height, weight, and brain size like Homo habilis, but eventually they may have been about 25% taller and 50% heavier, with a brain 75% larger.

Like Homo habilis, early Homo erectus used Oldowan tool-making techniques (Odell 2013). But by 1.8 MYA, their stone tools were noticeably sharper and had straighter edges than those made by Homo habilis. By 1.5 MYA, their skills had evolved sufficiently for their toolmaking to be given a new name, Acheulian, after Saint Acheul in southwest France, the site where these more advanced tools were first discovered. The basic Acheulian tool was a teardrop-shaped hand axe, made by chipping pieces off rock rather than just banging rocks together as was the practice for Oldowan tools. Unlike the hand axes made by Homo habilis and early Homo erectus, the Acheulian axes were sharpened on both sides. They were improved and widely used for the next million years before being replaced by finer, more precise flint tips. Homo erectus made choppers, cleavers, hammers, and flakes that were used as knives and scrapers.

Another important tool used by Homo erectus was fire. Initially, they may have taken advantage of fires caused naturally by lightning, and kept them burning by adding twigs, wood, and slow-burning dung. About 1 MYA, Homo erectus learned how to use flints to make fire, which they then used more systematically for cooking meat, providing light and warmth at night, and frightening off predators (Rightmire 1993).

Estimating the population of Homo erectus is complicated by their wide dispersion over Africa and Eurasia. The authors assume that in the early years, when their physical capabilities were more limited and they were confined to Africa, total population may have been about the same as or a little larger than that of Homo habilis – i.e., a few thousand. As their physical and mental abilities increased, their hunting skills improved, and they dispersed more widely, their population would have risen. Based on known locations and estimated land-carrying capacity, it has been suggested that the population may have expanded to 7,000-35,000 about 500 TYA. Since the Great Dispersion commenced quite soon after the appearance of the species, and for lack of better information, the authors use the mid-range of about 20,000 as their working estimate of the average population of Homo erectus.

The per capita income of Homo erectus would have been higher than that of Homo habilis. Their diet contained a wider range of plant foods, more root vegetables, and a larger supply of higher quality meat. According to William Calvin from the University of Washington in Seattle, "Homo erectus ate well". For perhaps a third of their existence they were able to cook their food, making it easier to digest and extract the nutriment. The quality and quantity of their tools were also greater. But their per capita income was still substantially below that of a low-income person in the CAR in 2020. They lacked anything but rudimentary clothing and accommodation, and had limited if any access to the education, medical, and entertainment services available to even low-income people in the CAR. The authors set the average annual per capita income of Homo erectus at 25 percent above that of Homo habilis – i.e., at $50.

The estimates made by the authors imply average annual GWP of Homo erectus of I$ 1 million, equivalent to about eight times the mean income of US households in 2020. Era GWP of Homo erectus is estimated at I$1.8 trillion, about the same as that of Canada in 2020.

Neanderthals

Fossil evidence of Homo neanderthalis (commonly referred to as Neanderthals) cover the period 430-40 TYA. However, it is believed that they may have evolved about 500 TYA. The Neanderthals derived their name from the place where their first fossils were discovered, a limestone cave in the Neander Valley near Dusseldorf in Germany. Their territory ranged from Wales and Portugal in the west to the Altai mountains of Siberia in the east and the eastern boundary of the Mediterranean Sea in the south.

The Neanderthals were not as tall as Homo erectus – their height ranged from 1.5 to 1.75 m (4'11" to 5'9"); however, reflecting the colder climates in which they lived, they were heavier, weighing 64-82 kg (140-180 lb). The global climate was also cooling during the Neanderthal period, and later members of the species were generally shorter, but weighed about the same, as the early members. The brain size of late Neanderthals, of 1,200 to 1,750 cm^3, was on average about 50% larger than that of Homo erectus. It was also larger than the brains of modern humans, but not as complex.

The diet of the Neanderthals varied with their location. Studies using nitrogen isotype ratios found in the collagen of fossil bones indicate that, in steppe habitats, they consumed about 80% meat and 20% vegetables. They may have been apex predators (i.e., at the top of the food chain) for much of their existence, hunting, killing, and consuming large plant eaters such as mammoths, straight-tusked elephants, woolly rhinoceroses, and aurochs (the ancestors of domestic cattle) as well as smaller animals like red deer, reindeer, ibex (a species of wild goat), and sheep. In forest areas, their diet included plants such as hard seeds and nuts, and vegetables like mushrooms, pine nuts, and moss. DNA from dental calculus in a fossil found in Spain contained evidence of poplar bark, a basic ingredient of aspirin. However, the fossil also indicated the presence of an extremely unpleasant gastro parasite, and it is plausible that the bark was used as a medicine rather than as food.

Tools made by the Neanderthals included spear points and knives that most likely were set in wooden handles, scrapers, and pronged harpoons. They also used scrapers to make simple fur cloaks, draping the skin of a single animal around their body to keep warm. The Neanderthals lived during the last Ice Age, and it has been suggested that they would have had to cover about 70 percent of their bodies to survive the colder weather. They were skilled hunters, with impact scars on stone-tipped spears and on fossil bones suggesting that these were used to hunt medium- and large-sized animals. The spears appear to have been used at close-range for thrusting or throwing short distances as part of an ambush hunting strategy (there is no evidence that the Neanderthals used long-distance projectiles, which may have put them at a disadvantage in confrontations with Homo sapiens).

The average Neanderthal population is estimated to have been within the range of 7,000 – 9,000. This low number may reflect the "Boserupian Trap", where the size of a population is limited by its ability to obtain food, which in turn is constrained by the population's technology. If the population is small, innovation is limited, and the population remains small. This is consistent with the apparent lack of development of Neanderthal lithic technology. The authors assume that the average population size was the midpoint of the above range – i.e., 8,000.

The Neanderthals would have had a higher per capita income than Homo erectus. They had much more meat in their diet, they had clothing, and they possessed cultural goods. Child mortality was lower (about 45% compared to 60% for Homo erectus) and life expectancy for adults was longer, with many Neanderthals living past 40 years. There is evidence that they took care of their sick and elderly, and that they buried their dead, leaving bone shards as grave goods. Nevertheless, they would still have been much poorer than those in the lowest-income decile of the CAR, with considerable shortfalls in clothing, accommodation, education, health, and entertainment. The authors set their average per capita income at I$ 60, or 20% above that of Homo erectus and half the per capita income of those in the lowest decile in the CAR in 2020. Again, looking backward from today, that might seem to be a pitiful amount. But looking forward from the standpoint of Australopithecus africanus, Homo habilis, or Homo erectus it would have appeared quite luxurious.

The estimates for average population and average per capita income imply average annual GWP for the Neanderthals of I$ 480,000, equivalent to the annual income of three middle-income American families today. With duration of 350,000 years, the GWP of all Neanderthals would have been almost I$ 170 billion, or about the same as that of Bulgaria in 2020.

Homo Sapiens

About 400 TYA years ago, when the Neanderthals were spreading throughout Eurasia, Homo sapiens (wise man, also referred to as modern man) was emerging in Africa. It is believed that the two species had a common ancestor, possibly Homo erectus or Homo heidelbergensis (the latter named after Heidelberg in Germany, where their first fossils were found). Mitochondrial evidence suggests that about 200 TYA an evolutionary bottleneck (a time when most of the population suddenly dies out, perhaps due to some natural disaster) occurred for modern humans, since the DNA of all later modern humans can be traced back to a single woman, often referred to as "Mitochondrial Eve". Soon afterwards, some modern humans migrated out of Africa. The early migrations were not successful; however, about 60,000 years ago another, more clearly successful wave of migration occurred, with modern humans moving into Australia about 50 TYA, Europe 40 TYA, and the Americas 15 TYA. DNA evidence shows that they met and engaged in some interbreeding with other human species, such as the Neanderthals and the Denisovans (there is a 1-3% mixing of genes) but by about 45 TYA Homo sapiens was the only surviving human species.

Fossils from the Jebel Irhoud archaeological site in Morocco indicate that the diet of early humans contained meat from gazelles, wildebeests, zebras, hartebeests, buffalos, porcupines, tortoises, and snakes as well as freshwater molluscs and seasonal ostrich eggs. Unlike the Neanderthals, who adapted their diet to changes in location and other conditions, modern humans were more assertive and innovative, adapting their technology to maintain a largely plant-based diet by, for example, digging deeper in the soil for roots.

Early modern humans used Mousterian-like Levallois flake and core tools. However, by about 90 TYA, they were making tools that were significantly different to those used by the Neanderthals, including blade tools, sharper scrapers (some of which were used for smoothing and scraping leather), and fine bone tools. They also formed coordinated groups for killing large herd animals and developed specialised hunting weapons. Toward the end of the hunting era, when the mega-fauna was disappearing and acquiring meat was becoming more difficult, there was a burst of hunting innovations. These included the atlatl (spear thrower), which increased the range and impact of spears by effectively lengthening the thrower's arm. They also included the toggle-headed harpoon (a bone or antler harpoon with a hole for attaching a rope; after being thrown, the harpoon would stay in the body of the prey while the hunter kept hold of the rope).

The hunting era ended with the advent of a new dominant technology, farming. Most historians date the start of the farming era between 14 TYA and 10 TYA. For these calculations, the authors take the mid-range, of 12 TYA (10 BCE). The duration of the Homo sapiens hunting era, which overlapped with that of Homo neanderthalis until only a few thousand years after the two groups met, is therefore estimated at about 200,000 years.

Estimates of the global population at the end of the hunting era range from 1 to 10 million. The authors adopt the intermediate estimate by McEvedy and Jones (1978), of 4 million. This implies an average annual growth rate of about 0.004 over the 200,000-year period. If this rate were to be uniform, the average population would be 600,000. Population growth was not smooth; however, it is as likely that the level was above as below trend in any one year, and the authors adopt the rounded figure of 600,000 as the estimate of average population size for modern humans during the hunting era.

The per capita income of modern humans during this era would have been higher than that of the Neanderthals. They were superior hunters, they were also accomplished fishers, and they would have had a more varied and secure diet. Also, during the last part of the hunting era they made much-improved clothing and produced more complex works of art. It is assumed that average per capital income was I$ 90, or 50% higher than that for the Neanderthals.

Average annual GWP of modern humans during the hunting era is therefore estimated at I$ 5.4 million. The combined GWP of Homo erectus, the Neanderthals and Homo sapiens during the hunting era is estimated at a little more than I$ 3 trillion. Other members of the human species such as Heidelbergensis and the Denisovans also contributed to the GWP of humans; however, their numbers were small and they had disappeared by the last part of the hunting era, which appears to have been the highest growth part. Their contribution is set at 5% of the total of the other groups, increasing the GWP of humans during the hunting era to I$ 3.2 trillion, equivalent to 95% of the GDP of Great Britain in 2020.

Table 2. GWP During the Hunting Era

	Homo erectus	Homo neanderthalis	Homo sapiens	Others	Total
Av. population	20,000	8,000	60,000		24,000
Av. PCY (I$)	50	60	90		67
Annual GWP (I$)	1,000,000	480,000	5,400,000		1,600,000
Duration (years)	1,800,000	350,000	200,000		2,000,000
Era GWP (I$ b)	**1,800**	**168**	**1,080**	**152**	**3,200**

Source. Various, including estimates by the authors.

The Farming Era

For the farming era, there is only one species of human – Homo sapiens. At the start of the era, they were concentrated in four major, disconnected regions – the so-called Old World of Europe and Western Asia; Africa; East Asia; and the New World of the Americas and Australasia. The transition from hunting to farming as the dominant source of growth in GWP appears to have taken place independently in each region. Many causes have been suggested. First, the end of the 100,000-year glacial period both released previously frozen land and provided more moisture in the air, creating abundant resources for farming. Second, elimination of the megafauna during the last 10,000 years of the hunting era, due possibly to overhunting by increasingly skilled humans, the warming of the climate, and/or a 4-5 km wide comet that exploded over southern Canada, increased the need to find new sources of food and other materials. Third, slow evolution might only then have produced the right mix of mental, social, and cultural development needed for the development of farming.

In the Old World, farming emerged first in the fertile crescent, a boomerang-shaped region in the Middle East that includes parts of what are now Iraq, Syria, Lebanon, Palestine, Israel, Egypt, Turkey, and Iran. The main reason for this location was the abundance of fertile soils deposited by the regular flooding of the Tigris, Euphrates, and Nile rivers. Once agriculture was established, development was enhanced by the suitability of the area for irrigation and by the exchange of ideas as traders passed through the crescent in their travels between Europe and Asia.

There were several bursts of technological development over the course of the farming era. During the Neolithic period (12 TYA -5.5 TYA), these developments included domestication of plants and animals, irrigation, and invention of the plough, polished stone axes, pottery, and textiles. Plants were domesticated by selecting the larger and healthier seeds to grow the next season's crop. Eight so-called Neolithic founder crops – emmer wheat, einkorn wheat, hulled barley, peas, lentils, bitter vetch, chickpeas, and flax – were cultivated in the fertile crescent. Rice, mung, soy, and azuki beans were domesticated in China. Dogs were domesticated in Mesopotamia by raising wolf cubs (whose mothers may have been killed) and successively keeping the more friendly ones for breeding purposes. Later, they were used for managing other animals. Pigs were domesticated around 13 TYA, followed by sheep about 12 TYA and cattle around 10.5 TYA. In South America potatoes, beans, cocoa, llamas, and alpacas were domesticated 10-7 TYA and cotton was being grown in Peru 5.6 TYA. Irrigation was developed in Mesopotamia as population growth moved farmers into drier regions, where agriculture was only possible with the support of irrigated canals. In some areas, such as Babylonia and Assyria, the soil was prone to dry up, harden and crack, and ploughs were invented to keep the land arable. Farming also led to the privatisation of property, encouraging effort and innovation by ensuring that those responsible for producing goods and services would retain a substantial proportion of their output. Pottery was used for storing grain and cloth for providing warmth. New tools were made by grinding and polishing harder stones, rather than chipping away at softer stones as was done during the scavenging and hunting eras. Polished stone axes were used to clear land and fashion materials for buildings as well as to defend and appropriate land.

Another burst of innovation occurred during the European agricultural revolution, which commenced toward the end of the farming era. It included the introduction of higher-yielding crops, crop rotation, improved agricultural equipment, selective breeding, and a more efficient pattern of land ownership. Crops with low yields such as rye were replaced with higher-yielding crops such as wheat and barley. During the late Middle Ages, farmers used a three-year, three field crop rotation routine, with a different crop being grown in each of two fields, and the third field lying fallow. From about the mid-1600s, they increasingly moved to a four-field rotation system, such as wheat, barley, turnips (which are deep-rooted and can absorb minerals not available to shallow-rooted plants), and clover (which turns nitrogen from the air into fertiliser). Robert Bakewell and Thomas Coke introduced selective breeding of animals, mating together those with desirable characteristics and using inbreeding to stabilise these qualities. Bakewell also produced hornless sheep with square, meaty bodies and longer, more lustrous hair. He practised selective breeding with cattle, which traditionally had been used primarily as oxen for pulling ploughs and as dairy cows for producing milk. As a result, the weight of bulls for slaughter rose from about 170 kg (370 lb) in 1700 to 380 kg (840 lb) a century later.

The farming era gave way to the industrial era around the end of the 18th century. Eric Hobsbawm (historian at the University of London) suggested that the Industrial era began in the 1780s but was not fully felt until the 1830s. T. S. Ashton (economic historian at the University of London) held that it occurred between 1760 and 1830. For this chapter, the authors selected 1800 as a round-figure date for the end of the farming era and the start of the industrial era.

Compared with the scavenging and hunting eras, global population increased rapidly during the farming era. According to McEvedy & Jones, it rose from 4 million 12 TYA (10 000 BCE) to 7 million 6 TYA, and then much more quickly to 170 million in AD 1. It required the first 1.5 millennia of the

Christian era for population to double again, but it then accelerated to 900 million in AD 1800. These data suggest that the average annual population for the whole era would have been about 60 million.

There is some evidence (reduced height and weight) that per capita income may have fallen in the early part of the farming era. The likelihood is greater for the per capita surplus (which uses welfare as the target variable and subtracts the welfare costs of producing the goods and services that generate the economic surplus) since women especially had to spend long periods in the fields under hot sun, bending over and straining their backs to plant and tend crops. Diseases were also more likely to be contracted from animals, famines resulted when crops failed, and pandemics more than decimated (reduced by a tenth) the global population on several occasions. However, for the entire farming era, average per capita income would have been at least several times higher than in the hunting era. For most of the period and for most of the population, food was more varied and could be stored longer, and clothing, housing (fixed houses rather than collapsible tents), medicines, entertainment, and art were wider in range and better in quality. But while people enjoyed more material welfare than in the hunting era, living standards would not have been substantially higher than that of the lowest-income decile in a least-developed country today. Those living in the farming era did not have electricity – or even kerosene lamps – and health support was so poor that few lived much beyond 30 years. The authors set average per capita income during the farming era at I$ 250, about twice that of people in the lowest-income decile in the CAR in 2020.

Table 3. GWP in the Farming Era

Av. population (m)	60
Av. PCY (I$)	250
Annual GWP (I$ m)	15,000
Duration (years)	11,800
Era GWP (I$ b)	177,000

Source. Various, including estimates by the authors.

Average annual GWP during the farming era is therefore estimated at I$ 15 billion, a little less than the GDP of Mauritania in 2020. For the whole era, GWP is estimated at I$ 177 trillion, or about 10 percent more than GWP in 2020.

The Industrial Era

The industrial era was partly a consequence of the success of the European agricultural revolution. The latter increased labour productivity on farms, releasing thousands of workers (many of whom were children), who then migrated to the towns and cities in search of employment. It increased the incomes of landlords, providing them with financial resources to invest in the emerging manufacturing and service businesses and a starting market for the products of these new businesses (Ashton 1998). Imperialism played a part by providing Europe with access to the raw materials needed to manufacture the new products as well as the larger market needed to exploit economies of scale.

However, the availability of resources and markets is not a sufficient set of conditions for the emergence of a new technological era. There must also be new breakthrough technologies and, as emphasised by Snooks (1989,) the drive of economic agents to exploit these technologies. The latter was provided by the intense competition that prevailed among about 20 European states as well as by entrenchment of the principle of private property and free enterprise, and the incentives these gave entrepreneurs to create value from the new technologies.

Among the most important of the early technologies was the spinning jenny, which was patented by James Hargreaves in 1770. The original version could work eight spindles at a time and could be operated by unskilled labour. It was instrumental in changing cloth weaving from a cottage to a factory industry. In 1771, Richard Arkwright constructed what may have been the first modern factory, the Cromford Mill in the Derwent Valley, Derbyshire. The mill was water-powered, initially employed 200 workers, and operated two 12-hour shifts. Factories allowed specialisation and economies of scale, with consequent increases in labour productivity.

In 1712, Thomas Newcomen invented the steam engine; however, his product had few practical applications and roused little interest. In 1763, Thomas Watt introduced changes that made the engine commercially attractive. and it soon became the main source of power for British industry. In 1804, Richard Trevithik built a locomotive that made the first recorded railway journey, carrying five wagons loaded with 10 tons of steel and 70 men about 10 miles in just over four hours. Then, in 1829, a locomotive made by George and Robert Stephenson completed a journey over a one-mile track in a test of the claim that locomotives provided the best propulsion for a railway between Liverpool and Manchester. In 1837, Sir William Fothergill Cooke and Charles Wheatstone successfully demonstrated their electrical telegraph by sending a message from Euston to Camden in London. In 1838, the first commercial service was established in the United Kingdom (between Paddington and West Drayton) and in 1844, the first telegraph was installed in America (between Baltimore and Washington DC). In 1838, Samuel Morse and his assistant, Alfred Vail, developed a code, known as the Morse Code, that allowed easier transmission of messages across telegraph lines. In the 1860s, Alfred Nobel developed an explosive that he called 'dynamite', after the ancient Greek word 'dunamis" meaning power. It proved to be much stronger and safer for shattering rocks than the traditional gun powder (known then as 'black powder'). Nobel asked that his invention be used only for peaceful purposes; however, inventors have little power over the longer term uses of their discoveries, and dynamite quickly became the explosive of choice for military purposes. In 1826, Joseph Niépce created the first permanent photograph (of his estate in Burgundy, France), using an image obtained from a primitive camera and a pewter plate. In 1829, William Burt patented the first typewriter. Producing documents with his machine was slower than writing by hand, and Burt's invention had limited commercial value. Then, in 1867, Christopher Sholes developed a much more efficient typewriter. Initially, Scholes arranged the keys in alphabetical order, which had two major disadvantages: first, it placed some of the most-used keys in positions where they were not easy to reach; and second, frequent hitting of neighbouring keys in rapid succession caused the keys to stick. In 1882, to overcome these difficulties, Scholes developed the QWERTY keyboard (named after the first six letters in the first line of the keyboard). In 1831, Michael Faraday invented the Faraday Disk, the world's first electric generator. Although his machine was ineffective, his discovery of electromagnetic induction quickly led to the dynamo, the first generator able to deliver power to industry.

The industrial era was replaced by the information era. Again, there is no consensus about the commencement date of the latter era. The authors adopt a popular position that development of the World Wide Web in 1993 and the high penetration of personal computers and email by the late 1990s (50% of US households) make the end of the 20th century an appropriate demarcation date between the industrial and the information eras.

To estimate GWP for the industrial era, the authors divided the 200-year period into two parts: the first 150 years and the last 50 years. The justification for this is that reasonably firm data are available for the GDP of most countries, and thus for GWP, from 1950; however, only scattered estimates are available for GDP, and only then for a small number of countries, for the earlier period.

During 1801-1950, global population rose from 900 million to 2.5 billion and averaged 1.6 billion (Table 4). Average per capita income is assumed to have increased from about I$ 1,000 in 1800 to I$ 4,300 in 1950 (an average annual rate of about 1%) and to have averaged I$ 2,400. GWP for the 150-year period is therefore estimated at I$ 576 trillion.

Over the succeeding 50-year period, global population rose to 6.2 billion and averaged 4.1 billion. Per capita income increased to I$ 12,000 (an average annual rate of 2.1%) and averaged I$ 7,300. GWP therefore increased from I$ 11 trillion to I$ 74 trillion and averaged just under I$ 30 trillion. For the 50-year period, GWP was about I$ 1,500 trillion. For the whole era, GWP is estimated at a little under I$2,100 trillion.

Table 4. GWP During the Industrial Era

	1801-1950	1951-2000	1801-2000
Av. population (m)	1,600	4,100	2,225
Av. PCY (I$)	2,400	7,300	5
Av. Annual GWP (I$ b)	3,840	29,930	10
Duration (years)	150	50	200
Era GWP (I$ t)	**576**	**1,497**	**2,073**

Source. Various World Bank and IMF publications and estimates by the authors.

The Information Era

The information era is also referred to as the 'digital', 'computer' and 'new media' era. It is based on the use of systems for storing, retrieving, sending, and receiving information. The era has produced new products such as the computer, the internet, and digital TV; new processes such as internet banking, online commerce, and automated inventory control; and new distribution systems such as cable and wireless TV.

Unlike for the earlier eras, Malthusian-type arguments have not been needed or used to explain the fall of the industrial era and the rise of the information era. Rather, new eras seem to be an almost inevitable result of the integration of research into the economic and business systems. While there had been some largely uncoordinated research taking place in small universities for several hundred years before the industrial era, its embedment in higher education institutions, national and regional government bodies, and large private companies increased rapidly in the latter half of the 19th century and exploded during the Great Wars and in the second half of the 20th century. The stick of demanding KPIs for researchers and the carrot of huge profits and exciting lifestyles for entrepreneurs have led to an acceleration in both the generation and exploitation of new technologies. That it was the information era that came next, and not the biotech, nanotech, or robotics era, seems to be due mainly to the general-purpose nature of information technology – it supports innovation in a wide range of fields.

The first computer may have been the abacus, whose invention has been attributed variously to the ancient Greeks, the Romans, and the Chinese, suggesting that it may have been developed independently in several places. The abacus was followed by the slide rule, which was invented by William Oughtred in 1622. The next major step toward the modern computer was taken by Charles Babbage, who designed the so-called analytical engine, a proposed general-purpose computer, between 1833 and 1871. However, due to conflicts with his engineer, inadequate funding and weak business skills, Babbage never completed the construction of his machine. In 1890, Herman Hollerith designed a punch card system that was used to process the 1880 US census, taking only 3 years compared to 7 years with manual processing. (Hollerith later formed IBM.) In 1941, the first general-purpose computer, the S3, was built. In 1943-44, John Mauchly and J. Presper Eckert built the Electronic Numerical Integrator and Calculator (ENIAC), considered the grandfather of digital computers, and in 1946 they built the UNIVAC, the first commercial computer for business and government applications. In 1947, William Shockley, John Bardeen and Walter Brattain of Bell Laboratories invented the transistor, and in 1958 Jack Kilby and Robert Noyce unveiled the integrated circuit (known as the computer chip). In the early 1970s, a team at IBM led by Alan Shugart invented the 'floppy disk', which allowed data to be shared among computers, and Robert Metcalfe at Xerox developed the Ethernet, which led to the

connection of multiple computers. In 1981, IBM released its first personal computer (PC). In 1990, Tim Berners-Lee created the Hyper Text Mark-up Language (HTML), which gave rise to the World Wide Web. In 1996, Sergey Brin and Larry Page developed the Google search engine, and in the late 1990s Wi-Fi allowed computer users to connect to the Internet without wires. In 2004, Mark Zuckerberg incorporated Facebook. In 2005, the video-sharing device, YouTube, was introduced. In 2007, Apple added many computer functions to the iPhone, changing it from a mobile phone to a smart phone and producing a sea-change in cultural and social behaviour. In 2016, the first programmable quantum computer was released.

Comprehensive and reasonably reliable data on GWP are available through 2019. Data for 2020 were not available when this chapter was being written. Also, they will have been considerably affected by measures taken to restrict the spread of COVID-19. To produce a "normalised" estimate of GWP for 2020, the authors projected the latest GDP figures forward to that year on the assumptions that (normalised) real GDP for all countries would have continued to increase at the actual average annual rate of the preceding four years, and that inflation was zero in 2020.

During 2001-20, global population rose from 6.2 billion to 7.7 billion (an average annual rate of 1.1%, or half the corresponding rate in the 1960s) and, for comparability with the analysis for the earlier eras, averaged 6.9 billion. Per capita GWP increased from I$ 12,000 to I$ 19,000 and averaged I$ 15,400 a year. Total GWP therefore rose from I$ 74 trillion to I$ 146 trillion and average I$108 trillion. Total GWP for the era-to-date was I$ 2,142 trillion.

Historical Gross World Product

Historical GWP is estimated at I$4,394 trillion. The scavenging and hunting eras together produced less than a tenth of 1%, the farming era generated 4%, and the industrial era accounted for 47%. Almost half (49%) was produced in the already-completed part of the information era, which covers only 20 years of the 3-million-year history of humankind.

Future Gross World Product

The projections in this section are derived from a World Income Model (WIM) developed initially by

Table 5. Historical GWP

	Scavenging	Hunting	Farming	Industrial	Information	Total
Av. population (m)	0.002	0.024	60	2,225	6,942	0.211
Av. PCY (I$)	40	65	250	4,657	15,425	6,654
Av. Annual GWP (I$ b)	0.0001	0.0016	15	10,363	107,083	1.406
Duration (years)	1,000,000	2,000,000	11,800	200	20	3,000,000
Era GWP (I$ t)	**0.1**	**3**	**177**	**2,073**	**2,142**	**4,394.33**
% shares	0.002	0.071	4.028	47.163	48.737	100.000

Source. Various World Bank and IMF publications, World Factbook, author estimates.

Figure 1.

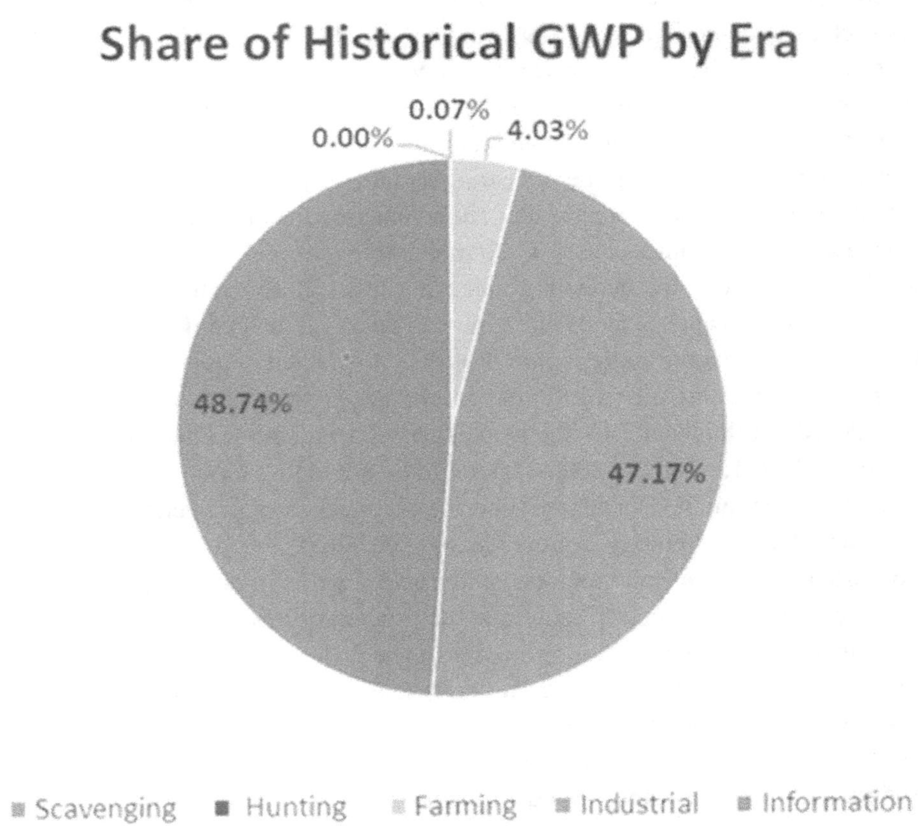

Hooke (2019) and updated by Hooke and Alati for this chapter. The model includes data and projections for 178 countries and protectorates that comprise more than 99% of GWP. it is used by the authors to provide context for more in-depth country and regional analyses, where workshops and questionnaires are used to produce refinements for individual economies on an as-need basis – i.e., when a request has been made for an in-depth analysis and/or a projection for a specific economy or group of economies.

Different to the historical analysis which uses a per capita income multiplied by population framework, WIM adopts a labour productivity times work force approach to forecasting GDP in each country. Country projections are merged to form 21 regional totals, and the latter are summed to obtain projections for GWP.

The workforce is defined as the population between the ages of 15 and 64 inclusive that are willing and able to work. This is shown in the model as the product of the proportion of the population that is of working age (a1) and the proportion of the working age population that is in the work force (a2). Population projections are from data published by the UN Population Division. For a1 and a2, the authors take the average of the 2020 rates for Germany and Japan, two high-income countries with relatively stationary populations, as the norm for all countries when their populations stabilise. Thus a1 = 0.60 and a2 = 0.84. The product of the two (a3 in the model) is 0.50. In the scene-setting run of the model, the authors assume that the actual value for a3 for each country outside Africa will converge smoothly on this rate by 2060 and will do so in African countries by 2080.

For labour productivity (GDP divided by the work force), the assumptions are that the United States will continue to be the technology leader, with annual growth continuing at the average rate realised during 2001-20 (1.0%), plus an acceleration factor. The latter considers the persuasive evidence produced by researchers such as Graeme Snooks (1989), Ray Kurzweil (2005) and Peter Diamandis (2012) that while entropy obviously applies to the universe as a whole, there are components within the universe where order is increasing and is doing so at an increasing rate. The time taken to double per capita GWP was millions of years during the scavenging and hunting eras but was only a few thousand years during the farming era, about 70 years during the industrial era and 30 years at the growth rate achieved in the completed portion of the information era. Further, the actual growth rate during the information era to date has probably been substantially understated. GDP methodology is well designed for measuring changes in the volume of production of homogeneous goods but is less well suited to encapsulating growth in the value of services, where the value added is often more in quality than quantity. It has been estimated that that in the IT sector, official figures may capture less than 20 percent of the increase in value. Reflecting the measurement and forecasting difficulties, the accelerator factor used in this chapter is set at the low rate of 0.1% per decade. It thus increases the annual growth of labour productivity in the United States from 1.1% in the 2020s to 1.8% in the 2090s. It should be noted that faster -than-expected increases in the acceleration factor can quickly shrink the timeline of longer-term projections.

For countries within the technology frontier (set by the United States), it is assumed that the globalisation of factor, product, and financial markets, with their positive implications for international trade, investment, and technology transfer, will continue the process of labour productivity convergence that was led by the four Asian tigers in the third quarter of the 20th century and is now gaining traction in lower-income economies. The trend away from goods to services and continued technical progress in the transport and communication sectors are likely to see further acceleration in the productivity convergence process. The assumption made is that labour productivity of all countries except those in Africa will merge to within 5% of that of the United States by 2060 and that of African countries will merge to within 5% of that of the United States by 2080. It is assumed that the United States, with its highly innovative culture, will then maintain this small productivity difference until the end of the century.

The model projects that GWP will be 16 times higher in 2100 than in 2020. The value of the goods and services produced during the remainder of the 21st century will be 18 times greater than the value produced during the history of humankind up to 2020. The model also suggests that there will be significant changes in the regional distribution of GWP. The share of northeast Asia, which rose from 3% in 1950 to 26% in 2020, is projected to fall to 20% by 2040 and 10% in 2100 due mainly to demographic factors, especially affecting China and Japan. The share of Europe, which was 45% in 1950 and 22% in 2020, is expected to fall to 13% by the end of the century, reflecting demographic changes as well as its already quite high relative labour productivity. The major regions whose shares will increase are South Asia and Africa. The share of South Asia rose from 6% in 1950 to 12% in 2020 and is projected to increase further to 27% in 2060 before falling back to 20% in 2100. The share of Africa is projected to rise from 5% in 2020 to 7% in 2040 and 40% in 2100, reflecting both strong income convergence and demographic factors.

Figure 2.

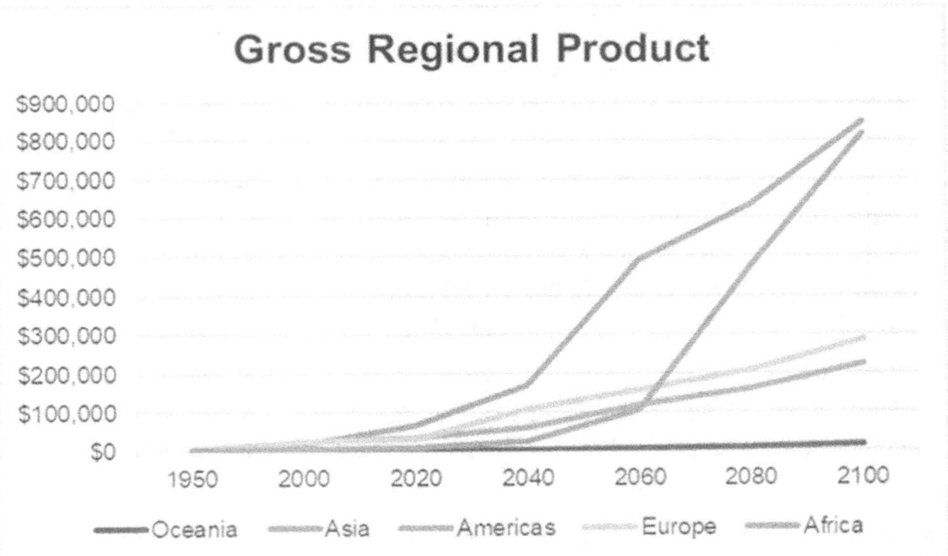

The World's Largest Economies

During the scavenging and hunting eras humans were nomadic, moving from place to place in search of food. Groups ranged in size from extended families to tribes of about 30 people during the scavenging era (based on the behaviour of chimpanzees today) and up to 100 people during the hunting era.

The first village (a small community in a rural area), the first town (a populated area with fixed boundaries and a local government), the first city (similar to a town but with more people), and the first economy

Table 6. GWP, GRPs, and GDPs of China, India & USA, 1950-2100 (2020 I$)

	1950	2000	2020	2040	2060	2080	2100
Oceania	120	936	1,656	2,128	5,962	8,968	14,313
Asia	**1,621**	**19,786**	**65,777**	**169,283**	**490,104**	**635,165**	**850,662**
Northeast	325	10,082	36,473	72,761	148,929	181,195	229,299
Southeast	470	3,261	9,555	26,512	78,733	109,683	151,323
South	650	4,837	16,666	62,148	239,960	312,556	423,453
West	96	1,033	2,061	4,804	12,022	16,833	24,516
Central	81	572	1,021	3,060	10,460	14,898	22,071
Americas	**3,277**	**23,236**	**33,886**	**59,773**	**118,559**	**159,516**	**223,923**
North	2,709	16,824	22,743	31,639	45,656	65,656	98,207
Central	127	1,806	3,374	8,890	24,744	32,378	44,076
South	441	4,605	7,768	19,244	48,159	61,481	81,640
Europe	**5,313**	**23,552**	**30,523**	**107,448**	**155,541**	**207,291**	**287,729**
Eastern	1,264	5,936	8,029	13,899	24,856	31,628	41,861
Northern	933	3,876	5,557	70,625	99,330	135,107	191,146
Southern	1,124	4,999	5,908	8,437	12,280	14,814	18,586
Western	1,993	8,741	11,029	14,487	19,075	25,743	36,136
Africa	**1,375**	**4,616**	**7,129**	**26,028**	**105,534**	**473,120**	**819,263**
Northern	440	1,999	2,772	7,645	21,395	60,970	96,372
Western	446	1,249	1,691	7,539	34,229	159,009	283,237
Southern	237	664	910	2,205	5,358	13,077	18,913
Eastern	143	401	1,134	5,752	29,807	157,243	270,641
Middle	108	303	622	2,887	14,745	82,821	150,100
GWP	**11,707**	**72,126**	**138,970**	**364,659**	**875,702**	**1,484,060**	**2,195,890**
China	171	2,892	28,483	60,983	130,208	158,418	200,475
India	471	2,786	11,893	45,053	170,667	222,299	301,173
USA	2,520	15,383	20,866	28,746	41,198	59,246	88,618

Source. World Factbook 2020 and projections by the authors

(a large set of inter-related production and consumption activities) emerged in Mesopotamia during the early and middle parts of the farming era. The first village would have emerged near the beginning of the farming era, possibly on the banks of the Euphrates or Tigris rivers. The first town may have been Jericho, built around 11 TYA. It covered 2.5 hectares, housed about 4,000 people (plus their animals,) and was surrounded by a wall of undressed stone. The first city with supporting physical evidence such as documentation was Uruk, in Iraq. Cuneiform texts indicate that Gilgamesh, King of the city's first dynasty and hero of the famous Epic of Gilgamesh, built the walls of the city about 6,700 years ago.

The first modern economy, in what is now southern Iraq, was that of Sumer, which emerged about 4,000 BCE. It was based on grain and other crops grown in the valleys of the Euphrates and Tigris rivers, which generated surpluses that allowed part of the population to specialise in manufacturing (pottery and

Figure 3.

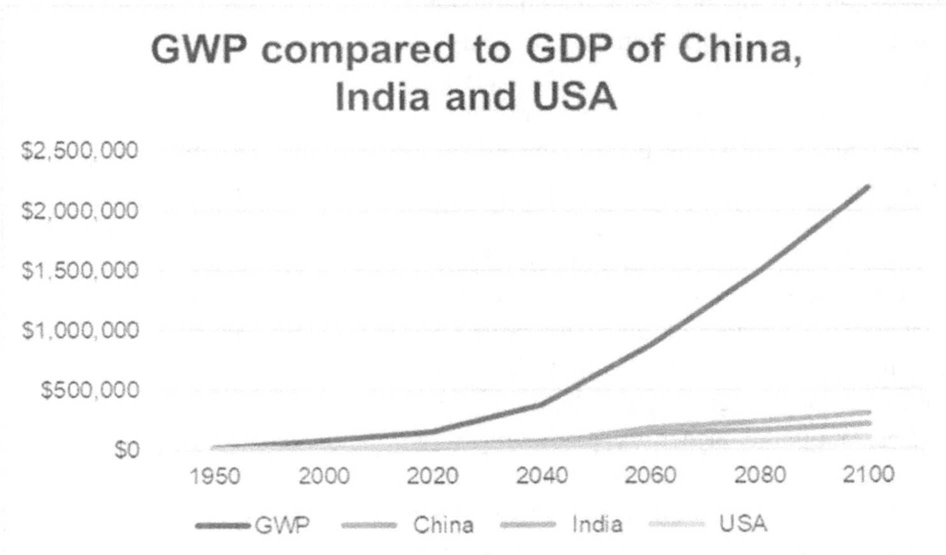

textiles) and services (including religion, defence, and public administration). The next major economy, and the first empire, was that of Akkadia, which dominated Mesopotamia during 2350-2150 BCE. Although based primarily on farming, the Akkadian empire (especially under its first king, Sargon) used military conquest as an important secondary strategy for increasing income. Sargon did not invent modern conflict – the dozen or so city states of Mesopotamia were frequently engaged in battle. However, once the outcome of a battle had been decided and some tribute paid, earlier victors had returned home, only to have to the loser recover and again become a viable enemy. Sargon added destruction of the enemy's towns and productive capacity to his conquest strategy, making his victories more permanent. When the empire declined, due mainly to power struggles for the throne, the economies of Assyria and then Babylonia may have become the largest in the world.

For about 500 years commencing 2000 BCE, the Indus economy was dominant. It covered about 500,000 square miles and its two major cities - Harappa and Mohenjo-Daro - each supported about 30,000 people. The Indus economy was built on farming, with trade being an important but subsidiary strategy (Hooke 2019). Then, for about 500 years commencing 1550 BCE, and especially during the reign of Amenhotep III, the Egyptian economy was probably the world's largest. It was based primarily on agriculture, practised along the banks of the Nile, but was also supported by profits from trade and tribute from conquest. In the Old World, the economy of ancient Greece may have been the largest in the early part of the first millennia BCE. Greece did not have much land that was suitable for agriculture and based its economic strategy firmly on trade. The Roman economy may then have become the largest. With trade being dominated by the Greeks in the east and Carthage in the west, Rome adopted conquest as its main strategy for generating wealth above what could be created from the primary technology of farming.

On at least three occasions, India appears to have had the largest economy. The first was the Mauryan period (325-185 BC) following the formation of the first Indian empire by Candra Gupta (who abdicated in 297 BC and died in the traditional Jaina manner - committing suicide by slow starvation). Farming, trade, and sound government were the basis of the economy's prosperity. Second was the period from AD 320 to 540, when the Gupta dynasty ruled the country. This was a time of comparative stability, and both the economy and classical Indian culture flourished. Third was the Mughal period, which commenced in AD 1526. Economic prosperity was based on farming, which was augmented by trade and supported by an efficient, centralised administration.

However, the dominant economy historically has been that of China. For two of the last four millennia, it may have been the biggest in the world, reflecting its relatively large population, advanced technology, and stable government. Hooke (2019) suggests that China held the top position on the global economic ladder during the latter half of the Chou period (1111-21 BC), during the Ch'in period (221-206 BC) and during much of the Han period (206 BC – AD 220). Its decline from the early 15th century to the late 20th century can be attributed primarily to its withdrawal (by imperial decree) from the world economy and thus also from the competitive forces that were driving innovation abroad, especially in Western Europe.

The other two economies completing the list of 10 largest economies are the United Kingdom and the United States. While the former was among the most dynamic economies from about 1500 AD, its GDP did not pass that of China until it incorporated India into its empire in 1858. It remained the biggest economy for only about 60 years, with the United States becoming the largest from the end of World War I and occupying this position until it was in turn passed by China in 2014.

China is expected to remain the largest economy until the middle of the 21st century. Its potential to benefit further from per capita income convergence will ensure high growth in the coming decades and its large work force will keep its GDP permanently above that of the United States. However, it will not remain long above India on the global economic ladder. India's population is projected to pass that of China in the mid-2020s and to be more than 50% larger by 2100. During the coming decades, India's work force will grow faster than its population (the work force to population ratio is currently 40%, compared to the global average of 46%) while China's work force will grow more slowly (its work force to population ratio is 60%). China's loss of its "demographic gift", and India's shedding of its "demographic burden" will halve the current difference in their per capita incomes. There are further reasons for concern about the Chinese economy. First, its high growth during 1990-2010 was supported by exceptionally high investment (its total investment to GDP ratio was about twice the global average, and much higher than that of any other major economy). The investment ratio is now declining as China rebalances its economy in favour of higher consumption. There is also the possibility that the slowdown in the growth of GDP during the second decade of this century may reflect structural weaknesses. When the four tigers moved into convergence mode, they enjoyed double-digit growth until their per capita income was close to that of the OECD countries. However, China's growth rate appears to have fallen well below 10% even though its per capita income is only 30% of that of OECD countries. The opportunity cost of centrally imposed political stability in terms of GDP foregone may be quite high.

CONCLUDING COMMENTS

The projected strong future growth of GWP gives rise to both challenges and opportunities. The challenges relate mainly to the need for additional natural resources to produce the larger volume of products. Since the writings of Thomas Malthus, there have been concerns that we are running out of resources and that we need to slow down and perhaps even reverse economic growth to prevent catastrophic shortages, particularly of food and energy. To date, these fears have not materialised, mainly because breakthrough

Table 7. Three Major Economies, 1950-2100

	1950	2000	2020	2040	2060	2080	2100
GDP (I$ b)							
India	471	2,786	11,893	45,053	170,667	222,299	301,173
China	171	2,892	28,483	60,983	130,208	158,418	200,475
USA	2,520	15,383	20,866	28,746	41,198	59,246	88,618
Work force (m)							
India	147	413	543	683	859	823	788
China	309	733	814	731	655	586	524
USA	79	139	164	180	197	208	220
Productivity (I$)							
India	3,206	6,749	21,908	65,971	198,660	270,214	382,291
China	555	3,946	34,992	83,375	198,660	270,214	382,291
USA	32,095	110,315	127,232	159,923	209,115	284,435	402,412

Source. World Factbook 2020 and projections by the authors

technologies have solved the immediate problems and beyond, allowing humanity to continue to prosper and dream of an even brighter future. Further, most of the projected growth of GWP is in the less-resource intensive services sectors of the global economy, and many new technologies are being developed that will reduce considerably the demand for natural resources per unit of GWP (such as vertical farming, 3D printing of meat and other foods, and small-scale atomic energy plants). Also, as is indicated in chapter 20, we will not for much longer be limited by the resources available on Earth.

The opportunities lie in the additional resources that a larger world economy makes available to address global problems such as those associated with pandemics, natural Earth-bound disasters like fires and volcanic eruptions, and collisions with near-earth objects. The advent of farming allowed Mesopotamia and other economies to generate savings that led to the accumulation of capital, rising labour productivity, widespread increases in living standards, and vastly increased absolute and health-adjusted life expectancy. The economic growth provided by the new technologies discussed in other chapters of this book will allow even more resources to be allocated to achieving such goals. They promise to underpin the opportunity of an increasingly exciting and fulfilling future for humankind.

REFERENCES

Ashton, T. S. (1998). *The Industrial Revolution, 1760-1830*. Oxford University Press.

Calvin, W. (2004). *A Brief History of the Mind: From Apes to Intellect and Beyond*. Oxford University Press.

Castells, M. (2010). *The Rise of the Network Society*. Wiley-Blackwell.

Coolidge, F., & Wynn, T. (2011). *The Rise of Homo sapiens: The Evolution of Modern Thinking*. Wiley-Blackwell.

Diamandis, P., & Kotler, S. (2012). Abundance: The Future is Better than You Think. Free Press (Simon & Schuster).

Dominguez-Rodrigo, M. (2002). Hunting and Scavenging by Early Humans: The State of the Debate. *Journal of World Prehistory*, *16*(1), 1–54. doi:10.1023/A:1014507129795

Facchini, F. (2003). *A Day with Homo habilis: Life 2,000,000 Years Ago*. Amazon.

Factbook.World (2021). CIA.

Grine, F., Fleagle, J., & Leakey, R. (2006). *The First Humans – Origin and Early Evolution of the Genus Homo*. Springer Science & Business Media.

Hanson, R. (2000). *Long Term Growth as a Sequence of Exponential Models*. Carow Hall.

Hooke, A. (2019). *Global Economic Development: Past, Present and Future*. Lakeland House.

Kurzweil, R. (2005). *The Singularity is Near*. Penguin Putnam Inc.

Mazoyer, M., & Roudart, L. (2005). *A History of World Agriculture*. New York Press.

McEvedy, C., & Jones, R. (1978). *Atlas of World Population History*. Puffin.

Odell, G. (2013). *Stone Tools: Insights into Pre-History*. Springer-Verlag.

Preston, S. (2017). *Homo Erectus as a Man*. CreateSpace Independent Publishing Platform.

Rightmire, G. (1993). *The Evolution of Homo erectus*. Cambridge University Press.

Snooks, G. D. (1989). *The Dynamic Society*. Routledge.

Chapter 2
Technological Convergence:
From Biological Evolution to Technological Evolution

Jose Luis Cordeiro
The Millennium Project, Spain

ABSTRACT

Technological convergence is accelerating and allowing humanity to move from slow and erratic biological evolution to fast and precise technological evolution. The expression "emerging technologies" is used to cover new and potentially powerful fields such as biotechnology, artificial intelligence, and nanotechnology. Although the expression might be somewhat ambiguous, several clusters of different technologies are advancing exponentially and will be critical to humanity's future. NBIC is a common abbreviation that stands for nanotechnology, biotechnology, information technology, and cognitive science. Other technologies like robotics, quantum computing, and space technologies can be added towards an accelerating "technological convergence" that might lead to a "technological singularity" as proposed by US engineer and futurist Ray Kurzweil. According to Kurzweil, we will reach a "technological singularity" by 2045, when we will be able to transcend many of our current limitations and move from biological humans to technological transhumans, both on planet Earth and beyond.

INTRODUCTION

Since the Big Bang, the universe has been in constant evolution and continuous transformation. First there were physical interactions, followed by chemical reactions, then biological processes, and finally now technological evolution. As we begin to ride the wave into human redesign, the destination is still largely unknown, and the opportunities and threats are also growing.

Biological evolution continues but it is just too slow to achieve the possibilities available today thanks to technological evolution. Natural selection with trial and error can now be substituted by technical selection with engineering design. Humanity's monopoly as the only advanced sentient life form on the planet is coming to an end, and will be supplemented by a number of posthuman incarnations, including

DOI: 10.4018/978-1-7998-6772-2.ch002

enhanced humans, transhumans, robots and cyborgs, as we approach the "technological singularity". Moreover, how we re-engineer ourselves could fundamentally change the ways in which our civilization functions and raise crucial questions about our identities and moral status as human beings.

The famous astronomer and astrobiologist Carl Sagan popularized the concept of a Cosmic Calendar about three decades ago. In his 1977 book, *The Dragons of Eden: Speculations on the Evolution of Human Intelligence*, Sagan wrote a timeline for the universe, starting with the Big Bang about 15 billion years ago. Today, we think that it all started about 13.7 billion years back, and we keep updating and improving our knowledge of life, the universe and everything. In his Cosmic Calendar, with each month representing slightly over one billion years, Sagan dated the major events during the first 11 months of the cosmic year (see Table 1).

Table 1. Cosmic Calendar: January – November

Big Bang	January 1
Origin of Milky Way Galaxy	May 1
Origin of the solar system	September 9
Formation of the Earth	September 14
Origin of life on Earth	~ September 25
Formation of the oldest rocks known on Earth	October 2
Date of oldest fossils (bacteria and blue-green algae)	October 9
Invention of sex (by microorganisms)	~ November 1
Oldest fossil photosynthetic plants	November 12
Eukaryotes (first cells with nuclei) flourish	November 15

Source: J.L. Cordeiro based on C. Sagan (1977)

Most of what we study in biological evolution happened in the last month. In fact, Sagan wrote that the first worms appeared on December 16, the invertebrates began to flourish on the 17th, the trilobites boomed on the 18th, the first fish and vertebrates appeared on the 19th, the plants colonized the land on the 20th, the animals colonized the land on the 21st, the first amphibians and first winged insects appeared on the 22nd, the first trees and first reptiles evolved on the 23rd, the first dinosaurs appeared on the 24th, the first mammals evolved on the 26th, the first birds emerged on the 27th, the dinosaurs became extinct on the 28th, the first primates appeared on the 29th and the frontal lobes evolved in the brains of primates and the first hominids appeared on the 30th. Basically, humans are just the new kids in the block, and only evolved late at night on the last day of this Cosmic Calendar (see Table 2).

The previous Cosmic Calendar is an excellent way to visualize the acceleration of change and the continuous evolution of the universe. Other authors have developed similar ideas to try to show the rise of complexity in nature. For example, in 2005, astrophysicist Eric Chaisson published his latest book, *Epic of Evolution: Seven Ages of the Cosmos*, where he describes the formation of the universe through the development of seven ages: matter, galaxies, stars, heavy elements, planets, life, complex life, and society. Chaisson presents a valuable survey of these fields and shows how combinations of simpler systems transform into more complex systems, and he thus gives a glimpse of what the future might bring.

Table 2. Cosmic Calendar: December 31

Origin of *Proconsul* and *Ramapithecus*, probable ancestors of apes and men	~ 1:30 p.m.
First humans	~ 10:30 p.m.
Widespread use of stone tools	11:00 p.m.
Domestication of fire by Peking man	11:46 p.m.
Beginning of most recent glacial period	11:56 p.m.
Seafarers settle Australia	11:58 p.m.
Extensive cave painting in Europe	11:59 p.m.
Invention of agriculture	11:59:20 p.m.
Neolithic civilization; first cities	11:59:35 p.m.
First dynasties in Sumer, Ebla, and Egypt; development of astronomy	11:59:50 p.m.
Invention of the alphabet; Akkadian Empire	11:59:51 p.m.
Hammurabi legal codes in Babylon; Middle Kingdom in Egypt	11:59:52 p.m.
Bronze metallurgy; Mycenaean culture; Trojan War; Olmec culture; invention of the compass	11:59:53 p.m.
Iron metallurgy; First Assyrian Empire; Kingdom of Israel; founding of Carthage by Phoenicia	11:59:54 p.m.
Ashokan India; Ch'in Dynasty China; Periclean Athens; birth of Buddha	11:59:55 p.m.
Euclidean geometry; Archimedean physics; Ptolemaic astronomy; Roman Empire; birth of Christ	11:59:56 p.m.
Zero and decimals invented in Indian arithmetic; Rome falls; Moslem conquests	11:59:57 p.m.
Mayan civilization; Sung Dynasty China; Byzantine empire; Mongol invasion; Crusades	11:59:58 p.m.
Renaissance in Europe; voyages of discovery from Europe and from Ming Dynasty China; emergence of the experimental method in science	11:59:59 p.m.
Widespread development of science and technology; emergence of global culture; acquisition of the means of self-destruction of the human species; first steps in spacecraft planetary exploration and the search of extraterrestrial intelligence	Now: The first second of New Year's Day

Source: J.L. Cordeiro based on C. Sagan (1977)

Both Sagan and Chaisson have written excellent overviews about evolution, from its cosmic beginnings to the recent emergence of humans and technology. This way of looking into the long-term pass to the present is a new academic discipline called Big History, examining history from the Big Bang to the present. Big History resists specialization, and searches for universal patterns or trends.

A more futuristic look from the past to the future is given by engineer and inventor Ray Kurzweil in his 2005 book: *The Singularity is Near: When Humans Transcend Biology*. Kurzweil wrote about six epochs with increasing complexity and accumulated information processing (see Table 3).

According to Kurzweil, humanity is entering Epoch 5 with an accelerating rate of change. The major event of this merger of technology and human intelligence will be the emergence of a "technological singularity". Kurzweil believes that within a quarter century, non-biological intelligence will match the range and subtlety of human intelligence. It will then soar past it because of the continuing acceleration of information-based technologies, as well as the ability of machines to instantly share their knowledge. Eventually, intelligent nanorobots will be deeply integrated in our bodies, our brains, and our environment, overcoming pollution and poverty, providing vastly extended longevity, full-immersion virtual reality incorporating all of the senses, and vastly enhanced human intelligence. The result will be an intimate merger between the technology-creating species and the technological evolutionary process it

Table 3. The Six Epochs of the Universe according to Kurzweil

Epoch 1	Physics and chemistry (information in atomic structures)
Epoch 2	Biology (information in DNA)
Epoch 3	Brains (information in neural patterns)
Epoch 4	Technology (information in hardware and software designs)
Epoch 5	Merger of technology and human intelligence (the methods of biology, including human intelligence, are integrated into the exponentially expanding human technology base)
Epoch 6	The universe wakes up (patterns of matter and energy in the universe become saturated with intelligent processes and knowledge)

Source: J.L. Cordeiro based on R. Kurzweil (2005)

spawned. Appendix I shows some of the possible future developments until the end of this century, based on the previous events, and remembering that the rate of technological change is accelerating thanks to the emerging and exponential technologies.

Computer scientist and science fiction writer Vernor Vinge first discussed this idea of a technological singularity in a now classic 1993 paper, where he predicted:

Within thirty years, we will have the technological means to create superhuman intelligence. Shortly after, the human era will be ended.

Other authors talk about such technological singularity as the moment in time when artificial intelligence will overtake human intelligence. Kurzweil has also proposed the *Law of Accelerating Returns*, as a generalization of Moore's law to describe an exponential growth of technological progress. Moore's law deals with an exponential growth pattern in the complexity of integrated semiconductor circuits (see Figure 1).

Kurzweil extends Moore's law to include technologies from far before the integrated circuit to future forms of computation. Whenever a technology approaches some kind of a barrier, he writes, a new technology will be invented to allow us to cross that barrier. He predicts that such paradigm shifts will become increasingly common, leading to "technological change so rapid and profound it represents a rupture in the fabric of human history." He believes the *Law of Accelerating Returns* implies that a technological singularity will occur around 2045:

An analysis of the history of technology shows that technological change is exponential, contrary to the common-sense 'intuitive linear' view. So we won't experience 100 years of progress in the 21st century—it will be more like 20,000 years of progress (at today's rate). The 'returns,' such as chip speed and cost-effectiveness, also increase exponentially. There's even exponential growth in the rate of exponential growth. Within a few decades, machine intelligence will surpass human intelligence, leading to the Singularity —technological change so rapid and profound it represents a rupture in the fabric of human history. The implications include the merger of biological and non-biological intelligence, immortal software-based humans, and ultra-high levels of intelligence that expand outward in the universe at the speed of light.

Figure 1. Moore's Law
Source: Our World in Data

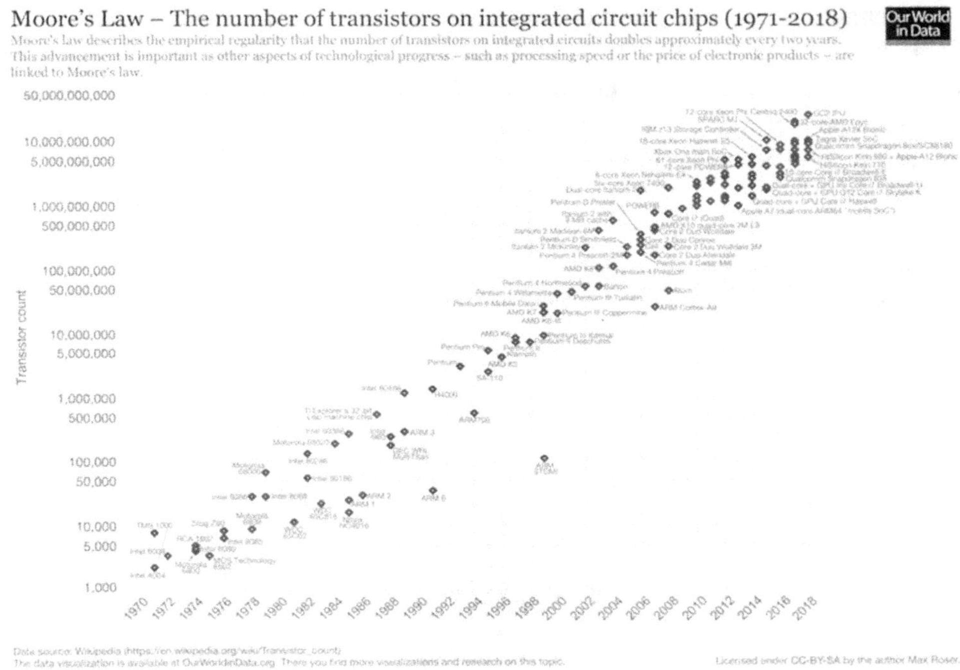

TECHNOLOGICAL CONVERGENCE

Futurists today have diverging views about the singularity: some see it as a very likely scenario, while others believe that it is more probable that there will never be any sudden and dramatic changes due to progress in artificial intelligence. However, most futurists and scientists agree that there is an increasing rate of technological change. In fact, the rapid emergence of new technologies has generated scientific developments never dreamed of before, like the possibility of biological rejuvenation.

The expression "emerging technologies" is used to cover such new and potentially powerful technologies as genetic engineering, artificial intelligence, and nanotechnology. Although the exact denotation of the expression is vague, various writers have identified clusters of such technologies that they consider critical to humanity's future. These proposed technology clusters are typically abbreviated by such combinations of letters as NBIC, which stands for Nanotechnology, Biotechnology, Information technology and Cognitive science. Various other acronyms have been offered for essentially the same concept, such as GNR (Genetics, Nanotechnology and Robotics) used by Kurzweil, while others prefer NRG because it sounds like "energy." Journalist Joel Garreau in *Radical Evolution* uses GRIN, for Genetic, Robotic, Information, and Nano processes, while author Douglas Mulhall in *Our Molecular Future* uses GRAIN, for Genetics, Robotics, Artificial Intelligence, and Nanotechnology. Another acronym is BANG for Bits, Atoms, Neurons, and Genes.

The inaugural NBIC Conference for Improving Human Performance was organized in 2003 by the NSF (National Science Foundation) and the DOC (Department of Commerce). Since then, there have been many similar gatherings, in the USA and overseas. The European Union has been working on its own strategy towards converging technologies, and so have been other countries in Asia, starting with Japan and later South Korea and China.

The idea of technological convergence is based on the merger of different scientific disciplines thanks to the acceleration of change on all NBIC fields. Nanotechnology deals with atoms and molecules, biotechnology with genes and cells, infotechnology with bits and bytes, and cognitive science with neurons and brains. These four fields are converging thanks to the larger and faster information processing of ever more powerful computers (see Figure 2).

Figure 2. Technological Convergence NBIC
Source: J.L. Cordeiro based on M.C. Roco and W.S. Bainbridge (2003)

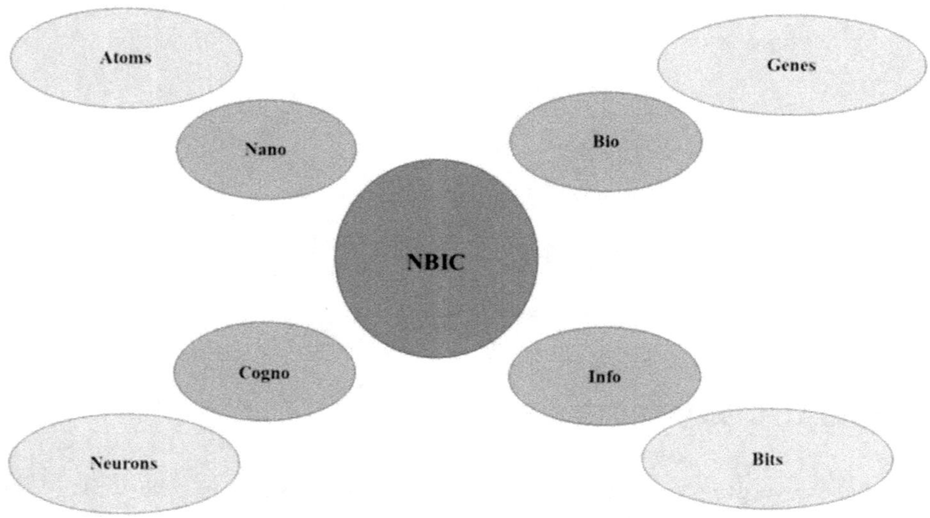

Experts from the four NBIC fields agree about the incredible potential of technological evolution finally overtaking and directing biological evolution. Bill Gates of Microsoft has stated that:

I expect to see breathtaking advances in medicine over the next two decades, and biotechnology researchers and companies will be at the center of that progress. I'm a big believer in information technology... but it is hard to argue that the emerging medical revolution, spearheaded by the biotechnology industry, is any less important to the future of humankind. It, too, will empower people and raise the standard of living.

Larry Ellison of Oracle agrees: "If I were 21 years old, I probably wouldn't go into computing. The computing industry is about to become boring". He explains that: "I would go into genetic engineering." Biologist Craig Venter has said that he spent 10 years reading the human genome, and now he is planning to write new genomes. He wants to create completely new forms of life, from scratch. Indeed, Venter and his team created in 2010 the first synthetic bacterium, and they call it *Synthia*. Scientist Gregory Stock also believes that cloning, even though a fundamental step in biotechnology, is just too simple and unexciting: "why copy old life forms when we can now create new ones?"

Biological evolution allowed the appearance of human beings, and many other species, through millions of years of natural selection based on trials and errors. Now we can control biological evolution, direct it and go beyond it. In fact, why stop evolution with carbon-based life forms? Why not move into silicon-based life, among many other possibilities? Robotics and artificial intelligence will allow us to do just that.

Scientist Marvin Minsky, one of the fathers of artificial intelligence at MIT, wrote a very famous 1994 article "Will robots inherit the Earth?" in *Scientific American*, where he concludes: "Yes, but they will be our children. We owe our minds to the deaths and lives of all the creatures that were ever engaged in the struggle called Evolution. Our job is to see that all this work shall not end up in meaningless waste." Robotics expert Hans Moravec has written two books about robots and our (their) future: *Mind Children* in 1988 and *Robot* in 1998. Moravec argues that robots will be our rightful descendants and he explains several ways to "upload" a mind into a robot. In England, cybernetics professor Kevin Warwick has been implanting his own body with several microchip devices and published in 2003 a book explaining his experiments: *I, Cyborg*. Warwick is a cybernetics pioneer who claims that:

I was born human. But this was an accident of fate – a condition merely of time and place. I believe it's something we have the power to change... The future is out there; I am eager to see what it holds. I want to do something with my life: I want to be a cyborg.

As these authors and thinkers suggest, we need to start preparing ourselves for the coming NBIC realities of technological convergence, including robotics, artificial intelligence, quantum computing and even space technologies. Thanks to technological evolution, humans will transcend our biological limitations to become transhumans and eventually posthumans. To ease this transition into a posthuman condition, we must ready ourselves for the distinct possibility that the Earth, and other planets, will be inherited by not just one but several forms of highly intelligent and sentient life forms. Thus, the philosophy of humanism is not enough for a world, and a universe, where future life forms will continue evolving.

FROM HUMANISM TO TRANSHUMANISM

A new philosophy has been proposed to continue the ideas of humanism in a new world where science and technology are the major drivers of change. Julian Huxley, the English evolutionary biologist and humanist that became the first director-general of UNESCO and founder of the World Wildlife Fund, wrote that:

The human species can, if it wishes, transcend itself—not just sporadically, an individual here in one way, an individual there in another way, but in its entirety, as humanity. We need a name for this new belief. Perhaps transhumanism will serve: man remaining man, but transcending himself, by realizing new possibilities of and for his human nature.

"I believe in transhumanism": once there are enough people who can truly say that, the human species will be on the threshold of a new kind of existence, as different from ours as ours is from that of Pekin man. It will at last be consciously fulfilling its real destiny.

Huxley originally published those words in his essay *Religion Without Revelation* (1927), which was later reprinted in his book *New Bottles for New Wine* (1957). Other scientists and philosophers discussed similar ideas in the first half of the 20th century, and these ideas slowly helped to create new philosophical movements considering nature and humanity in a continuous state of flux and evolution. English scientist John Burdon Sanderson Haldane and French philosopher Pierre Teilhard de Chardin helped to identify new trends in the future evolution of humanity. Thanks to them and many others, the philosophy of transhumanism has greatly advanced since Huxley first used that word. The idea of Extropy was mainly developed by British philosopher Max More (see Appendix 2) and transhumanism by Swedish philosopher Nick Bostrom and other members of the World Transhumanist Association, now called HumanityPlus (see Appendix 3). Extropy and transhumanism explore the boundless possibilities for future generations, while we approach the possible technological singularity.

"Humans" can no longer be regarded as a stable category let alone one which occupies a privileged position in relation to all that is subsumed under the category of the non-human. On the contrary, humans must be understood as a tenuous entity which is related to the animal, the "natural" and indeed other humans as well. Humans are at a crossroads like other natural species which are reclassified in the face of new relational dynamics and shifting epistemological paradigms. Moreover, such dynamics and interpolation serve to reveal the boundaries of humans as a corporal, cognitive, and agency-laden construct. Discovering such boundaries, one may glean where humans end, where humans are called into question, and where humans stand to augment themselves or become more than human.

Our understanding about ourselves and about our relationships with nature around us has increased significantly due to the continuous advances in science and technology. Reality is not static since humans and the rest of nature are dynamic, indeed, and both are changing constantly. Transhumanism transcends such static ideas of humanism as humans themselves evolve at an accelerating rate. In the beginning of the 21st century, it is now clear than humans are not the end of evolution, but just the beginning of a conscious and technological evolution. In general, transhumanism considers super intelligence, super longevity, and super wellbeing.

Thanks to the accelerating rate of technological change, humans are transcending biological limitations. Human beings are crossing the traditional boundaries of what being human meant. Such radical changes have profound philosophical implications in what it is to be human and the interactions between humans with an increasingly modified environment. What is natural and what is not natural has a new meaning in transhumanism, since everything is changing and evolving, but not just biologically but also technologically.

THE HUMAN SEED

Since English naturalist Charles Darwin first published his ideas about evolution on *The Origin of Species* in 1859, it has become clear to the scientific community that species evolve according to interactions among them and with their environment. Species are not static entities but dynamic biological systems in constant evolution. Humans are not the end of evolution in any way, but just the beginning of a better, conscious and technological evolution. The human body is a good beginning, but we can certainly improve it, upgrade it, and transcend it. Biological evolution through natural selection might be ending, but technological evolution is only accelerating now. Technology, which started to show dominance over biological processes some years ago, is finally overtaking biology as the science of life.

As fuzzy logic theorist Bart Kosko has said: "biology is not destiny. It was never more than tendency. It was just nature's first quick and dirty way to compute with meat. Chips are destiny." Photo-qubits might also come after standard silicon-based chips, but even that is only an intermediate means for augmented intelligent life in the universe.

Homo sapiens sapiens is the first species in our planet which is conscious of its own evolution and limitations, and humans will eventually transcend these constraints to become enhanced humans, trans-humans and posthumans. It might be a rapid process like caterpillars becoming butterflies, as opposed to the slow evolutionary passage from apes to humans. Future intelligent life forms might not even resemble human beings at all, and carbon-based organisms will mix with a plethora of other organisms. These posthumans will depend not only on carbon-based systems but also on silicon and other "platforms" which might be more convenient for different environments, like traveling in outer space.

Eventually, all these new sentient life forms might be connected to become a global brain, a large interplanetary brain, and even a larger intergalactic brain. The ultimate scientific and philosophical queries will continue to be tackled by these posthuman life forms. Intelligence will keep on evolving and will try to answer the old-age questions of life, the universe and everything. With ethics and wisdom, humans will become posthumans, as science fiction writer David Zindell suggested:

"What is a human being, then?"

"A seed."

"A... seed?"

"An acorn that is unafraid to destroy itself in growing into a tree."

FROM THE EARTH TO OUTER SPACE

We will finally begin colonizing outer space in the following decades, as physicist Michio Kaku writes in his best-selling book *The Future of Humanity: Terraforming Mars, Interstellar Travel, Immortality, and Our Destiny Beyond Earth*, and entrepreneur Elon Musk explains in his many presentations. In fact, Musk emphasizes the critical importance of becoming a multi-planetary civilization to guarantee the survival of humanity, and later a possible transhumanity or posthumanity, as we keep evolving technologically, even outside our original home planet Earth. He plans to start with a colony on Mars, that will begin with a possible manned trip projected to leave Earth on 2024 (or 2026, if it has to be delayed).

Currently, not just national space agencies are planning space missions, like NASA from the USA, ESA from Europe, CNSA from China, Roscosmos from Russia and JAXA from Japan, but also private entrepreneurs like Elon Musk with Space X, Richard Branson with Virgin Galactic and Jeff Bezos with Blue Origin, are joining the new space race. And this race is truly global, full of public and private initiative, with both competition and collaboration.

Nikolai Semenovich Kardashev was a Russian astrophysicist who in 1964 proposed a scale to measure the level of technological progress in an advanced civilization. His scale is only theoretical and highly speculative in terms of an actual civilization; however, it puts the energy and power consumptions of an entire civilization in a cosmic perspective. His original scale had three designated categories called Type

I, Type II, and Type III. These are based on the amount of usable energy that a civilization has available at its disposal, and the degree of space colonization as well. In general terms, a Type I civilization has achieved mastery of the resources of its home planet, Type II of its solar system, and Type III of its galaxy.

Table 4 shows the Kardashev scale within the context of different power levels from the smallest to the highest values. In fact, the numbers correspond to power levels instead of energy levels, but remember that power is the amount of energy per unit of time: one Watt (or W, the standard SI unit of power) is defined as one Joule (or J, the standard SI unit of energy) per second. Therefore, if we are clear about the time span being considered, there should be no problem using power or energy values consistently.

A Type I civilization is one that can harness all the power in a single planet (in our case, planet Earth has about 174×10^{15} W in available power). A Type II civilization is one that can harness all of the power available from a solar system (approximately 385×10^{24} W for our Sun), and a Type III civilization is able to harness all of the power available from a single galaxy (approximately 5×10^{36} W for the Milky Way). These figures are extremely variable since planets, solar systems and galaxies vary widely in luminosity, size, and many other important parameters. As a general reference, and using very round numbers, we could say that a Type I civilization can harness about 10^{16} W, a Type II about 10^{26} W, and a Type III about 10^{36} W, all of these figures plus or minus one or two orders of magnitude. In fact, astrophysicist Carl Sagan preferred to use a logarithmic scale instead of the orders of magnitude proposed by Kardashev.[1] Thus, our human civilization is currently at about 0.72 in a logarithmic scale before reaching 1.0 corresponding to Type I status. Such a logarithmic scale indicates an exponential growth for energy consumption. Our prehuman ancestors started harnessing fire about half a million years ago, and we might reach Type I status in the following decades. This exponential growth is also explained by Kaku, who writes about different propulsion systems available to different types of civilizations:

- Type 0:
    - Chemical rockets
    - Ionic engines
    - Fission power
    - EM propulsion (railguns)
- Type I:
    - Ram-jet fusion engines
    - Photonic drive
- Type II:
    - Antimatter drive
    - Von Neumann nano probes
- Type III:
    - Planck energy propulsion

Space advocates have realized that there are almost unlimited amounts of energy available in outer space. That is why it is so important to become a Type I civilization that can better explore and colonize the universe, beginning with the Moon and Mars in the next few years, and eventually moving to other parts of our own solar system and galaxy.

Some authors have considered even higher types of civilizations than the three originally defined by Kardashev. For example, a Type IV civilization could have control of the energy output of a galactic supercluster (approximately 10^{42} W in our case) and a Type V civilization could control the energy of

Table 4. Energy Scale and Kardashev Civilization Types (Power in Watts)

Example	Power	Scientific notation
Power of Galileo space probe's radio signal from Jupiter	10 zW	10×10^{-21} watt
Minimum discernable signal of FM antenna radio	2.5 fW	2.5×10^{-15} watt
Average power consumption of a human cell	1 pW	1×10^{-12} watt
Approximate consumption of a quartz wristwatch	1 μW	1×10^{-6} watt
Laser in a CD-ROM drive	5 mW	5×10^{-3} watt
Approximate power consumption of the human brain	30 W	30×10^{0} watt
Power of the typical household incandescent light bulb	60 W	60×10^{0} watt
Average power used by an adult human body	100 W	100×10^{0} watt
Peak power consumption of a Pentium 4 CPU	130 W	130×10^{0} watt
Power output (work plus heat) of a person working hard	500 W	500×10^{0} watt
Power of a typical microwave oven	1.1 kW	1.1×10^{3} watt
Power received from the Sun at the Earth's orbit per m^2	1.366 kW	1.366×10^{3} watt
2015 world average power use per person	2.3 kW	2.3×10^{03} watt
Average photosynthetic power output per km^2 in ocean	3.3 - 6.6 kW	$3.3 - 6.6 \times 10^{3}$ watt
2015 US average power use per person	12 kW	12×10^{3} watt
Average photosynthetic power output per km^2 in land	16 - 32 kW	$16 - 32 \times 10^{3}$ watt
Approximate range of power output of typical cars	40 - 200 kW	$40 - 200 \times 10^{3}$ watt
Peak power output of a blue whale	2.5 MW	2.5×10^{6} watt
Mechanical power output of a diesel locomotive	3 MW	3×10^{6} watt
Average power consumption of a Boeing 747 aircraft	140 MW	140×10^{6} watt
Peak power output of largest class aircraft carrier	190 MW	190×10^{6} watt
Electric power output of a typical nuclear plant	1 GW	1×10^{9} watt
Power received from the Sun at the Earth's orbit by km^2	1.4 GW	1.4×10^{9} watt
Electrical generation of the Three Gorges Dam in China	18 GW	18×10^{9} watt
Power consumption of the first stage of Saturn V rocket	190 GW	190×10^{9} watt
2015 US electrical power consumption	0.5 TW	0.5×10^{12} watt
2015 world electrical power consumption	2.0 TW	2.0×10^{12} watt
2015 US total power consumption	3.7 TW	3.7×10^{12} watt
2015 world total power consumption	16 TW	16×10^{12} watt
Average total heat (geothermal) flux from Earth's interior	44 TW	44×10^{12} watt
World total photosynthetic energy production	75 TW	75×10^{12} watt
Heat energy released by a hurricane	50 - 200 TW	$50 - 200 \times 10^{12}$ watt
Estimated total available wind energy	870 TW	870×10^{12} watt
World´s most powerful laser pulses	1.2 PW	1.2×10^{15} watt
Estimated heat flux transported by the Gulf Stream	1.4 PW	1.4×10^{15} watt
Total power received by the Earth from the Sun (Type I)	174 PW	174×10^{15} watt
Luminosity of the Sun (Type II)	385 YW	385×10^{24} watt
Approximate luminosity of the Milky Way galaxy (Type III)	5×10^{36} W	5×10^{36} watt
Approximate luminosity of a Quasar	1×10^{40} W	1×10^{40} watt
Approximate luminosity of the Local Supercluster	1×10^{42} W	1×10^{42} watt
Approximate luminosity of a Gamma Ray burst	1×10^{45} W	1×10^{45} watt
Approximate luminosity of all the stars in the known universe	2×10^{49} W	2×10^{49} watt
The Planck Power (basic unit of power in the Planck units)	3.63×10^{52} W	3.63×10^{52} watt

Source: J.L Cordeiro based on J.L. Cordeiro (2006) and M. Kaku (2018)

the entire universe. Such an advanced civilization would approach or surpass the limits of speculation based on our current scientific understanding, and it may not be possible. Finally, some science fiction authors have written about a Type VI civilization that could control the energy over multiple universes (a power level which could technically be infinite) and also about a Type VII civilization that could have the hypothetical status of a deity (able to create universes at will, using them as an energy source).

Kardashev´s civilization types are originally defined according to the total energy available to a planet. Indeed, our Sun continuously delivers to Earth over 10,000 times the power consumed by humanity today: 174 PW (1.74×10^{17} W) from our Sun versus 16 TW (1.6×10^{13} W) used currently by all humans alive now. Indeed, solar energy is by far the largest external source of energy available to our civilization. However, beyond solar energy, we still have plenty of energy sources available on our planet to move us towards a Type I civilization.

Table 5 shows the different energy contents (specific energy measured in terms of Mega Joules per kilogram, MJ/kg) available in several energy sources. Hydro power was one of the first energy sources used by humans, but its energy content is extremely low: only 0.001 MJ/kg for water stored at a height of 100 meters. Bagasse, animal dung, manure and wood fuels are relatively better, ranging from 10 to 16 MJ/kg. Humanity then moved to coal, whose energy content goes from about 22 to 30 MJ/kg, depending on the type and quality of coal. Now hydrocarbon fuels are the main energy source, with 22 to 55 MJ/kg from methanol to methane, for example. Additionally, since the middle of the 20th century, several countries have also started using nuclear fission, and some fast breeder reactors produce 86,000,000 MJ/kg with Uranium.

Humanity has continuously increased its energy sources throughout history, starting from different types of biomass and hydropower in the distant past, to coal and hydrocarbons constituting the largest energy sources today. Nonetheless, nuclear energy, first fission and later fusion, might also become major energy sources in the near future, while we keep moving towards renewable energies to finally become a Type I civilization powered mostly by solar power.

We are advancing exponentially in our energy uses from harnessing fire about half a million years ago to developing sustainable nuclear fusion in the coming decades. Eventually, we should be able to harness the full energy content of matter and convert matter directly into energy, according to Einstein´s famous equation, and even annihilate matter and antimatter to produce more energy. To have an idea of the incredible energy potential represented just by our planet, the Earth has an estimated mass of 5.98×10^{24} kg, which represents a theoretical energy content of 5.37×10^{41} J. Therefore, just our oceans have enough water to power humanity beyond Type I status over very long geological time scales. Additionally, in the case of our Sun, its mass is estimated to be 1.99×10^{30} kg, which is theoretically equivalent to 1.79×10^{47} J. Those numbers are really enormous, and they represent more than enough energy for billions and billions of years.

Considering the visible and known universe, its total mass-energy is currently estimated at about 4×10^{69} J. In a few words, there is certainly no lack of mass-energy in the universe. Moreover, ordinary matter is now considered to be only about 4% of the total matter-energy density in the observable universe, which also includes 22% dark matter and 74% dark energy as well. Since matter and energy can't be destroyed but only converted from one type to another, as scientists believe, there are almost unlimited amounts of energy for our civilization to keep expanding throughout the universe after reaching the "Energularity."

The Earth, the Sun, the Milky Way galaxy and the visible universe have more than enough energy to power our advancing human-machine civilization for the following decades, centuries, millennia, and even billions of years into the future. It is thus possible to convert the immense energy supplies available

Table 5. Approximate Energy Content of Different Materials (MJ/Kg)

Fuel type	Energy content (MJ/kg)
Pumped stored water at 100m dam height (hydro power)	0.001
Battery, lead acid	0.14
Battery, lithium-ion	0.7
Battery, lithium-ion nanowire	2.5
Bagasse (cane stalks)	10
Animal dung, manure	12 - 14
Wood fuel $(C_6H_{10}O_5)_n$	14 - 16
Sugar ($C_6H_{12}O_6$, glucose)	16
Methanol (CH_3-OH)	22
Coal (anthracite, lignite, etc.)	22 - 30
Ethanol (CH_3-CH_2-OH)	30
LPG (liquefied petroleum gas)	32 - 34
Butanol (CH_3-$(CH_2)_3$-OH)	36
Biodiesel	38
Olive oil ($C_{18}H_{34}O_2$)	40
Crude oil (medium petroleum averages)	42 - 44
Gasoline	46
Diesel	48
Methane (CH_4, gaseous fuel, compression dependent)	55
Hydrogen (H_2, gaseous fuel, compression dependent)	140
Nuclear isomers (Ta-180m isomer)	41,340
Nuclear isomers (Hf-178m2 isomer)	1,326,000
Nuclear fission (natural Uranium in fast breeder reactor)	86,000,000
Nuclear fusion (Hydrogen, H)	300,000,000
Binding energy of helium (He)	675,000,000
Mass-energy equivalence (Einstein's equation $E = mc^2$)	89,880,000,000
Annihilation of matter and antimatter (twice $E = mc^2$)	180,000,000,000

Source: J.L Cordeiro based on J.L. Cordeiro (2006) and M. Kaku (2018)

in the universe into usable power, but it will certainly take massive investments and lots of imagination, creativity, science, and technology.

Our civilization is still in its infancy, and barring any wild cards, geopolitical crises, nuclear wars, bio terrorism, nano grey goo, environmental disasters or extraterrestrial contacts, science and technology will keep uncovering the contours of the universe. As Russian rocket scientist Konstantin Tsiolkovsky, one of the "founding fathers" of astronautics, said over a century ago:

The Earth is the cradle of humanity, but one cannot stay in the cradle forever.

REFERENCES

Asimov, I. (1994). *I, Robot*. Bantam Books. (Original work published 1950)

Bostrom, N. (2005). *Superintelligence: Paths, Dangers, Strategies*. Oxford University Press.

Bostrom, N. (2005). A History of Transhumanist Thought. *Journal of Evolution and Technology / WTA, 14*(1).

British Telecom. (2005). *Technology Timeline*. London, UK: British Telecom. www.bt.com/technologytimeline

Chaisson, E. (2005). *Epic of Evolution: Seven Ages of the Cosmos*. Columbia University Press. doi:10.7312/chai13560

Clarke, A. C. (1984). *Profiles of the Future: An Inquiry into the Limits of the Possible*. Henry Holt and Company.

Condorcet, Marie-Jean-Antoine-Nicolas de Caritat. (1979). *Sketch for a historical picture of the progress of the human mind*. Westport, CT: Greenwood Press.

Cordeiro, J. L. (2006). *Energy 2020: A Vision of the Future in Creating Global Strategies for Humanity's Future*. World Future Society.

Cordeiro, J. L. (2010). *The Future of Energy and the Energy of the Future*. Singularity Institute for Artificial Intelligence.

Cordeiro, J. L. (2010). *Telephones and Economic Development: A Worldwide Long-Term Comparison*. Lambert Academic Publishing.

Cordeiro, J. L. (2018). *La Muerte de la Muerte*. Editorial Planeta, Deusto.

Darwin, C. (2003). *The Origin of the Species*. Fine Creative Media. (Original work published 1859)

de Chardin, T. P. (1964). The Future of Man. New York, NY: Harper & Row.

Drexler, K. E. (1987). *Engines of Creation*. Anchor Books. www.e-drexler.com/d/06/00/EOC/EOC_Cover.html

Drexler, K. E. (2013). *Radical Abundance: How a Revolution in Nanotechnology Will Change Civilization*. Public Affairs.

Foundation for the Future. (2002). *The Next Thousand Years*. Bellevue, WA: Foundation for the Future. www.futurefoundation.org/documents/nty_projdesc.pdf

Fumento, M. (2003). *BioEvolution: How Biotechnology is Changing the World*. Encounter Books.

Garreau, J. (2005). *Radical Evolution: The Promise and Peril of Enhancing Our Minds, Our Bodies—and What It Means to Be Human*. Doubleday.

Glenn, J. (2019). *State of the Future 19.1*. The Millennium Project. www.StateOfTheFuture.org

Haldane, J. B. S. (1924). *Daedalus; or, Science and the Future*. K. Paul, Trench, Trubner & Co.

Haraway, D. (1991). A Cyborg Manifesto. In *Simians, Cyborgs and Women: The Reinvention of Nature*. Routledge.

Hawking, S. (2002). *The Theory of Everything: The Origin and Fate of the Universe*. New Millennium Press.

Hughes, J. (2004). *Citizen Cyborg: Why Democratic Societies Must Respond to the Redesigned Human of the Future*. Westview Press.

Huxley, J. (1957). Transhumanism. In *New Bottles for New Wine*. Chatto & Windus. (Original work published 1927)

Kaku, M. (2005). *Parallel Worlds: The Science of Alternative Universes and Our Future in the Cosmos*. Doubleday.

Kaku, M. (2012). *Physics of the Future: How Science Will Shape Human Destiny and Our Daily Lives by the Year 2100*. Anchor.

Kaku, M. (2014). *The Future of the Mind: The Scientific Quest to Understand, Enhance, and Empower the Mind*. Doubleday.

Kaku, M. (2018). *The Future of Humanity: Terraforming Mars, Interstellar Travel, Immortality, and Our Destiny Beyond Earth*. Doubleday.

Kurian, G. T., & Molitor, G. T. T. (1996). *Encyclopedia of the Future*. Macmillan.

Kurzweil, R. (1999). *The Age of Spiritual Machines*. Penguin Books., www.kurzweilai.net

Kurzweil, R. (2005). *The Singularity is Near: When Humans Transcend Biology*. Viking., www.singularity.com

Kurzweil, R. (2012). *How to Create a Mind: The Secret of Human Thought Revealed*. Penguin Books.

Minsky, M. (1987). *The Society of Mind*. Simon and Schuster.

Minsky, M. (1994). Will robots inherit the Earth? *Scientific American*. www.ai.mit.edu/people/minsky/papers/sciam.inherit.txt

Moravec, H. (1988). *Mind Children*. Harvard University Press.

Moravec, H. (1998). *Robot: Mere Machine to Transcendent Mind*. Oxford University Press. www.frc.ri.cmu.edu/~hpm/book97

More, M. (2003). *The Principles of Extropy*. The Extropy Institute. www.extropy.org/principles.htm

Mulhall, D. (2002). *Our Molecular Future*. Prometheus Books.

Paul, G. S., & Cox, E. (1996). *Beyond Humanity: Cyberevolution and Future Minds*. Charles River Media.

Roco, M. C., & Bainbridge, W. S. (Eds.). (2003). *Converging Technologies for Improving Human Performance*. Kluwer. doi:10.1007/978-94-017-0359-8

Roco, M. C., & Montemagno, C. D. (Eds.). (2004). The Coevolution of Human Potential and Converging Technologies. New York, NY: New York Academy of Sciences.

Sagan, C. (1973). *The Cosmic Connection*. Doubleday.

Sagan, C. (1977). *The Dragons of Eden: Speculations on the Evolution of Human Intelligence*. Random House.

Stock, G. (2002). *Redesigning Humans: Our Inevitable Genetic Future*. Houghton Mifflin Company.

Tegmark, M. (2017). *Life 3.0: Being Human in the Age of Artificial Intelligence*. Knopf.

Venter, J. C. (2008). *A Life Decoded: My Genome: My Life*. Penguin.

Venter, J. C. (2013). *Life at the Speed of Light: From the Double Helix to the Dawn of Digital Life*. Viking Adult.

Vinge, V. (1993, Winter). The Coming Technological Singularity. *Whole Earth Review*.

Warwick, K. (2003). *I, Cyborg*. Garnder's.

Wells, H. G. (1902). The Discovery of the Future. *Nature, 65*. www.geocities.com/yokelcraig/hgwells1.html

World Transhumanist Association (Humanity+). (2002). *The Transhumanist Declaration*. World Transhumanist Association. www.transhumanism.org/declaration.htm

Zindell, D. (1994). *The Broken God*. Acacia Press.

Zubrin, R. (2019). *The Case for Space: How the Revolution in Spaceflight Opens Up a Future of Limitless Possibility*. Prometheus.

APPENDIX 1.

Brief Chronology of Possible Future Events Until 2099

2020s: Second Bridge towards indefinite lifespans using biotechnology (Ray Kurzweil)
2020s: Worldwide eradication of poliomyelitis and measles
2020s: Vaccines against malaria and HIV
2020s: 3D bioprinting of simple human organs
2020s: Commercial cloning of human organs with own cells from patients
2020s: Beginning of commercial rejuvenation treatments with stem cells and telomerase
2020s: AI and robot doctors complement and supplement human doctors
2020s: Telemedicine spreads worldwide
2023: Global population reaches 8,000,000,000 people according to the United Nations
2024: First landing on the Moon by a man and a woman (NASA Artemia program)
2024: First manned trips to Mars (Elon Musk)
2025: Molecular assemblers (nanotechnology) are possible (Ray Kurzweil)
2026: Global population reaches 8,000,000,000 people according to the US Census Bureau
2029: Longevity escape velocity is reached, also called the Methuselarity (Ray Kurzweil)
2029: An advanced AI finally passes the Alan Turing's test (Ray Kurzweil)
2030s: Third Bridge towards indefinite lifespans using nanotechnology (Ray Kurzweil)
2030s: Neural implants allow connections of human brains and Internet (Elon Musk)
2030s: Consolidation of the first human colony in Mars (Elon Musk)
2030s: Cure for Parkinson´s and Alzheimer's disease
2030s: Cure for the majority of cancers
2030s: Worldwide eradication of malaria and HIV
2037: Global population reaches 9,000,000,000 people according to the United Nations
2039: Mental transfer from brain to brain becomes possible (Ray Kurzweil)
2040s: Fourth Bridge towards indefinite lifespans and immortality using AI (Ray Kurzweil)
2040s: Interplanetary Internet connects to Earth, Moon, Mars, and spaceships
2042: Global population reaches 9,000,000,000 people according to the US Census Bureau
2045: Aging is cured, and death becomes optional (Ray Kurzweil)
2045: The Singularity: AI surpasses all human intelligence (Ray Kurzweil)
2049: Distinction between reality and virtual reality disappears (Ray Kurzweil)
2050: Humanoid robots win English football cup (British Telecom)
2050s: Humans and machines merge to advance civilization on Earth and beyond
2050s: First reanimations of cryopreserved patients (Ray Kurzweil)
2060s: The human-machine civilization reaches Type I in the Kardashev Scale
2072: Picotechnology starts (pico is one thousand times smaller than nano, Ray Kurzweil)
2099: Femtotechnology starts (femto is one thousand times smaller than pico, Ray Kurzweil)
2099: Lifespan becomes irrelevant in a world of "amortality"

APPENDIX 2

The Principles of Extropy

- **Perpetual Progress:** Extropy means seeking more intelligence, wisdom, and effectiveness, an open-ended lifespan, and the removal of political, cultural, biological, and psychological limits to continuing development. Perpetually overcoming constraints on our progress and possibilities as individuals, as organizations, and as a species. Growing in healthy directions without bound.
- **Self-Transformation:** Extropy means affirming continual ethical, intellectual, and physical self-improvement, through critical and creative thinking, perpetual learning, personal responsibility, proactivity, and experimentation. Using technology – in the widest sense to seek physiological and neurological augmentation along with emotional and psychological refinement.
- **Practical Optimism:** Extropy means fueling action with positive expectations – individuals and organizations being tirelessly proactive. Adopting a rational, action-based optimism or "proaction", in place of both blind faith and stagnant pessimism.
- **Intelligent Technology:** Extropy means designing and managing technologies not as ends in themselves but as effective means for improving life. Applying science and technology creatively and courageously to transcend "natural" but harmful, confining qualities derived from our biological heritage, culture, and environment.
- **Open Society – information and democracy:** Extropy means supporting social orders that foster freedom of communication, freedom of action, experimentation, innovation, questioning, and learning. Opposing authoritarian social control and unnecessary hierarchy and favoring the rule of law and decentralization of power and responsibility. Preferring bargaining over battling, exchange over extortion, and communication over compulsion. Openness to improvement rather than a static utopia. Extropia ("ever-receding stretch goals for society") over utopia ("no place").
- **Self-Direction:** Extropy means valuing independent thinking, individual freedom, personal responsibility, self-direction, self-respect, and a parallel respect for others.
- **Rational Thinking:** Extropy means favoring reason over blind faith and questioning over dogma. It means understanding, experimenting, learning, challenging, and innovating rather than clinging to beliefs.

APPENDIX 3

The Transhumanist Declaration

1. Humanity stands to be profoundly affected by science and technology in the future. We envision the possibility of broadening human potential by overcoming aging, cognitive shortcomings, involuntary suffering, and our confinement to planet Earth.
2. We believe that humanity's potential is still mostly unrealized. There are possible scenarios that lead to wonderful and exceedingly worthwhile enhanced human conditions.
3. We recognize that humanity faces serious risks, especially from the misuse of new technologies. There are possible realistic scenarios that lead to the loss of most, or even all, of what we hold

valuable. Some of these scenarios are drastic, others are subtle. Although all progress is change, not all change is progress.

4. Research effort needs to be invested into understanding these prospects. We need to carefully deliberate how best to reduce risks and expedite beneficial applications. We also need forums where people can constructively discuss what should be done, and a social order where responsible decisions can be implemented.
5. Reduction of existential risks, and development of means for the preservation of life and health, the alleviation of grave suffering, and the improvement of human foresight and wisdom should be pursued as urgent priorities, and heavily funded.
6. Policy making ought to be guided by responsible and inclusive moral vision, taking seriously both opportunities and risks, respecting autonomy, and individual rights, and showing solidarity with and concern for the interests and dignity of all people around the globe. We must also consider our moral responsibilities towards generations that will exist in the future.
7. We advocate the well-being of all sentience, including humans, non-human animals, and any future artificial intellects, modified life forms, or other intelligences to which technological and scientific advance may give rise.
8. We favour allowing individuals wide personal choice over how they enable their lives. This includes use of techniques that may be developed to assist memory, concentration, and mental energy; life extension therapies; reproductive choice technologies; cryonics procedures; and many other possible human modification and enhancement technologies.

Chapter 3
Technology Convergences to Enhance Education, Health, and Community Self-Sufficiency and Resilience

Don MacRae
HIAlba, Australia

ABSTRACT

This chapter considers existing and emerging technology convergences leading to advances in education and health as well as in other areas such as plant-based food production. Some advances will be enhanced by scientific research and technology directed to supporting life in outer space. A proposed revival of crofting and crofting communities throughout the highlands and islands of Scotland serves to illustrate how technological convergences and related growth of business could contribute to improved performance in the target areas as well as achieve high levels of self-sufficiency and post-COVID resilience of communities worldwide.

STAGING THE CONTEXT

Existing and emerging technological convergences in education and health make major contributions to achieving high levels of self-sufficiency and resilience of communities in the post-covid era. Further contributions to increasing self-sufficiency and resilience from advances in housing, employment and belonging through systems of governance that are politically decentralised, increasingly decarbonised both enabled by effective digitalisation are outlined by MacDonald & MacRae (2018) in *Maximising Scotland's Well-Being by Bravely Innovating*. Among other things research, education and training are achieving advances in plant-based food production resulting in benefits in health and healthcare and delivering myriad business opportunities. Some of these advances are resulting from scientific research and technology supporting life in outer space. Chapter 10 considers how existing AI advances could

DOI: 10.4018/978-1-7998-6772-2.ch003

lead to benevolent artificial general intelligence (AGI) through building trust-based social capital in education and health.

MacDonald & MacRae (2018) propose four types of communities which could have relevance worldwide:

1. Harnessing convergencies in technologies to sustain high levels of self-sufficiency and post-covid resilience in food production and nutrition, education, health, housing, employment and belonging of communities in remote, rural and regional areas.
2. "Imagination creates reality" (to quote Richard Wagner) when claiming the feasibility of establishing offshore communities:
    a. harnessing wind and wave power integrated with capacities to generate hydrogen-based power from greening offshore oil & gas installations at a scale to power the electricity supergrids of continents: supergrids relying on dispatchable electricity through breakthroughs in hydrogen storage and fuel-cell technology, supported by the emergence of graphene to enable superconducting of electricity for near zero transmission losses.
    b. integrating with the structures of power generation the growth and harvesting of seaweed at massive scales for the purposes considered later in this chapter.
    c. integrating with the structures of power generation massive scale applications of recently commercialised technologies for Direct Air Captur*e* of CO2 – see *IEA* (2020) – possibly powered by *Kite Power Systems* (2020) for bio-sequestered acceleration of seaweed growth for dually earned carbon credits.
    d. housing communities throughout the foregoing industrial infrastructure and in ocean-based "seascrapers" – *Shimizu Corp of Japan* (2020) has developed a concept for a subsea city for 5,000 inhabitants with claims of realisation by 2030.
3. Regional and suburban communities growing through the establishment and operation of a wide range of gigafactories to provide the machinery required by (1) and (2), including e-vehicles (cars, buses, trucks, trains and planes).
4. Cities focussed on:
    a. harnessing direct foreign investment to deliver the foregoing.
    b. manufacturing advanced products for local needs and exports, including ventilators, PPE, masks and other needs of ensuring effective healthcare resilience against pandemics, and renewable energy-based steel and metal making for, inter alia, specialised ship building to service (2).
    c. preparing to manage transitions to emerging revolutions in transportation, renewable energy generation, education, health and plant-based production from vertical farming and precision fermentation (both considered later) at scales which will increase markedly their self-sufficiency in nutrition and boosting post-covid resilience to deal with future pandemics.

The Case

A proposed revival of crofting – see *Crofting Commission* (2020) – and crofting communities throughout the Highlands & Islands of Scotland could foster the building of social capital supporting the creation of high levels of self-sufficiency and post-covid resilience of communities in remote, rural and regional areas worldwide. Crofting is characterised by its supportive communities, or "townships". There are over 20,000 individual crofts, typically established on 2–5 hectares for better quality forage, arable and vegetable production. Each community manages poorer quality hill ground as common grazing for cattle and sheep.

The key element of this revival is the establishment and operation of supporting eco-industrial parks (EIPs) fostering the symbiotic operation of SMEs and large companies integrated with the flow of products from nearby crofts and crofting communities. Sometimes referred to as industrial ecology or the circular economy, EIPs facilitate physical exchanges between enterprises, consisting of materials, energy, water, and by-products for collective competitive advantage.

In this case the enterprises of EIPs would primarily create software to facilitate exports of processed plant-based foods of communities in the remote, rural and regional target areas. These in turn lead to exports of enabling education, training and healthcare products and services from EIPs reflecting profound changes in the operation of education in the post-covid era. Experiences and attendant software for educational products would be drawn from sustainably growing speciality crops through convergences of technological breakthroughs. Advanced methods of producing and processing these crops for a wide range of novel, high-quality products and services could lead to myriad business opportunities including many start-up enterprises.

Technology Convergences in Education and Healthcare To Achieve Clean, High-Yield Crops

Growing high-yield speciality crops without pesticides, weed killers, and chemical fertilisers integrated with the processing operations of EIPs would stimulate many convergences of technologies. The global education reach, supported by experiences gained in the application of mixed reality technologies, AI and 5G, will be profound.

Cases are made for the growing of hemp, sugar beet, seaweed, and hazelnuts (harvested from extensive afforest developments on common land and in the provision of on-croft wind barriers). It is envisaged that crofts will also operate intensive plant-based permaculture providing nutrition for much larger numbers of on-crop workers, residents, and visitors. In a way permaculture operates at the micro-level of industrial ecology, facilitating physical exchanges between residuals from growing fruit, vegetables and crops and afforestation operations to increase collective productivity. Thus, lessons from plant-based permaculture could benefit the collective growth of crops. Other communities worldwide would focus on crops that could best be grown in their target areas.

A recent technology breakthrough has demonstrated that the application of rock dust to soils would markedly enhance these physical exchanges while carbonising, and re-mineralising the soil and promoting human health by providing mineral and nutrient dense food. Crofts certainly have the rocks. An EIP enterprise could reduce these to dust for on-croft dispersal by drones: an alternative to the use of drones for crop spraying – see Mazur (2016), "Six Ways Drones Are Revolutionizing Agricultur*e*". In many cases staff on-croft or in EIPs could be fulfilling, in whole or in part, the requirements of gaining professional and technical qualifications. The network of croft-based field stations could lead to the establishment and maintenance of databases served by sensors providing information on soil types and ameliorative responses, including carbonisation of varying levels and application quantities and frequencies of rock dust. Field station staff could not only operate drone delivery of rock dust but also contribute to initiatives to improve drone related technologies, including the optimisation of delivery programs.

A further recent advance in increasing soil health involves the application of fermented crop residues as organic matter while at the same time helping crofters in reducing their carbon footprint. Referred to as Bokashi, details of how to produce this potent natural fertiliser are provided by Gillbard (2020) in a recent issue of *Farmer's Weekly*. An extension of the proposed sensor based IoT network could be used to test for desired soil ameliorating outcomes throughout the process of fermentation. Databases for this as well as rock dust responses could cover all crops leading to the development of machine learning (ML) and deep learning (DL) algorithms to optimise treatments by both across soil types and conditions.

Add to the foregoing advances in robotics in agriculture, primarily for autonomous tillage, fertilising, planting, and harvesting of crops on massive scales. Just as there is a move to realising the many benefits of non-tillage cropping and regenerative farming – see *Regenerative International* (2016) article "Regenerative—Not 'Climate-Smart'—Agriculture Needed to Feed the World and Cool the Planet" - there is also a move to use robots in weed and pest control, obviating the myriad costs to human health and natural environments from the use of weed killers and pesticides. In the USA *Greenfield Robotics* (2020) have developed what they refer to as tiny autonomous mower robots which cause, after repeated mowing, even the most pernicious weeds give up. In the UK, the *Small Robot Company* (2020) "is reimagining farming with robotics and artificial intelligence". It has made significant progress in the development and commercialisation of three farmbots, "Tom", "Dick" and "Harry", to plant, feed and weed arable crops autonomously, with minimal waste. Weeding is done, not by mowing, but by electric zapping of the weed roots. The AI facility" Wilma" will oversight the interplay between these working robots, especially through efforts to collect and analyse data on soil conditions, terrain problems and equipment performance to optimise productivity.

All this foreshadows the increasing application of AI in agriculture and in the case of crofting to facilitate ever more cognitively enhanced dialogue through computation and augmented reality between field station staff and EIP professionals and technicians based in tertiary education campuses, SMEs, and companies. This could involve managing convergences of:

1. the databases arising from measurements made by sensor based IoTs managed by field stations.
2. ML to assist in optimising applications of rock sand and Bokashi and other crop management and permaculture responses under investigation.
3. networking with IoT sensor-based data collected from field operations with medium to long term prospects for using DL to probe for system-wide insights across many EIP-field stations operating worldwide.

Clearly there would be myriad opportunities for considerable beneficial interactions between education at all levels from primary schooling to post-doctoral endeavour with the challenge of achieving nutritional self-sufficiency in crofting communities and beyond to the realisation of similar outcomes worldwide. It would also become evident that fundamentally favourable health and healthcare outcomes would arise from the opportunities to enjoy plant-based foods free of pesticides, chemical fertilisers, and weed killers.

Technology Drivers To Maximise Benefits From Interplay of Speciality Crops

Strong cases exist for growing the specialty crofting crops of hemp, sugar beet, seaweed, plant-based permaculture of fruit and vegetables and afforestation of hazelnuts.

Hemp

Recent years have seen rapid increases in growing hemp for a wide range of purposes, including:

1. so called superfoods, including nutritious seeds, high protein flours, and oils.
2. medicinal products such as CBD from the root of hemp, used to treat arthritis, asthma, coughs, bloating, syphilis, and heart problems including valvular insufficiency, "senile heart," and weak heart muscles following pneumonia.
3. Bioplastics from the by-product of hemp-CBD operations.
4. hurds/shives making up to 70% of plant biomass as an input to hempcrete for 3D printing of buildings.
5. as an input with other crop residues such as sugar beet pulp and residues from on-croft plant-based permaculture for Bokashi fermentation into clean, high potency fertilisers.

Hemp is a superlative growth crop providing many farming and environmental benefits:

(1) produces nearly four times as much raw fibre as an equivalent sized tree plantation.
(2) trees take about 20 years to mature… hemp 14 weeks.
(3) needs no pesticides as it is unpalatable to insects.
(4) needs no herbicides because it grows too quickly for any weed to compete.
(5) grows in virtually any climate and soil condition, and is excellent for reclaiming otherwise unusable land, rejuvenating poor soils.
(6) between 1.7 – 1.9 tonnes of CO_2 are absorbed from the atmosphere for each tonne of hemp cellulose produced (typically 10 tonnes of hemp straw per hectare grown).

Business Opportunities From Hemp

1. Processing of hemp into products for use as a fertiliser.
2. Large range of food products from hemp seeds – seeds as addition to many dishes; high protein powders, hemp flours, hemp oil as healthy alternative to so-called vegetable oils (either low in nutrition or unhealthy), hemp milk, hemp cheese substitutes.
3. CBD – evidence of claimed benefits in treating a wide range of health concerns is strong only for the treatment of epilepsy in children – *FDA* (2018) recently approved the first ever cannabis-derived medicine. Research strengthening the evidence for treating other concerns will greatly increase the revenue generated by the wide range of treatments already in existence.

Sugar Beet

In relation to sugar beet, the processing of 70kg of mash from a tonne of plant to extract sugars could provide the feedstock for the revolutionary plant-based food technology of precision fermentation (PF) considered below. The Scottish company *Cellucomp Ltd* (https://www.cellucomp.com/) is developing new "wonder" materials using the efficient extraction of nanocellulose particles from sugar beet residuals. Of particular interest is the use in manufacturing superlight composites which could greatly reduce the cost of manufacturing windmill blades, the overall cost of the installation of windmill assemblies and marked reductions in the costs of their decommissioning. Similar to hempcrete, nanocellulose particles can be added to concrete to improve the effectiveness of 3D printing of buildings (Ahmat et al, 2019).

Business Opportunities From Sugar Beet

1. Processing of Sugar beet into products for use as a fertiliser, raw materials for EIP-based precision fermentation of a wide range of highly nutritious foods, as an additive to concrete for 3D house printing.

Seaweed

The harvesting of seaweed as a food and fertiliser has been realised for centuries if not millennia. The *FAO* (n.d.) has provided an *Introduction to Commercial Seaweeds*. The market revenue of commercial seaweed is projected to surpass US$ 85 billion by 2026, achieving a projected CGAR of more than 12% between 2020 and 2026. This will be mainly due to rising consumption as a food ingredient as well as increasing adoption projected for pharmaceutical and personal care products. The raw product is primarily from aquaculture as controlling quality from wild harvesting is less reliable. Rich in minerals and vitamins strengthens the case for using seaweed as a source of hydrocolloids, such as agar, alginate, and carrageenan. As these hydrocolloids provide several functional and health benefits, they are increasingly finding diversified applications in food & beverages, pharmaceuticals, dietary supplements, bio-refining, and wastewater treatment.

The advent of vast arrays of floating wind farms holds great potential to co-establish and operate seaweed farms to sustain industrial scale production of many non-food products such as proven biodegradable bioplastics (Israel 21c, 2019) for the same range of uses as fossil fuel based plastics, including food & drink packaging *Australian Seaweed Institute* (n.d.), leather (Sidhu, 2020), biofuels (Carrington, 2013), cosmetics (Jesimani et al, 2019), nanotechnology (PHS.ORG, 2018) to produce, *inter alia*, low cost solar cells, myriad fertiliser applications, and revolutionary feedstock for cattle to reduce to near zero their methane excretions (Kesteven, 2016).

For centuries Japan has cultivated various seaweed at massive scales as a food. The tonnage of nori alone amounts to over 350,000 tonnes per annum. With a value of about US$2B this is one of the world's most valuable crops produced by aquaculture.

Crofts and crofting communities and for that matter most of the target region communities with coastal access would be the front-line players in delivering the national benefits of a ten-fold growth of the seaweed industry to about $60B pa.

Business Opportunities From Seaweed

Plant-based food production and processing seaweed could result in an explosion of opportunities:

1. Processing of seaweed into products for use as a fertiliser.
2. Nori and myriad edible seaweed products packaged in seaweed plastic alternatives.
3. Caviar – several companies are producing black and red vegan pearls.
4. Seaweed-based craft beers, as direct ingredient and via seaweed fertilised barley.
5. Wide range of snack foods selling with craft beer.

6. Increasing range of vegan cheese alternatives using kelp-extracted agar and carrageenan.
7. Seaweed oil as a nutritious alternative to most of the suspect vegetable oils in a wide range of processed vegan foods.
8. Kelp – staple in many Asian cuisines as a source of vitamins, minerals, and antioxidants.
9. Cosmetics – research on kelp sources for natural cosmetics (Utroske, 2016).
   (10) Tech advances: boosting performance of superconductors, li-ion batteries, and fuel cells (*Nanowerk*, 2017).
   (11) Kelp extracts to produce low-cost solar cells (Anand & Suresh, 2015).

Intensive Permaculture of Plant-Based Foods

The *Vegan Organic Network* (n.d.) provides accounts of plant-based permaculture free of domestic animal husbandry. The literature of practitioners from their e-shop explicates the art and the science involved. Integration of the latest advances in grey and black water recycling from on-croft residences and associated composting with rock dust treatment and Bokashi fertilisers could further enhance outcomes. These include realising the benefits of on-croft accelerated fruit tree growing for food and wind shelters to maximise the growth of a very wide range of organic vegetables, berries, and herbs with greenhouses in the mix for colder seasons.

Afforestation

The common lands of crofting communities provide opportunities for afforestation: the establishment of a forest or stand of trees (on-croft for food and windbreaks to enhance the productivity of plant-based permaculture) in areas where there was no previous tree cover. Fertilisation could be based on Bokashi fermentation using cleared grasses among other inputs such as crop and permaculture residues and seaweed. Forest crop yields could be enhanced by drone application of rock dust prior to planting. UK company *Dendra Systems*, formerly *BioCarbon Engineering*, seeks to help restore the world's forests using seed-planting drones and AI enabling software (Graham, 2020). Drones are first used to determine the most suitable locations for planting by assessing the soil conditions of a specified terrain. They then fire into the ground biodegradable pods filled with a germinated seed and nutrients. Two operators of ten drones can plant 400,000 trees per day, a billion in less than a year.

It is likely that native trees providing a valuable crop such as hazelnut would be preferred as this would enhance the maintenance and enhance the carbon sequestration capacities of peatlands dispersed throughout plantations of orchard-woodlands. Writing in *Nature*, Leifeld & Menichetti (2018) report on the underappreciated potential of peatlands in global climate change mitigation strategies. The *European Commission* (2018) funded Pantheon project is developing a robotic system to increase the productivity as well as sustainability of hazelnut farming. The robots traverse ground and swarm in the air through hazelnut orchards gathering data on trees, including the presence of pests and diseases and water levels, which is transmitted to field stations. This facilitates automatic responses such as adjusting irrigation systems, and supporting decision-making by farmers and agronomists. Robots can also perform measurements of tree and crop growth. Precision farming techniques minimise the application of chemicals and maximise water conservation, thereby increasing the yield of hazelnut crops and deploying workers more efficiently. Truffles being most often found near hazelnut trees poses the question of the likely efficacy of adding truffle spores to the hazelnut seed pods for planting by drones. And for good measure adding spores to the Bokashi fertiliser mix? And the likely emulation of truffle sniffer dogs by robots? And extending the bots to truffle digging even where root structures pose difficulties? All on a 24x7 timetable? (Harrison, 2019).

Hazel nutshells have been processed to produce low-cost activated carbon for use in the pharmaceutical industry, primarily as an adsorbent to remove pollutants during the drug manufacturing process. Further, could hazelnut twigs, bark and other residues be compacted into workable timber materials such as the high-quality engineered products from forest residues by *TimTek* (n.d.)?

Convergences in Existing and Emerging Food Production Technologies

Advances in existing and emerging food production technologies extend beyond contributing to meeting the self-sufficiency needs of the target regions to not only urban and conurbation communities but also to envisaged communities in outer space-stations, the Moon, Mars and beyond. Technologies include vertical farming, plant-based meat substitutes, cell-based meat, 3D food printing, and precision fermentation.

Vertical Farming

The case for vertical farming (VF), throughout conurbations both on Earth and on other planets, is gathering momentum on several fronts. *AeroFarms* (www.aerofarms.com) claim that it has achieved plant growth that is 390 times greater per square foot than on a traditional field. *VertiCrop* (www.verticrop.com) has developed a hydroponic technology with the claim of 20 times more yield than open field agriculture.

Offshore living could also benefit from *in situ* food production of plant-based foods by extending initiatives of the Top 25 vertical farming companies (Tao, 2019). The Scottish company *Intelligent Growth Solutions* (www.intelligentgrowthsolutions.com) has developed some of the most advanced enabling technologies to date. The relevance to space exploration is spelt out by Cuello (2019) in the article "One Giant Leap: Space Programs Spin Off Vertical Farming".

Indeed, the more this space-related research advances in the target communities, probably centred in their eco-industrial parks, the better for its potential applications in outer space. The convergence of the VF industry and precision fermentation advances will likely enable NASA and other space programs to develop effective food production systems on the Moon, Mars and beyond.

Precision Fermentation

The claims for precision fermentation (PF) are that it has the potential to deliver affordable plant-based, pure foods of the highest nutritional value with desirable textures and structures for humans and ever decreasing costs of production (Tubb & Seba, 2019). It has the potential to heighten the reality of sustaining human life in outer space. In this case, the sugar feedstock for PF processes could be achieved from aeroponic growth of fruits with high sugar content. Aeroponics is also likely to be the principal means of producing fresh plant-based food in outer space, complementing wide-ranging PF food choices.

3D Printed Plant-Based Food

3D food printing is the process of manufacturing food products using a variety of additive manufacturing techniques. Most commonly, food grade syringes hold the printing material, which is then deposited through a food grade nozzle layer by layer. Currently there is a small number of professional-grade 3D food printers, mostly being used for catering, events, and restaurants. However, they are beginning to impact on food choices using a fraction of the resources that the food industry demands. One of the most interesting emerging trends revolves around meat-free steaks, burgers, poultry and pork products. Many ventures are undertaking the difficult task of printing plant-based foods that mimic the texture, colour, and taste of real meat products. Fish, however, has been difficult to imitate, especially its characteristic tender flaky texture, as well as its rather full, strong, sometimes salty flavour. A few start-ups companies have ventured into printed vegan sushi, but now, a new start-up in Europe has recently 3D printed plant-based salmon Listek (2020) and tuna fillets as part of a broader aim to reinvent plant-based seafood.

Technological Convergences Enabling Education With A Global Reach

The educational experiences and outcomes of crofting community residents could be enriched by the proposed on-croft activities and the interactions with eco-industrial parks (EIPs). This could traverse all levels of education from pre-school to post-doctoral research

In *Bravely Innovating* 25 new towns are proposed as extensions of existing towns throughout the Highlands & Islands, all comprising a major EIP operating to support crofts and crofting communities (MacDonald & MacRae, 2019). The question is asked: Could technology support the operation of a campus of the existing University of the Highlands (UHI) in each of the proposed 25 new towns as they reach residential populations of 5,000 to 10,000? Can we envisage a distributed university with a student body of 200,000 using distributed ledger technology (DLT) platforms designed to allocate and network resources to teaching at all campuses? This sounds like a massive increase to the present UHI student body of about 10,000 but would still be an order of magnitude less than the largest in-person student bodies in the world.

By 2030 UNESCO estimates that 69 million teachers will be needed, as 263 million children worldwide currently lack access to basic education.

Convergences in Spatial Technologies Delivering Educational and Training Outcomes

The increasing advent of the spatial technologies of Augmented Reality (AR) and Virtual Reality (VR) in education via head mounted displays could accelerate solutions to close the education and training gap between developing and developed nations. With AR integrating individuals with their physical world and VR immersing them in virtual worlds, *The Future Is Faster Than You Think* reports:

...when AI and VR converge with wireless 5G networks, our global education problem moves from the nearly impossible challenge of recruiting teachers and funding schools for the hundreds of millions in need, to the much more manageable puzzle of building a fantastic virtual educational system that we can give away for free to anyone with a headset. It's quality and quantity on demand.

The *Alliance for Affordable Internet* (www.a4ai.org) is working to achieve such outcomes. As a global coalition it is working to drive down the cost of internet access in low- and middle-income countries through policy and regulatory reform. *SpaceX* through its *Starlink* satellites program (www.starlink.com) intends to provide satellite internet connectivity to underserved areas of the planet funded by projected returns from large-scale commercial operations. *Starlink's* satellite internet constellation will consist of thousands of mass-produced small satellites working in combination with ground transceivers.

AR by overlaying digitally created content with real world observation and supporting users to interact with both the real world and digital elements or augmentations is resulting in many applications for industry and business, including boosting efficiency, enhancing training, and keeping workers safe. VR now fulfilling its hype with better, and less expensive, technology, provides business with a near limitless technology for exploring possible futures, products, and services. Applications include enhancing croft-based field work, effective training for surgery in target area communities, and in the case of military combat, effective where hazards and difficulties inhibit iterative 'live' training exercises. Increased retention outcomes are achieved, by learning 'through doing' to build skills such as dispute resolution, public speaking, and corporate communication. Mixed Reality (MR) seamlessly blends the user's real-world environment and digitally created content, through the coexistence and interaction of AR and VR environments. Holographic Reality (HR), while less developed, facilitates interactions with virtual characters and worlds and will likely surpass the cumbersome headset-bound reach of AR and VR.

Near term HR transformations are anticipated in communication, 3D design, advertising, the memories industry, and entertainment, as well as education, medical imaging, and drug design. The capacity to represent anything real or digital (objects, people, etc.) with extremely high three-dimensional fidelity will have to await markedly decreased costs of the immense computational power required to manipulate the vast data needs of very high pixel densities. Perhaps quantum computers will deliver. With such a cost-effective advance, HR will make it possible to conduct business meetings, conference gatherings, and group interactions where the holographic image of all participants, regardless of global location, are projected into a common conference room where all are holographically present. The ensuing HR disruptions will be much more profound than the considerable covid disruptions which are nearly zeroing international business travel, markedly reducing commuting to and from work, and video conferencing to connect those who can work from anywhere. A major step in this direction has been forged by *Nvidia* (2020) in its release of *Maxine* – a cloud-AI-based video-streaming platform better connecting millions working and studying remotely. *Maxine* provides developers with a suite of GPU-accelerated AI conferencing software to enhance video quality – the internet's main source of traffic. The platform is a "cloud-native" solution allowing service providers to harness new AI effects — including gaze correction, super-resolution, noise cancellation, face relighting, and more — for end users, including participants in the 30 million web meetings estimated to take place every day.

For the past five years *NASA* (2015) and Microsoft have been collaborating on enhancing the autonomy of astronauts on the International Space Station by developing the holographic capacities of HoloLens, a mixed reality headset. Referred to as "Sidekick" the technology is being used by astronauts to conduct ever more complex science and exploratory investigations. In the past year, the technology has advanced such that HoloLens 2 is helping to build NASA's Orion Spacecraft (Langston 2020). Technicians have no need to refer to computer screens and paper instructions in conducting their assembly tasks. The mixed reality headsets make visible everything from animations of fittings to engineering drawings to torque values with repetitive tasks completed 90% faster and all but eliminating assembly mistakes. Not only will the autonomy of astronauts be further enhanced in meeting the challenges of projected Lunar and Mars explorations and eventually settlements, but Microsoft is now applying HoloLens 2 to meet wide-ranging challenges of industry and business.

Replacing old hard disk drive technology still used for cloud storage by holographic storage is being investigated by Microsoft as an old yet new idea of using holograms to store data in a much smaller physical space (Thomsen et al, 2020). Recent research indicates that the black holes of outer space concentrate enormous amounts of information akin to a massive hard disk, likening black holes to holograms *SISSA* (2020). More recent research is now investigating the proposition that black holes are holograms within holograms.

The prospect of quantum computation (see Chapter 6) delivering cost-effective HR may be reciprocated by the prospect of a nanoscale holographic technique opening the way for quantum computation *ScienceDaily* (2019). From such nanoscale holography to the increasing reach of HR in business, education, health etc., science envisions no boundaries as increasing effort is expended on the question of whether humans exist in a hologram conceivably encompassing the universe Goodman (2018). XR technology-mediated experiences combine virtual and real-world environments and realities. Here the 'X' is a placeholder for V(R), A(R), M(R), and H(R) though it also represents an undefined or variable quality/quantity. *VRChat* (www.vrchat.com) apps support the spatial creation, publication, and exploration of virtual worlds with other individuals and communities of interest worldwide. Head mounted displays to deliver the foregoing include: Google Daydream *(standalone VR)*, HP Reverb *(tethered VR)*, HTC Vive, Magic Leap One, Microsoft HoloLens (MR), Oculus Quest, Oculus Rift, Samsung Gear VR, Samsung Odyssey, Sony Playstation VR, and Valve Index. The global AR and VR market size was US$ 11B in 2017. It is projected to reach US$ 571B by 2025, growing at a CAGR of 63% from 2018 to 2025.

Spatial Networks

There are no boundaries to the potential of spatial networks. While social X(R) pushes the boundaries of communications there is a significant difference between media engagement and real-time networking. Interacting with digital media feeds in virtual spaces is engaging with real people. In this sense social is the individual not the media. So spatial networking entails the engagement part of social media, in effect connecting with other people and communities. With X(R) technologies participants are spatially immersed in an experience. Such spatial networks unlock a whole new perspective on connected learning and real-world engagement.

The organisation *Educators in VR* (www.educatorsinvr.com) are focussed on preparing for a future rich in X(R) immersive technologies, particularly its integration into the workforce. It has conducted many wide-ranging projects, led by experts in their field of study and designed to bring together others with deep interest in attendant topics supported by the spatial technologies and industries represented. They have projects on language learning arts, virtual coaching and self-development, math, science, cultural heritage preservation, and much more.

Most spatial networks and immersive tools are cross-platform, thus available on several different X(R) devices. Many of the existing platforms run on traditional devices allowing participants, to join using PC/Mac computers and mobile phones (Android/iOS).

As immersive pioneers revolutionizing remote education with interactive experiences in social VR, *Educators in VR* are preparing teachers and academic institutions for future incorporation of virtual spaces and activities into the classroom. The community provides educators and learners with a virtual classroom, allowing them to connect with a wider market of students and their peers. Social X(R)s free and open access is ideal for schools, educators, consultants, trainers, and entrepreneurs. In schools there is great enthusiasm according to co-founders Daniel Dyboski-Bryant and Lorelle VanFossen:

VR keeps students locked in their chairs. They want to engage. They want to learn. They can't get enough of this, and it's fascinating and so exciting as a teacher to see this new passion for learning.

It's taken off in ways we never dreamed possible. We read the research. We see the studies. We talk to the educators integrating VR into the classroom and they are living the proof that VR is changing education.

Accessing Free World Class Courseware Through Spatial Networking

Consider now the development of spatial networks securely enabled by distributed ledger technology (DLT) platforms to access and engage creatively with free high-quality courses being developed for interactive e-learning and to facilitate efforts to contribute to the growing store of such free courses. Examples include the 2,400 online undergraduate and post-graduate courses from MIT's *OpenCourse-Ware* (www.ocw.mit.edu) with inspirational stories from Students, Educators and Self-Learners. OCW's *Unlocking Knowledge* program is a web-based publication of virtually all MIT course content. It is open and available to the world and is a permanent MIT activity. Through OCW's *Empowering Minds* program educators improve courses and curricula, making their schools more effective; students find additional resources to help them succeed; and independent learners enrich their lives and use the content to tackle some of the world's most difficult challenges, including sustainable development, climate change, and cancer eradication.

No prizes for guessing that of the top 10 or so MIT undergraduate courses most visited recently four were computer based – introductory courses in Computer Science and Programming, Algorithms, Artificial Intelligence, and Computational Thinking and Data Science – together with a course on the Principles of Microeconomics. In a similar vein *WooTube* (www.misterwootube.com) has posted an award-winning series of videos on high school certificate courses in mathematics and information technology for students and teachers. The IT courses include: Decision Support Systems, Communications, Databases, Multimedia, and Assorted IPT Video.

Also of growing interest are courses in GPU-accelerated computing. GPU computing is the use of a *graphics processing unit* as a co-processor to accelerate *central processing units* (CPUs) for general-purpose scientific and engineering computing. In this respect, *NVIDIA Teaching Kits* (www.developer.nvidia.com/teaching-kits) are complete course solutions for use by educators in a variety of academic disciplines that benefit from GPU-accelerated computing. The kits provide full curriculum design coupled with ease-of-use. As an example, educators can bridge academic theory with real-world application to empower next-generation innovators with critical artificial intelligence (AI) skillsets. Co-developed with leading university faculties, initiatives include:

1. The Deep Learning Institute – kit covers introductory and advanced deep learning topics.
2. Accelerated Computing – kit covers accelerated parallel computing topics.
3. Robotics Teaching – kit covers introductory and advanced multidisciplinary topics.

Another commercial education service worth noting is provided by the CGI, a global business with 78,000 professionals in hundreds of locations across the Americas, Asia-Pacific, and Europe. A provider of end-to-end IT and business process services facilitating the ongoing evolution of the businesses of their clients, CGI is also a global leader in creating digital content and tools for primary, secondary, and tertiary schools. *CGI* (n.d.), through their "Empowered Learning" program, offers every student access to powerful digital learning tools. Digital transformation in higher education is viewed as more than a technology solution, "it is a cultural change that encompasses people, process and technology". In effect students are provided with digital and innovative environments to develop skills and tools enriching the transition from education to working environment.

Technological Convergences Extending The Global Reach of Health and Healthcare

Digital health will progressively bring state-of-the-art healthcare and medicine to the crofts and crofting communities in the target regions from an eco-industrial park base. This will be accelerated as 5G extends the reach of telehealth to conducting remotely controlled robotised surgery. And a future where drones support the delivery of medicines and defibrillators (Wong 2017) by operating as full-sized drone ambulances *APCollege* (n.d.) with wearable MRI/CT scanning devices and 5G in healthcare (Stracuzzi, 2020) are all common place worldwide. New technologies utilising light and sound to track tumours and measure neural activity may replace the CT and MRI scanners with cheaper, more efficient and wearable systems (Jepsen, 2018). To further boost the self-sufficiency and self-reliance of communities recent advances in 3D printing increases the feasibility the of producing masks, PPE equipment and ventilators at local EIPs: 3D Printing Community Responds to Covid-19 and Coronavirus Resources (Petch, 2020).

Seminal Contribution of Distributed Ledger Technologies To Health and Healthcare

This section considers applications of Distributed Ledger Technology (DLT) to delivering various health and healthcare services as follows:

1. DLT Acceleration of Advances in Health Services and Genomics
2. DLT-Based Application to Delivering Ultra-Secure Personalised Healthcare
3. DLT and Destressing Medical Practitioners
4. DLT-Based Application to Assist in Coping with Depression and Suicide
5. Fast-tracking Training in Distributed Ledger Technology and Blockchain Coding
6. Other DLT Uses in Healthcare

DLT Acceleration of Advances in Health Services and Genomics

Distributed Ledger Technology (DLT) is providing opportunities to revolutionise health services by "de-siloising" operations for increased system efficiency and effectiveness while fulfilling the dream of enabling patients to access securely their own health records. This is being explored and tested by the UK National Health Service – *Can Blockchain save the NHS?* (Unblocked, 2018) – largely based on tech savvy doctors and the recommendations of the report of the *Government Office for Science* (2016): *Distributed Ledger Technology: beyond block chain.* At the leading edge of delivering core capacities stands MIT's research project MedRec (n.d.) trialling a records system to manage the complexities (including secure patient access) of multi-institutional, lifetime medical records.

Epidemiological campaigns at national and international scales and attendant drug design, production and timely delivery also stand to benefit from advances in DLT: Global Public Health: Blockchain for Contagious Disease Relief.

Central to the de-siloising advances in the worlds of medical research is 'Permissioned Blockchains' configured to control participant transactions, define their roles to access and contribute to maintaining their identity. Of even greater relevance is the capacity for participants to allow databases they control on a prescribed DLT platform operating in the public domain to be used for calculation and scenario simulation purposes whilst keeping the data private even in public networks, once again through an MIT-led advance – "SNARKs for C: Verifying Program Executions Succinctly and in Zero Knowledge" (Ben-Sasson et al, 2013). Given the availability of this capacity in the early 2000s, particularly DLT-enabled authentic IP ownership and hyper-secure protocols, it is likely that CRISPR as the leading advance in molecular cell engineering would be several years ahead of where it is today: in fact, by the time a panel of judges at the US Patent and Trademark Office gets around to finally deciding on patent ownership, CRISPR may have lost its preeminent title as the most precise tool for genetic engineering.

The SNARK enabled Blockchain as a Service (BaaS), specifically a scenario simulation service, has already been commercialised by the NASDAQ listed company *SimulationsPlus* with a market capital of US$ 1.27B. Of relevance here is the recent "Large-scale investigation of the reasons why potentially important genes are ignored" (Stoeger et al, 2018). With research to date concentrating on about 2,000 of the 19,000 genes in the human genome, gene-specific strategies resulting from this investigation may assist in identifying significant but ignored genes thereby probing novel directions of investigation. Clearly, this could be accelerated by applying AI (algorithms derived from gene-specific strategies) to Big Data sets being generated from machine/deep learning probes at multiple investigation sites connected by SNARK-based BaaS-DLT. This could also accelerate the vast number of research efforts in stem cell research directed to the discovery of commercial drug treatments. For example, the award-winning Scottish company *Cytochroma* (www.cytochroma.uk) confirms that "drug discovery is becoming more expensive and ineffective due to the lack of models and tests to predict toxicity".

DLT-Based Application To Delivering Ultra-Secure Personalised Healthcare

Effective access to comprehensive patient data through electronic health records (EHR) systems has had to deal with records fragmented across many healthcare stakeholders, often due to various providers using different EHR systems, attendant siloed data holdings, and concerns with the heightened risk of security breaches by centralising these systems.

Blockchain in Healthcare: A Data-Centric Perspective provides an account of the how the management of healthcare records can be greatly improved by applying three of the major strengths of distributed ledger technologies such as blockchain:

1. Ultra-secure and immutable store of information.
2. Decentralisation of transactions.
3. Built-in gamification – application of typical elements of game playing, typically as an online marketing technique to encourage engagement with a product or service.

Descriptions are provided of the application of these strengths to areas of great promise that include:

1. Personal Health Record (PHR).
2. Population Health.
3. Personalised Medicine.

Personal Health Record (PHR)

The PHR is seen as the key building block of the healthcare blockchain ecosystem with many efforts in existence to develop and popularize various versions since the early 2000s in the face of inadequate security protocols and interoperability problems curtailing its adoption. The *Blockchain in Healthcare* paper details the ways in which DLT is well placed to overcome both these drawbacks and accelerate the transition to patient-centric healthcare by putting the patient and the patient's data at the centre of the healthcare system.

Population Health

Population health data is medical information that is specific to a particular demographic. Blockchain provides strong solutions for population health management facing challenges that include low data security, share-ability, and interoperability due to the siloing of patient information in insecure, non-compatible EHR systems that cannot easily exchange information. As a result, there is a scarcity of usable population health data sets across diverse patient populations.

Personalized Medicine

DLT-blockchain can mainstream personalized medicine. Personalized medicine is tailored to an individual based on their health data including medical history and genetics. Because blockchain can facilitate ready access to a person's PHR and to relevant population health data, healthcare providers may soon provide routine individualised treatment.

Consider a future in which family doctors upload the results of using advanced scanning technology to diagnose heart failure problems with 100% accuracy instead of the reported *35% to 50% today without this equipment.* Further consider CT and MRI scanning technology that is wearable allowing the family doctor to upload vital information on the condition of all a patient's organs aided by AI-ML algorithms to produce 100% accurate diagnosis. It does this by utilising light and sound to track tumours and measure neural activity replacing CT and MRI scanners with cheaper, more efficient wearable systems, conceivably within the next five years. In time this wearable technology could be of sufficiently low cost that patients at high risk could wear it away from the surgery and be connected online to their family doctor.

Now add to this the information that is being ascertained from medical advances in genomics and we start to build data of immense value to the patient per se and possibly a financial return from releasing it to the population health data holding.

Destressing Medical Practitioners

Perhaps the key characteristic of best practice hospitals and their networks of associated general and specialist medical practitioners is a cohort of doctors that is supported in maintaining and strengthening their initial reasons and passions for a career in medicine.

It is conjectured that worldwide only the top 1-2% of school matriculants will gain entry to medical school with the top <1% for leading institutions. Does this typically lead to a significant cohort of under-utilisation of a nation's intellectual horsepower? Are too many medical practitioners insufficiently challenged intellectually unless they take the medical research pathway? Is this likely to intensify as auto-diagnosis proves superior to even the best of practitioners and specialists? Will this lead to increasing the prevalence of chronic stress – "Prevalence of chronic stress in general practitioners and practice assistants: Personal, practice and regional characteristics?" (Viehmann, 2017) – and burned-out doctors?

To existing countervailing strategies, consideration may be given to adding the following activities designed to engage medical practitioners individually and in groups located anywhere on the planet through a secure DLT-based network. Headlining this initiative could be the expression of vignettes on the history/emergence of medical based movements such as the heroic global reach of MSF, or the Flying Doctor Service in Australia, the near miraculous containment of potential pandemics (e.g., recent heroics in arresting spread of Ebola), case studies of the most effective practices in containing Covid-19 and so on.

Suggested activities include how doctors individually and collectively:

1. achieve the near impossible against intransigent bureaucracies.
2. specify medical research initiatives - with all manner of success stories from the past.
3. build websites, blogs, online fora to disseminate insights and success stories.
4. thrive in remote communities by communicating purposefully with peers (e.g., through initiatives like the Centre for Rural Emergency Medicine - www.ruralemergencymedicine.org) and communicating with others by participating in all manner of communities-of-interest such as indicated in the next point.
5. disseminate/share interests in books, travel, wine, nutritious food, music, art, photography, etc, again with case studies on how this has already been done to a level of considerable effect: e.g., in nutrition by Dr Gregor's NutritionFacts (www.nutritionfacts.org).
6. start up non-medically focussed charities and/or businesses (perhaps medical spin-off businesses) with examples of how this has been done by medical practitioners worldwide.
7. for the tech savvy, participate in the sorts of initiatives outlined earlier in the section on *Distributed Ledger Technology Acceleration of Advances in Health Services and Genomics.*

DLT-Based Application To Assist in Coping With Depression and Suicide

Worldwide far too many suicides are the ultimate expression of depression. Globally about 800,000 people take their lives every year, about twice the homicide rate. As reported in *The Scotsman* (2018) there were 784 suicides in Scotland in 2018, three times more for males than females. A change of policy is now proposed to examine cases beyond those who had been in contact with health and social services to identify treatment mistakes, including the examination of cases where no apparent blame could be attributed. Strategies to reduce suicides by 20% by 2022 are focussed on widening opportunities to enhance access by sufferers to opening up to their friends, family and confidential helplines such as the Samaritans who hold the view that most "who feel suicidal do not actually want to die; they do not want to live the life they have" and while this may seem to be a small distinction it is critical "and why talking through other options at the right time is so vital".

One outcome of such interventions could be the vast DLT-based accumulation of preventative case histories, released with the consent of individuals, by remote, regional and city locations within nations that will become available for AI-based Big Data analysis. Among other things this will increasingly enhance the sophistication of machine-based conversing.

Other DLT Uses In Healthcare

Other DLT uses for the healthcare industry include:

(1) Beneficial behaviour modification programs (fitness, diet, vice management); (2) Medical products supply chain tracking; (3) Medication tracking and provenance; (4) Medical treatment audit trail; (5) Medical billing audit trail; (6) Patient consent management; and (7) Virtual clinical trial management.

Fast-Tracking Training in Distributed Ledger Technology and Blockchain Coding

Since a great deal of the potential outlined in the following is based on applications of DLT the case can be made to fast-track training in DLT-blockchain coding. This article from *The Conversation* and reprinted in *Scientific American* indicates how far advanced Scandinavia and Estonia are in the health-care applications of DLT-blockchain (Rutschman, 2018). A case could be made for increased efforts to establish worldwide training centres in DLT-blockchain coding. Relevant MIT-OCW courses could be drawn on for this purpose, especially where e-learning components are necessary.

Spatial Networking in Health

The medical field already utilizes several technologies that provide valuable information to doctors, surgeons, and nurses, making it one of the best use cases for X(R) technological convergences. With X(R)-enabled devices, this information can be applied more effectively than ever before.

Mixed reality on HoloLens 2 is changing the game for the health industry – from planning and performing live surgeries, to seeing ultrasounds in 3D, to training doctors, surgeons, and nurses to learn and understand human anatomy in new ways. By blending digital objects as 3D holograms in the physical world, HoloLens 2 allows health professionals to experience content with the context, format and people that make it most relevant. HoloLens 2 combines an untethered device with apps and solutions that help the medical community learn, work, communicate, and collaborate more effectively to produce better patient outcomes. It is the culmination of breakthroughs in hardware design, AI, and mixed reality development. The convergence of cloud computing and AI on a 5G network is being heralded by the software giants with unprecedented collaborative forays into digital health. Microsoft with its state-of-the-art MR technology HoloLens 2 is partnering with 5G provider Verizon and Medivis a provider of real-time immersive visualization capacities for the exploration of volumetric medical data and images, such as computed tomography (CT) scans.

In speaking about 'new services', CEO Tim Cook says that Apple's 'greatest contribution to mankind' will be health-related. An estimated increase in services to 40% of their business that could lead to a massive investment in healthcare services in addition to Apple's foray into hardware for wearables currently oriented to building fitness and collecting data supporting the massive global drive into personalised healthcare. This is being enabled by wearables performing a wide range of scans, including CT (computed tomography), MRI (magnetic resonance imaging) and PET (positron emission tomography). This is attracting serious investment from major engineering enterprises such as GE Healthcare, Siemens Healthineer and Philips Healthcare. Partnering with such engineering powerhouses would likely result not only in further opportunities for Apple to design massively winning products but also to be at the forefront of contributing to the required technological advances, especially in harnessing the full potential of convergences with AI and 5G.

In China, *Alibaba* (2020) has developed "CT Imaging Analytics" under their "Fighting Covid-19 Technology Center", a global initiative to support disease control centres and genome research institutions worldwide in mitigating the risks associated with this pandemic. Given the typical scan time of 5-15 minutes, the world's CT imaging departments are struggling to meet the rising demands of the Covid-19 crisis. Add to this the fatigue and human errors caused by the long stressful hours worked by front-line staff dedicated to meeting escalating demands. Hence the need to conduct rapid CT imaging analytics. Using AI-enabled algorithms, the resulting technology can predict the probability of Covid-19 pneumonia with an accuracy of 96%. About 60 times faster than human detection methods, taking between three and four seconds to run each test and transmit the data, this technology can also estimate the proportion of pulmonary lesions.

A team of New York University scientists has developed a method using holographic imaging to detect both viruses and antibodies. The breakthrough has the potential to aid in medical diagnoses and, specifically, those related to the Covid-19 pandemic. Fully realized, the proposed tests – for the virus (current infection) and antibodies (immunity) – could be done in under 30 minutes, by minimally trained personnel. In assessing the best way to draw on these advances at the level of the target regions, lessons could be learnt from the experience in various countries curated by enterprises like the Centre for Rural Emergency Medicine. Its Director Dr Tim Baker, Chair of the Emergency Medicine Certificate and Diploma Committee of the Australasian College for Emergency Medicine, characterises their approach as follows:

Communities come up with great ideas to rural problems around the world. It is just tricky to know which community is succeeding in which area. I find that each small town has its own challenges and experts. Rather than provide direct advice, I usually try to let people know what others in similar situations have tried and what they have learned. Locals usually make the best decisions for locals if they are given adequate information.

This process could be enhanced by adding case studies on such decisions for AI-based Big Data analysis.

REFERENCES

Ahmat, H.O., Wilson, W., Sorelli, L., Tolnai, B., & Hamou, A. T. (2019). *Nanocellulose for improved concrete performance*. researchgate.net/publication/331134845_Nanocellulose_for_improved_concrete_performance_A_macro-to-micro_investigation_for_disclosing_the_effects_of_cellulose_filaments_on_strength_of_cement_systems

Alibaba. (2020). *Speeding Up CT Scans to Detect COVID-19*. https://www.alibabacloud.com/blog/speeding-up-ct-scans-to-detect-covid-19_596124

APCollege. (n.d.). Full-sized Drone Ambulances – the future vision. *Australian Paramedical College*. https://apcollege.edu.au/full-sized-drone-ambulances-future-vision/

Australian Seaweed Institute. (n.d.). *Seaweed Products R&D*. https://www.australianseaweedinstitute.com.au /about-us

Ben-Sasson, E., Chiesa, A., Genkin, G., Tromer, E., & Vizra, M. (2013). *SNARKs for C: Verifying Program Executions Succinctly and in Zero Knowledge*. https://eprint.iacr.org/2013/507.pdf

Carrington, D. (2013). Seaweed biofuels: a green alternative that might just save the planet. *The Guardian*. http://marineagronomy.org/sites/default/files/Seaweed%20biofuels_%20a%20green%20alternative%20that%20might%20just%20save%20the%20planet%20_%20Environment%20_%20The%20Guardian.pdf

CGI. (n.d.). *Empowered Learning by CGI*. https://www.cgi.com/uk/en-gb/brochure/education /empowered-learning

Crofting Commission. (2020). *Facts & Figures*. https://www.crofting.scotland.gov.uk/facts-and-figures

Cuello, J. L. (2019). *One Giant Leap: Space Programs Spin Off Vertical Farming*. https:// urbanagnews.com/blog/news/one-giant-leap-space-programs-spin-off-vertical-farming/

European Commission. (2018). *Pantheon: robots giving a helping hand to hazelnut farmers*. https://ec.europa.eu/digital-single-market/en/news/pantheon-robots-giving-helping-hand-hazelnut-farmers

FAO. (n.d.). *Introduction to Commercial Seaweeds*. http://www.fao.org/3/y4765e/y4765e04.htm

FDA. (2018). *FDA Approves First Drug Comprised of an Active Ingredient Derived from Marijuana to Treat Rare, Severe Forms of Epilepsy*. https://www.fda.gov/NewsEvents/ Newsroom/PressAnnouncements/ucm611046.htm

Gillbard, E. (2020). *Farmer's Weekly*. https://www.fwi.co.uk/arable/land-preparation/ soils/fermenting-organic-matter-better-for-soil-health-than-composting

Goodman, L. (2018). *The universe is a hologram and other mind-blowing theories in theoretical physics*. https://phys.org/news/2018-03-universe-hologram-mind-blowing theories-theoretical.html

Government Office for Science. (2016). *Distributed Ledger Technology: beyond block chain*. https://assets.publishing.service.gov.uk/government/uploads/system/uploads/attachment_data/file/492972/gs-16-1-distributed-ledger-technology.pdf

Graham, S. (2020). *Dendra Systems*. https://www.dendra.io/post/if-we-want-a-future-on-earth-were-going-to-have-to-restore-2-billion-hectares-of-land-lets-get-on-with-it

Greenfield Robotics. (2020). https://www.greenfieldrobotics.com/

IEA. (2020). *Direct Air Capture*. https://www.iea.org/reports/direct-air-capture

Jepsen, M. L. (2018). *How we can use light to see deep inside our bodies and brains*. https://www.ted.com/talks/mary_lou_jepsen_how_we_can_use_light_to_see_deep_inside_our_bodies_and_brains#t-7676

Jesimani, V., Du, H., Aslam, M., Pei, P., & Huang, N. (2019). *Potential Use of Seaweed Bioactive Compounds in Skincare—A Review*. https://www.ncbi.nlm.nih.gov/pmc/ articles/PMC6950024/

Kesteven, S. (2016). *Feeding cows seaweed could slash global greenhouse gas emissions*. https://www.abc.net.au/news/2016-10-19/environmental-concerns-cows-eating-seaweed/7946630

Kite Power Systems. (2020). http://www.kps.energy/

Kumar, N. (2018). *Global Public Health: Blockchain for Contagious Disease Relief*. https:// medium.com/swlh/global-public-health-blockchain-for-contagious-disease-relief-453df74f38da

Langston, J. (2020). *To the moon and beyond: How HoloLens 2 is helping build NASA's Orion spacecraft*. https://news.microsoft.com/innovation-stories/hololens-2-nasa-orion-artemis/

Leifeld, J., & Menichetti, L. (2018). The underappreciated potential of peatlands in global climate change mitigation strategies. *Nature Communications*, *9*(1), 1071. doi:10.103841467-018-03406-6 PMID:29540695

Listek, V. (2020). *3D Printed Vegan Salmon to be Commercialised with Help of FELIXprinters*. https://3dprint.com/270667/3d-printed-vegan-salmon-to-be-commercialized-with-help-of-felixprinters/

MacDonald, R., & MacRae, D. (2019). *Maximising Scotland's Well-Being by Bravely Innovating*. http://box5817.temp.domains/~hialbaor/wp-content/uploads/2019/04/HIAlba-IDEA-Overview-16a.pdf

Mazur, M. (2016). *Six Ways Drones Are Revolutionizing Agriculture*. https://www. technologyreview.com/2016/07/20/158748/six-ways-drones-are-revolutionizing-agriculture/

MedRec. (n.d.). *What is Medrec?* https://medrec.media.mit.edu/

Nanowerk. (2017). *Seaweed: From superfood to superconductor*. https://www.nanowerk.com/nanotechnology-news/newsid=46330.php

NASA. (2015). *NASA, Microsoft Collaborate to Bring Science Fiction to Science Fact*. https://www.nasa.gov/press-release/nasa-microsoft-collaborate-to-bring-science-fiction-to-science-fact

NVIDIA. (2020). *NVIDIA Announces Cloud-AI Video-Streaming Platform to Better Connect Millions Working and Studying Remotely*. https://nvidianews.nvidia.com/news/nvidia-announces-cloud-ai-video-streaming-platform-to-better-connect-millions-working-and-studying-remotely

Petch, M. (2020). 3D Printing Responds to Covid-19 and Coronavirus Resources. *3D Printing Industry*. https://3dprintingindustry.com/news/3d-printing-community-responds -to-covid-19-and-coronavirus-resources-169143/

PHYS.ORG. (2018). *New type of opal formed by common seaweed discovered*. https://phys.org/news/2018-04-opal-common-seaweed.html#jCp

Regenerative International. (2016). https://regenerationinternational.org/regenerative-not-climate-smart-agriculture-needed-to-feed-the-world-and-cool-the-planet

Rutschman, A. S. (2018). *U.S. Health Care Companies Begin Exploring Blockchain Technologies*. https://www.scientificamerican.com/article/u-s-health-care-companies-begin-exploring-blockchain-technologies/?

ScienceDaily. (2019). *New holographic technique opens the way for quantum computation*. https://www.sciencedaily.com/releases/2019/05/190504130303.htm

Shimizu Corp of Japan. (2020). https://www.shimz.co.jp/en/topics/dream/content01/

Sidhu, M. (2020). *WVTOX, This Is Why Kelp Leather Is the Best Vegan Leather Alternative To Animal Skin*. https://wtvox.com/fashion/kelp-leather/

SISSA. (2020). *Black holes? They are like a hologram*. https://www.sissa.it/news/black-holes-they-are-hologram

Small Robot Company. (2020). https://www.smallrobotcompany.com/

Stoeger T., Gerlach M., Morimoto R.I., & Amaral L.A.N (2018). *Large-scale investigation of the reasons why potentially important genes are ignored*. doi:10.1371/journal.pbio.2006643

Stracuzzi, M. (2020). *4 Revolutionary Use Cases of 5G in Healthcare*. https://www.telit.com/blog/4-revolutionary-use-cases-5g-healthcare/

Tao, M. (2019). *Top 25 vertical farming companies*. https://roboticsandautomationnews.com/2019/05/03/top-25-vertical-farming-companies/22181/

The Scotsman. (2018). *Leader comment: Learning the lessons of Scottish suicides*. https://www.scotsman.com/news/politics/leader-comment-learning-lesson-scottish-suicides-267281

Thomsen, B., Narayanan, D., & Rowstron, A. (2020). *In search for future of cloud storage, researchers look to holographic storage solutions*. https://www.microsoft.com/en-us/research/blog/in-search-for-future-of-cloud-storage-researchers-look-to-holographic-storage-solutions/

TimTek. (n.d.). *Redefining the Engineered Lumber Industry*. http://timtekllc.com/about.html

Tubb, C., & Seba, T. (2019). *Rethinking Food & Agriculture*. https://www.rethinkx.com/food-and-agriculture

Unblocked. (2018). *Can Blockchain save the NHS?* https://un-blocked.co.uk/events/ can-blockchain-save-nhs/

Utroske, D. (2016). *Researchers at work on new kelp source for natural cosmetics*. https://www.cosmeticsdesign.com/Article/2016/02/10/Researchers-at-work-on-new-kelp-source-for-natural-cosmetics

Vegan Organic Network. (n.d.). https://veganorganic.net/plant-based-permaculture/

Viehmann A., Kersting C., Thielmann A., & Weltermann B. (2017). *Prevalence of chronic stress in general practitioners and practice assistants: Personal, practice and regional characteristics.* . doi:10.1371/journal.pone.0176658

Wong, S. (2017). *Defibrillator drones could save lives before ambulance arrives.* https://www.newscientist.com/article/2134473-defibrillator-drones-could-save-lives-before-ambulance-arrives/

Wood, C. (2019). *Hologram Within a Hologram Hints at Fate of Black Holes.* https://www.quantamagazine.org/hologram-within-a-hologram-hints-at-solution-to-black-hole-information-paradox-20191119/

Chapter 4
Technological Beyond:
Merging of Man and Machine

Enric Bas
University of Alicante, Spain

ABSTRACT

In our condition as human beings, we have experienced a love-and-hate relationship with technology since the very beginning: from primitive humans (probably scared the very first time they saw fire) to 19th century Luddites to the recent conspiracy theories that link 5G to COVID-19, human beings have dreaded technology. But, for bad or for good, machines—which could be loosely defined from an etymological point of view as "structures of any kind created by humans"—have been, are, and probably will be an essential part in the history of mankind, thus playing a fundamental in the social evolution of human. This chapter attempts to shed some light and provide some insights into plausible conjectures by exploring future developments, either in the short and medium or in the long term, which might result from a synergistic context where exponentially-increasing technological innovation may lead to a radical change, an evolutionary paradigm shift in human life and civilization, by reaching toward a post-technological era characterized by consciousness of the noosphere.

INTRODUCTION

The unstoppable evolution: towards the next leap forward of the human species?

Man machine, pseudo human being

Man machine, super human being

The man machine, machine…

("The Man Machine", Kraftwerk, 1978)

DOI: 10.4018/978-1-7998-6772-2.ch004

In November 1999, more than 20 years ago, the inventor, entrepreneur, and visionary Ray Kurzweil (1999) wrote a controversial paper entitled "Spiritual Machines, the merging of man and machine" for *The Futurist* magazine where he literally predicted that "by 2009, computers will be embedded in our clothes. By 2019, they'll be hidden in our bodies. By 2099, human and machine intelligence will have merged" (sic). Few years later, during a TED talk (Kurzweil, 2005), the same author explained in abundant, well-grounded detail how the human brain was going to be reverse-engineered and nanobots would be operating our consciousness by the 2020s.

In turn, when Chris Oakes (2000), reporting for *Wired* in the "The Next 20 years" series interviewed Stanley Williams, the latter stated that three technologies born in the 2000s —nanotechnology; biotechnology; and information technology— would have transformed life by 2020. Even though each of these technologies can actually represent an industrial revolution in its own right, combining the three of them would result in an exponential transformation of our lives. Oakes even ventured to predict that computers were eventually going to develop the capabilities of human brains.

While none of these predictions has come true yet, this should not be regarded as a failure but rather merely as a timing problem. After all, everything that has been described above is on its way: the exponential growth of human interaction with technological devices and their dependence on those devices both in the private and in the social spheres of life can hardly be denied; and the progressive merging of man and machine emerges as a plausible future reality when considering the three technologies mentioned by Williams (in Oakes, 2000). It all simply forms part of the foreseeable —and thus plausible and even probable— human evolution in the coming years. And mankind's decision to follow this specific path clearly relates to the proactive and creative nature of human beings, who permanently push forward and pursue new challenges.

Needless to say, such predictions should not be taken as revealed truths but simply as inspiring visions. Kurzweil himself has often been strongly criticized by AI thinkers as well as cognitive and computing scientists who deeply questioned both his utopian optimism (most of these technological impacts are highly controversial) and the scientific foundations of his visions. Amongst these critics stand out the pessimistic cofounder of Sun Microsystems Bill Joy (Joy, 2000) and the biologist P.Z. Myers, in whose opinion Kurzweil only provides a "God for geeks" (Daniel, 2009).

Despite his polemical predictions, Kurzweil works as an advisor for Google (where he has been Director of Engineering and Machine Intelligence since 2012) and is a mentor of the NASA-supported Singularity University, amongst other successful innovation-oriented initiatives (including the invention of a wide variety of technology-based devices and the management of related companies) all through his lifetime career. Therefore, he has long been objectively influencing the future not only with his inventions but also through his inspiring visions and predictions; in short, by adopting a proactive attitude.

Kurzweil somehow represents —alongside a number of other influential future thinkers who may be referred to as "visionaries" or "futurists" such as Isaac Asimov, Arthur C. Clarke or Jules Verne— the very human desire to understand (the origin and meaning of life as well as our place and role in the universe) and transcend (death or the end of physical existence) which should act as the meeting point between technology and philosophy, or expressed differently, the place where Edison meets Nietzsche.

It is scientifically proven that technology has played a decisive role in the biological and cultural evolution of humans as a species (Ambrose, 2001; Taylor, 2010) by causing major developments at a behavioral and cognitive level (Stout et al., 2008, 2011). Along the same lines, according to Rothenberg (1993) we never stop trying to reconceive nature, either consciously or unconsciously rethinking it by means of our imagination and creative ability. Technology could not exist without the human vision and intent that drives it.

Furthermore, every technological innovation throughout human history has always meant a challenge that the people living at each specific period had to address, insofar as it changed their social, economic, and cultural landscapes. Humans have been discussing the proper place of machines in their lives, entangled in a never-ending—formerly moral and subsequently metaphysical— debate over centuries (Byrne, 1979). William F. Ogburn's "Cultural Lag Thesis" set up the conflict inherent to any technological advance and its roots: "cultural lag is seen as a critical ethical issue because the failure to develop broad social consensus on appropriate applications of modern technology may lead to breakdowns in social solidarity and (to) the rise of social conflict. Reasons for the cultural lag between technology and ethics include the social structural and market conditions under which each one of them develop " (Marshall, 1999).

The public debate concerning this moral dilemma still remains alive despite the inescapable arrival of the 4th Industrial Revolution that will follow the First (steam power and manufacturing from the 1850s onwards), the Second (electricity, synthetic materials, and mass production beginning in the 1900s), and the Third (Computing and electronic communications since the 1960s) (Daemmrich, 2017). That 4th industrial revolution appears as a brand-new stage where hyper-connection constitutes the norm and technology is both understood and assumed to be more pervasive and ubiquitous than ever (Bas, 2015), with an unprecedented transformative potential —thanks to its radical, disruptive, and breakthrough-favoring nature— since biological, digital, and physical systems synergistically merge to shape not only the future human self but also society as a whole.

Or, perhaps, this debate is still a current issue precisely because of the cultural lag between the immaterial world (ethics) and the material universe (technology) which move at different speeds. The exponential, diverse, and synergic nature of the transformations in progress —which lead to an increased merging of machines and *Homo Sapiens*— poses a big challenge as far as social change management is concerned. A "big step" in evolutionary terms for the human and social fabric where the changes are so profound that there has never been a time of greater promise and/or potential peril from the perspective of human history (Schawb, 2017).

As recently stated (Bejan, 2020), "we are [arguably] the human & machine species" (sic), since humans have always lived attached to their 'machines' –which, according to the etymological of this word, could be loosely defined as "structures of any kind created by humans." Those 'machines' (e.g. artifacts, contrivances, devices, gizmos…) which have progressively evolved to incorporate more powerful add-ons across human history can be considered an inextricable and essential driver for either the physical or the emotional and/or cognitive evolution of our species.

Consequently, for Adrian Bejan (2020), *Homo Sapiens* was never the simple representation of a naked body sketched in an anatomy book. Instead, it resulted from two mutually-feeding intertwined processes: a biological one (the continuous evolution of the human body); and a technological one (the human creations or `machines´ which, despite constantly evolving too, have done so at an incomparably smaller time scale than the human body). In a nutshell: we cannot dissociate human evolution from human creations, the so-called 'machines.'

There is no human evolution (biologically and sociologically speaking) without technological progress. The reason lies in the fact that technological progress contributes to human evolution: the search for ways to improve or make more effective all the processes concerning human life either individually or on a social basis. In other words, no plausible future scenario exists where humans reach a stage in which they can decide to stop —or even slow down— evolution and technological development; that would mean the disappearance of human civilization. Kurzweil's promise thus seems more likely to come true: "we will transcend all the limitations of our biology (…), that is, what it means to be human: to extend who we are'' (Vance, 2010)

The human/machine dialectics lies at the core of the evolution undergone by human beings as a species. It is in humans' DNA to move forward, to create, and to experiment; to try, to fail, and to try again; and also to generate —and accumulate— knowledge through trial and error. When something works in evolution, it remains; and regarding technological development, we are ready for the next step.

Following this logic, on the basis of the three types of knowledge proposed by Jürgen Habermas (technical; practical or hermeneutical; and emancipatory) and their contextualization within History, we created a typology with five "ways to approach future," including firstly a supernatural perspective and later a participatory one (Bas&Guilló, 2015):

Table 1. Ways to approach futures

	Supernatural	Hermeneutic	Technical	Emancipatory	Participatory
Origin	Primitive societies	Secular society	Industrial society	Post-industrial society	Liquid society
Paradigm shift	Old regime	Nation-state	Mass-capitalism	Globalization	Social networks
Key point	Religion	Ideology	Product	Knowledge	People
Dominant subsystem	Culture	Politics	Economics	Technology	Ecosystem
Foundation	Emotional (Rh)	Emotional(Rh)	Rational(Lh)	Rational(Lh)	Emotional(Rh)+ rational (Lh)
Method	Qualitative God´s orders	Qualitative Utopias	Quantitative Trend forecasting	Qualitative + quantitative Foresight	Creative futures
Attitude	Alienation	Alienation	Adaptation	Anticipation	Innovation

These five "ways to approach future" were presented in a sort of time arrow using a table format and placing them within a historical context for a better understanding of the different methods and attitudes which serve to envision the future as a product of its time —or, expressed differently, as a part of the Zeitgeist.

The table above could be considered a kind of evolutionary-based manner of understanding how human beings have addressed the future throughout history, insofar as it links each one of their approaches to a particular type of social system and, although all five formulas described still coexist nowadays, the very origin of each one is associated with a specific historical context, additionally showing a somewhat growing complexity in those approaches.

The crux of the matter is that the way to approach future has kept evolving across human history, exactly the same as the species —and its creations (machines). This happens because our attitude towards future, which closely relates to the fears and expectations aroused by uncertainty and change, basically depends on knowledge and on the ability to envision the future. Therefore, we may approach future differently, but something always remains present in every human being and society: the promise of a better future (induced by nature, religion, politics, science or technology…to quote but a few possible dimensions) in the search to endow our own existence with meaning.

The futurist Jim Dator (1996)wrote: "*Aiglatson* is 'Nostalgia' spelled backwards and is a word told to me by Gabriel Fackre to symbolize the yearning for things to come; revering the future; without being disrespectful to the past (remembering that once it was all that was humanly possible), preferring the dreams of the future to the experiences of the …past; always desiring to try something new; to go where no one has ever gone before in all areas of human —and non-human, and soon post-human— experience."

Thus, if craving for the future is inextricably human, just as the willingness to move ahead, looking towards the future can be considered essential to shape humans both as individuals and as social beings. If evolution stems from the search for a purpose and a meaning, should we call the foreseen technology-induced merging of man and machine "posthumanism"? Or perhaps only "humanism"?

In other words, why should we refer to something as "post-human" when it is nothing but a part of human evolution, a logical consequence of human nature? Is it because that might mean "closing the circle": human beings being 'devoured' by their own creations? Or could we possibly find ourselves at an intermediate stage, one more step along a huge evolutionary chain, inside something that will end up becoming bigger than us? Does the possibility exist to stop or control the evolutionary process when it comes to technological innovations? Since technology is an intrinsically human outcome, could we draw a distinction between human and technological evolution? And should we? Would human beings have a chance to survive as a species in that case? How does technology affect society? These are some of the first great questions that need to be addressed in relation to the moral and ethical issues involved in the merging of man and machine.

The "Extended Mind Thesis" or the man machine as a natural consequence of the human evolutionary process: from the liquid self to the extended self

"To be immortal is trivial; Except for man, all creatures are,

because they ignore death; the divine, the terrible,

the incomprehensible, is knowing oneself immortal"

-José Luis Borges

Human evolution reflects the story about how this species has adapted to the environment. Nevertheless, unlike other living creatures similarly impelled by a survival instinct, human beings have managed to develop complex tools for the purpose of shaping the environment for their own benefit.

The "ways to approach future" table implicitly shows this evolution by linking every philosophical and/or attitudinal change regarding the future to major societal changes respectively. All those societal changes, which mainly have to do with politics and the economy, can likewise be connected to technological change as a major driver of human progress from primitive societies to liquid modernity.

Only about 15,000 years ago, human beings hardly survived by collecting vegetables which periodically grew on a spontaneous basis in nature, and by chasing animals after walking huge distances. Each and every technological innovation, from fire to the wheel, to the fridge, to antibiotics, and to the internet, has caused radical social changes which transformed everything from life expectancy to lifestyles at the time, shaping human life for better or worse. Paraphrasing Marshall McLuhan, we shape our tools, and then our tools shape us:

"Physiologically, man in the normal use of technology (or his variously extended body) is perpetually modified by it and in turn finds ever-new ways of modifying his technology. Man becomes, as it were, the sex organs of the machine world, as the bee of the plant world, enabling it to fecundate and to evolve ever-new forms. The machine world reciprocates man's love by expediting his wishes and desires, namely, in providing him with wealth" (McLuhan, 1994, p. 46)

According to McLuhan, one technology —or medium— leads to another within an unstoppable dialectic between man and his very creation (the machine). This dynamic leads to a stage where it progressively becomes more and more difficult to distinguish both of them, to the extent that, by continuously embracing technologies, we relate ourselves to them through what he describes as "servo-mechanisms." In the words of Robert K. Logan, who has exhaustively analyzed the evolution of media and technologies as extensions of man: "at first, technology serves as an extension of humankind and then suddenly a flip occurs and humankind is transformed into an extension of its technology" (Logan, 2013) .

To put it in another way: although we create these tools —and induced environments— they create us too, thus leading us to a paradox: perhaps posthumanism or transhumanism should not be understood as the end of humanism —or the human species— but simply as a natural extension of humanism itself.

Our condition as humans has always pushed us to seek ways to enhance our biological, mental, and physical capacities by means of technology; and human nature is shaped by the specific technology and culture through which it finds an expression. Therefore, cyborgs (cybernetic organisms) —a recurrent paradigm for the merging of man and machine in popular culture— should not be seen as a mere mixture of flesh and wires. Instead, they would represent a less simplistic and more profound merging based on a technological extension of human cognition, i.e. a total symbiosis through technology (Clark, 2003).

We humans are and have always been tool-making creatures, this being an essential characteristic of our species —and of our evolution, since the ability to master a tool allows us to perform tasks that would be impossible otherwise. From this point of view, a mention could be made about the existence of a reflexive type of dialectics between human beings and their creation (the machine) where both feed each other in a never-ending process: machines are human creations the existence of which, inevitably shapes and redefines human existence not only by solving existing challenges but also by creating new ones that demand new solutions and tools. Consequently, according to Andy Clark, the merging of technology and biology becomes inevitable due to the fact that we human beings are creatures whose minds are special precisely because they have been tailor-made for multiple mergers and coalitions. Our creations —the machines— are by no means alien to us; in fact, they constitute an extension of ourselves, an indissoluble part of who we are.

Clark himself co-authored with the philosopher David Chalmers the 1998 seminal paper for the EMT (Extended Mind Thesis) where they argued that the mind does not reside exclusively in the brain or even the body, but extends to the physical world integrating all the artifacts which have an influence on cognitive processes. They paid special attention to those artifacts or tools which serve to store information (from a drawing in a cave to a symbol or a flag; from a notebook to a computer). However, in our view, those which may help access (eyeglasses, auditory prostheses, the Internet …) and/or manage (software programs/apps, cardiac pacemakers, insulin smart patches…) information broadly speaking should be considered as well (Clark & Chalmers, 1998).

McLuhan (1994) firstly argued that every machine (prostheses, cars, planes, or refrigerators to quote but a few) could be depicted as an extension of the human body crafted to turn complex processes into affordable ones. Going even further, the proponents of the EMT argue that, since they are extensions which go beyond the brain, such artifacts or tools created by humans combine with the traditional mind to create a cognitive coupled system. In this regard, the EMT emphasizes the complementariness of the internal and external elements that shape cognitive systems or processes, suggesting that the next cyborg technology generation will go one step beyond and fix imperfections or limitations in our body, eventually making it possible to add completely new functions to humans.

Nonetheless, the way in which EMT questions the "self" as some kind of central cognitive essence that makes me who and what I am is very challenging in ethical terms, since it somehow considers human transcendence and spirituality —which are linked to human values and conscience— somewhat irrelevant as opposed to the environment, which appears as a determining factor for our personal identity. The "extended self" as a concept related to EMT has been analyzed in depth from a philosophical point of view (Olson, 2011) and its obviously complex and controversial nature does not prevent it from being a very challenging and promising proposition fueled by a highly powerful transformative idea.

Numerous experts, scientists, and entrepreneurs working in AI —or in related fields— have explicitly or implicitly aligned themselves with the —"pragmatic and soul-less"— EMT approach, according to which the merging of man and machines (artifacts, devices ...and so on) and even the "extended self" are nothing but a natural part of human evolution. Thus, being optimists, they never worry about the moral or ethical implications and/or consequences that might derive or arise from prioritizing technological innovation with no restrictions whatsoever.

Others are extremely skeptical about the possibility of a machine becoming truly intelligent and argue that intelligence cannot be emulated by means of mathematical algorithms, which only execute the programmer's orders unaware of what they are doing and why (its purpose). This point of view relies on the assumption that a machine cannot be intelligent without previously becoming aware, which inevitably has some moral implications. By way of example, when Roger Penrose (who inspired Stephen Hawking) was interviewed by a popular Spanish weekly magazine (XLSemanal, 2020) he stated that a few years before winning the Nobel prize for Physics, the Oxford University the following: "I am not saying that it is impossible to build a conscious apparatus in a laboratory. Perhaps, but that device will not be a computer. You can only be intelligent if you are aware. In other words, there is no intelligence without awareness."

In any case, regardless of whether we agree or not on the limitations and possibilities when it comes to the merging of man and machine, the big question remains in the air: should humans face it without any kind of restriction, simply embracing technology in a deterministic manner (as something unstoppable which is intrinsic to human evolution) and —if the answer is "yes"— what implications might that have for human life, identity, and society? And even more importantly, would this mean the end of human civilization —as we know it— and the beginning of a new (post-human) civilization or simply humans continuing to be humans?

Therefore, a key point which needs to be addressed while analyzing or discussing the future of merging of man and machine would be human beings' identity (their self), which is in turn closely linked to another two concepts mentioned above: awareness and intelligence. This evidence could lead us to formulate some further questions related to those posed in the previous paragraph: Will machines have an identity? Will machines be self-conscious? Will they be intelligent? And —should the answer be "yes"— what implications might that have for human life, identity, and society? How will this merger happen? This is something we are going to explore in the coming pages.

The authors of a paper entitled "Towards a Liquid Self: how time, geography, and life experiences reshape the biological identity" (Grinoglio et al., 2014) proposed an inspiring hypothesis which is summarized in its abstract:

"(...) Here we present the new hypothesis of "liquid self" which integrates and extends the danger/damage theory. The main novelty of the "liquid self" hypothesis lies in the full integration of the immune response mechanisms into the host body's ecosystems, i.e. in adding the temporal, as well as the geographical/evolutionary and environmental dimensions, (((((((...which are synthetically expressed by what we suggested calling (sic) "immunological biography." Our hypothesis takes into account the important biological changes occurring with time (age) in the immune system (including immunosenescence and inflammaging), as well as changes in the organismal context related to nutrition, lifestyle, and geography (populations)...))))))). We argue that such temporal and geographical dimensions impinge upon, and continuously reshape, the antigenicity of physical entities (molecules, cells, bacteria, viruses), making them switch between "self" and "non-self" states in a dynamic (sic), "liquid" fashion (...)"

This idea of the "liquid self" stresses how change (time) and complexity (environment) jointly and continuously reshape human nature when it comes to immunology —or the natural ability to protect oneself? from danger. It thus basically suggests that, at a biological level, we humans are ever-changing entities continuously adapting in a struggle for survival.

This line of reasoning could be easily extrapolated from the biological arena to sociology when dealing with the Liquid Society or Liquid Modernity (Bauman, 2006), where (flexible and changing) networks are gradually replacing the (solid and recurrent) structure as the main conceptual reference issue when it comes to mapping social behavior and interactions. Such a Liquid Society could not exist without the presence of ICTs as key facilitators to generate, analyze, and transmit information. The same holds true for the Liquid Self —the changing man— which can was defined broadly by Fizz as a new stage where the main shift has led towards apps that let you introduce yourself to different people in different ways and at different times (Emerging Trends in Social Media, 2014)

It has already been suggested (Bas, 2015) that the Liquid Self —as a concept related to Bauman's Liquid Modernity and the increasingly prominent role of technology in social life— forms part of the current Zeitgeist: ICTs are more omnipresent, ubiquitous, and pervasive than any other previous technology. Besides, in conjunction with some other technologies (e.g. VR, AI, or Biotechnology), they are generating a technological environment with an enormous and unprecedented transformative potential for human beings and societies alike.

If, as mentioned above, McLuhan said that every technology created by humans, from a wheel to a chip to a prosthesis, is just an extension of themselves, a connatural part of them —despite being artificial by nature— the Liquid Self could serve as a metaphor of the step forward that ICTs imply: the true merging of man and machine as a continuous process where both feed each other and modify the body and mind of the former to the extent of probably meaning the prelude to a plausible civilizational shift. The big step beyond. Will something like the "extended self" —a sort of artificial consciousness, a truly autonomous and independent entity and identity— be the ultimate humanist creation?

Matryoshkas in a timeline; three steps for the upcoming evolutionary paradigm shift

"You are not a human being in search of a spiritual experience.

You are a spiritual being immersed in the human experience"

-Pierre Teilhard de Chardin

The man/machine or creator/creation dialectics is as old as human beings themselves. We have repeatedly emphasized in the preceding pages that the evolution of human beings cannot be distinguished from their own creations —the machines— which, far from being something strange or alien to humans, are only an expression of human nature and its creativity.

The reciprocity involved in this dialectics between man/machine has already been extensively analyzed: humans shape machines, but machines also shape humans in a never-ending, interactive dynamic process. Arguably, what we call "the merging of man and machine" is nothing but the logical consequence of the continuous feedback inherent to this reciprocal —as well as cumulative— process.

The merging of man and machine is accordingly an old story. Nothing new under the sun. The radical difference between ours 2020 and other previous stages of this historical process, though, lies in the existence (because of the exponential nature of technological innovation) of an unprecedently high multiplicity of merging, potentially interconnected drivers which are exponentially increasing complexity and making it harder to clearly envision plausible future scenarios. All these drivers converge to create the suitable conditions, at least technologically speaking, for a kind of "perfect storm" —a radical breakthrough, a challenging tipping point, an unprecedented disruption in the social fabric— with regard to the merging of man and machine.

The primary intention of this paper consisted in identifying the major drivers of change when it comes to the merging of man and machine, which could subsequently help build plausible future alternative scenarios according to the methodology of future scenarios (first designing the most probable, then the contingent, and finally the preferred scenario). However, thorough research and the contrast between objective data as well as subjective estimates and proposals made us change the original approach in the light of the findings listed below:

1. Technological progress is unstoppable owing to the quintessential character of creativity as an element inherent to human nature and evolution. Creativity can be neither banned nor limited, as that would mean the end of human evolution.
2. The merging of man and his own creations —let us call them as 'machines'— is nothing new or alien, but only an undifferentiated part of our civilizational progress as a human species, from the very beginning; from the origin.
3. Most of the identified technological drivers are basically ongoing ones whose further consolidation and developments basically depend on legal (regulatory framework), economic (large-scale investment), and cultural (moral acceptance) issues.
4. It was consequently deemed appropriate to focus the central debate not on what is possible —or even plausible— but on when it will be possible, and depending on whom, what and how. Unless the current technological structure (devices and systems) totally collapses due to a black swan, the technological route has already been drawn.
5. Therefore, in accordance with this approach, it makes no sense to cover alternative simultaneous plausible scenarios, as the Future Studies methodology suggests. It seems more convenient to explore the potential future possibilities consecutively rather than on an alternative basis. Three steps in the stairway to the future are proposed within a matryoshka-like conception.

These are three consecutive scenarios considered:

SCENARIO #1 (2030): THE CLOUD / GLOBAL-TOTAL NETWORK / HYPERCONNECTIVITY
SCENARIO #2 (2050): COLLECTIVE CONSCIOUSNESS/BRAIN / PLANETARY SMART SKIN
SCENARIO #3 (+2050): TRANSCENDENCE / SINGULARITY / A NEW HUMAN SPECIES

SCENARIO #1: HYPERCONNECTIVITY - IN THE 2030 HORIZON (SHORT TERM)

Thanks to the rapid advances made in increasing transfer speed, accessibility, and connectivity, the IoT (Internet of Things) is already here, slowly entering and becoming an integral part of our everyday life —as exemplified by the presence of such household devices or appliances as smart fridges, wireless printers, or cleaners like Roomba. And these technological breakthroughs are regarded as potentially useful to improve our quality of life, since they can help us replace routines or deal with them in a more efficient, accurate and coordinated way, as well as faster.

Being perceived as the introduction of technologies meant to rid humans of boring and repetitive tasks, these changes are initially welcomed by individuals, which in turn transforms the understanding of social capital.

IoT arises as a distinctive entity when more inanimate objects than human users are connected to the Internet (Evans, 2011), i.e. it has existed since about 2008. Therefore, IoT can no longer be seen as a promise now in 2020. It has already become a reality, even if it may take some time until it completely permeates social and economic life (homes, offices…), becoming fully mainstream and "visible" —the same as computers or mobile devices did not so long ago.

Connecting things through a widespread use of ICTs to effectively systematize routines under the supervision of humans did not suffice, though. Kevin Ashton, who works at MIT (Massachusetts Institute of Technology) and coined the term "Internet of Things (IoT)" during a presentation for Procter & Gamble back in 1999, drew on the D2D (device-to-device) communication experience to state the following over a decade ago:

If we had computers that knew everything there was to know about things —using data they gathered without any help from us— we would be able to track and count everything, and greatly reduce waste, loss, and cost … We need to empower computers with their own means of gathering information, so they can see, hear, and smell the world for themselves, in all its random glory … (Ashton 2009)

Therefore, the original M2M concept is moving ahead to wards a new stage: things, people, processes, and data are being increasingly integrated into the Internet of Things (IoT). If they reach a point where they become networked via the Internet and can cooperate with one another (embedded in an intelligent, dynamic, and widely distributed virtual system), that would be the Internet of Everything (IoE).

According to CISCO (2015), IoE means bringing things, processes, data, and people together to make networked connections more relevant and valuable than ever before —(thus) turning information into actions that create new capabilities, richer experiences, and unprecedented economic opportunity for businesses, individuals, and countries. At Cisco Live, CEO John Chambers said: "The simple concept, as you move forward with IoE, is that you have to get the right information at the right time to the right device to the right person to make the right decision" (Chambers, 2014).

Despite the existing open debate on whether IoT and IoE are the same thing or not (Weissberger, 2014), this IoE proposal seems of interest because it explicitly incorporates people into the equation. Under this perspective, IoT (connecting machines with machines) could be seen as a kind of preliminary step to the wider vision of IoE (connecting machines with people, data, and processes, by combining M2M with P2P and P2M).

IoE consequently emerges as a new stage where machines and humans (and potentially other living beings too) merge while integrating as nodes of a whole network: nodes with unique identifiable embedded devices, wirelessly connected to the ubiquitous Internet, with full access to data (cloud), and autonomous enough to send or receive information without the need for human intervention. The concept of IoE is being extended to its fullest by the implementation of the Internet of Nano Things (IoNT), which takes place through the incorporation of nano-sensors into diverse objects and the utilization of nano-networks.

Looking at the IoNT as a logical consequence —another wave— of the rationale behind the development of the Internet as "fixed computing," "mobility," "IoT," and "IoE" sounds like an interesting approach (Miraz et al., 2015) which perfectly fits within the philosophy underpinning this chapter, which revolves around the cumulative and "human" dimension of technological innovation, however far it might go.

This is accordingly what hyperconnectivity means as a scenario for 2030: everything and everybody being networked anywhere and anytime, probably without the choice of voluntary disconnection (if hyperconnectivity becomes the norm, probably refusing it will be unfair in social terms and also prosecuted and punished by the law). An exciting but simultaneously frightening scenario that is very likely to come true within the next ten years. Hyperconnectivity will entail a total, real, globalization, along with the culmination of the information age initiated by the Internet halfway through the 20th century.

SCENARIO #2: COLLECTIVE CONSCIOUSNESS - IN THE 2050 HORIZON (MEDIUM TERM)

Twenty-five years ago, Jennifer Kreisberg wrote an essay entitled "A globe, clothing itself with a brain" for Wired Magazine (Kreisberg, 1995) where she explained the vision developed by the philosopher of evolution Pierre Teilhard De Chardin in the 1940s. De Chardin, who influenced thinkers such as Marshall McLuhan (in *The Gutenberg Galaxy*) and Al Gore (in *Earth in the Balance*), described a global sphere of thought, a living unity containing all our collective thoughts, experiences, and consciousness which he called "noosphere" (mental world, derived from the Greek word "noös": mind).

Vladimir Vernadsky, a professor of geochemistry at La Sorbonne and coeval with De Chardin and Einstein, adopted the term ("noosphere") to outline the upcoming—and in his view inexorable— preponderant activity of man. For Vernadsky, all the technological and scientific advances made throughout the twentieth century, including Einstein's discovery of relativity and the mastery of atomic energy, meant that mankind was in the process of coming closer to noosphere development by enabling human beings to extend their reach into mother earth's whole ecosystem and even beyond it —the surrounding universe.

Therefore, according to Vernadsky (1998), the noosphere could be understood as the third Earth development layer, after the biosphere and the technosphere; the last of many stages in the evolution of the biosphere across geological history (Lahoz-Beltra, 2014). A third evolutionary stage for humans, characterized by the prevalence of mind and consciousness over the physical world.

Outstanding philosophers such as David Chalmers and neuroscientists like Christof Koch, who works as Chief Scientific Officer at the Allen Institute for Brain Science, have tried to approach the problem of consciousness (learning process, storage of memories, perceptions, and feelings) both in humans and computers, by exploring the notion that all things in the universe might be —or are potentially— conscious, as long as the information that they contain is sufficiently interconnected and organized. If so, could we then find ourselves halfway through a path towards a conscious web where the Internet becomes artificially intelligent? Towards a kind of 'global thinking layer,' an emerging Noosphere? (this takes us back to Teilhard de Chardin). Perhaps, in that case, we would be standing on the threshold of a breakthrough not only in the field of technology (AI, IoT, IoE…) but also in our understanding of consciousness itself. A substantially bigger jump indeed.

However, the almost widespread optimism about the advent of IoT (or even its more sophisticated sequels: IoE, IoNT…) is not shared when it comes to the noosphere or, expressed differently, to reaching a sort of global consciousness with a relevant autonomy concerning human action. Some world leading scientists among whom stands out Andrew Ng (expert in AI, entrepreneur, and a researcher as well as professor at Stanford), seem to be a little more skeptical than futurists such as Kurzweil (to name but a few authors already mentioned above) about the current expectations and timelines for developments related to expanding the human brain and being able to replicate it by building truly intelligent (self-conscious) machines (Ng, 2017).

Controversial initiatives like the Global Consciousness Project (GCP) and others —sometimes close to paranormal, new age, or parapsychology approaches— are not so misguided. GCP's IT, conceived in 1979 (and carried out in 1998) by the dean of the School of Engineering and Applied Science at Princeton University, Robert G. Jahn, had as its aim to undertake a scientific study about the interaction of human consciousness with mechanical and physical instruments. Numerous universities and scientists from all over the world have collaborated in this project.

As GCP pursues to detect the relationship that may exist between collective consciousness (or the collective unconscious) and physical reality using a global computer network, it has been deeply criticized. The people working at GCP have tried to establish some kind of action/reaction mechanism between various dramatic worldwide events (e.g. the death of Diana, Princess of Wales, the 9/11 attacks…) by assessing or trying to determine the extent to which global consciousness was affected by such shocks; and, at the same time, how these variations in global consciousness (reflected in sadness, hope, or any other states of mind) impacted on the real world.

This actually raises a very interesting question: Has mankind —as a whole— acquired self-consciousness yet? If it still has not, could that aim be reached soon by technological means?

Nonetheless, science mostly remains Newtonian/deterministic to a certain extent and relies on accumulated knowledge (including methods and tools to gather information, analyzing it, and eventually drawing conclusions therefrom). Consequently, speculative/creative thinking is still not very welcome if it does not fit within the established knowledge framework and/or challenges the current paradigm. For this reason, any approach in the boundaries of knowledge tends to be rejected, despite the fact that such "alternative" approaches or interpretative models are developed by prominent scientists on most occasions. Together with the Noosphere/Global Consciousness approach advocated in these highly controversial proposals rejected by most of the academia would be the Gaia Hypothesis, developed by the English chemist James Lovelock in the late 1970s and based on the work of the US oceanographer Alfred C. Redfield during the mid-1950s.

According to Lovelock, Gaia —or "mother nature"— is a complex entity that includes the biosphere, the atmosphere, the oceans, and the land; and constitutes in its entirety a cybernetic or feedback system that seeks an optimal physical and chemical environment for life on the planet (Lovelock and Margulis, 1974). The mathematician Norbert Wiener (Wiener, 1965) defined "cybernetics" —to which Lovelock refers in his description of Gaia— as "the scientific study of control and communication in the animal and in the machine."

Therefore, what Lovelock was suggesting is that Gaia (mother earth) functions as a self-regulatory system operating in an intelligent way; like any other living organism, as if it were in an automated process based on machine learning and controlled by a human… or, perhaps, as a self-conscious entity, with feelings, memories, and expectations. And that is the point where Gaia meets Global Consciousness.

Although those two approaches are separately very interesting and challenging at an intellectual level, they acquire a new meaning if we consider them together today (in 2020). A promising meaning because, while in the late 1970s (when the Gaian Hypothesis was formulated) or at the end of the 1990s (when Global Consciousness arose) we lacked the tools that we currently have at our disposal to assess how the planet —and/or its inhabitants— feel. Now, thanks to the drivers mentioned above (the Internet, AI, cloud computing…), we can map, identify, and measure collective behaviors and feelings (mood, reactions…) in real time all over the world and permanently; and we have the chance to predict them too (expectations?).

At present, such groundbreaking technologies are shaping a smart grid based on nodes, a much bigger and more sensitive one than the GCP network used to assess "global consciousness" and much more complex too, since it relies on the total convergence of IoE (things and humans are nodes connected to processes and data, anywhere and anytime) with Metadata, AI, Cloud Computing, Nanotechnology, Virtual Reality, and so and so forth. This would be the biggest leap forward in terms of knowledge generation and storage by humans (a sign of civilization), from cuneiform scripts and hieroglyphs to the achievement of phonetic languages through the invention of Gutenberg's printing press, to the systematization of European medieval knowledge with Diderot's encyclopedia, and to the world wide web/network of computers made possible through the Internet —commonly accepted as the medium which brings life to the noosphere.

In our perception, a gradual synergistic alignment of all these technologies is already taking place, even though we still find ourselves at the beginning of the road. Therefore, the effective configuration of a global brain or global consciousness, a kind of planetary smart skin (or thinking layer) fitting like a glove —i.e. reaching harmony and an acceptable level of sustainability to ensure its resilience (this is what the UN SDGs seek, is it not?)— with Gaia that would somehow end up validating both controversial hypotheses, could be closer than we think… *ma non troppo*. Our subjective appreciation is that it will not happen before 2050.

SCENARIO #3: TRANSCENDENCE - IN THE HORIZON OF 2050 AND BEYOND (LONG TERM)

Both Koch (2011) and Chalmers (2010) —mentioned above in relation to Scenario #2— have been linking computer consciousness to the idea of singularity: a hypothetical point in time when technological growth becomes uncontrollable and irreversible, ultimately resulting in unforeseeable changes for human civilization (Nazaretyan, 2014). Scientists working in this arena usually link the arrival of the singularity era to the previous achievement of a collective consciousness —or noosphere. According to Vinge and Kurzweil's original technological singularity hypothesis, in the future, the noosphere will be the natural environment where the 'human-machine superintelligence' emerges after reaching the point of technological singularity (Lahoz-Beltra, 2014).

More precisely, it was almost 40 years ago when the computer scientist Vernor Vinge coined the term "singularity," and explained what it meant, at the "VISION-21 Symposium" sponsored by the NASA

Lewis Research Center and the Ohio Aerospace Institute. In addition to stating that we find ourselves on the verge of a change comparable to the rise of human life on Earth, he predicted that, most probably between 2005 and 2030, we would have the technological means required to create superhuman intelligence and, shortly after, the human era would have come to an end (Vinge, 1993).

Vinge's prediction has not come true yet, but that does not necessarily mean that it is obsolete, at least in its formulation of a plausible scenario for the long term. The forecast made by Vinge does not explicitly imply a value judgement either. Nevertheless, the mere possibility of having intelligent, self-conscious supercomputers has obviously triggered a big debate where dystopian as well as utopian visions arise. If machines become intelligent and make their own decisions, which role will humans play? Will we be able to control these machines anyway? Should humans merge with them perhaps? Will this human/machine fusion take place? If so, are we already immersed in such an integrative process?

According to Kurzweil, the Singularity is such a rapid and profound technological change that it entails a rupture in the human history fabric which is taking place at this very moment —we still cannot comprehend it, though, at least with our current level of understanding. And he assigns so much relevance to Singularity because that emerging intelligence will continue to represent human civilization —a human-machine civilization already. This will be the next step in evolution, the next high-level "paradigm shift" (Kurzweil, 2001).

Back in April 2000, Bill Joy, Founder and Chief Scientist of Sun Microsystems, issued a warning about the real possibility of twenty-first-century technologies threatening to turn humans into an endangered species (Joy, 2000). It may sound too gloomy in the future, but the feasibility of this proposed scenario is defended even by the most optimistic scientists and visionaries, including Kurzweil (2001). The fear of humans losing control, which would result in our actual disappearance as a species (as it is known to us), logically stems from the plausibility of reaching a level of technological change that humans could not possibly assume. Scientific and technological exponential progress is undeniably on its way to reaching a significant peak; and this will probably lead to a crossroads where mankind will have to face its own destiny, dealing with unknown and even unexpected consequences.

Based on De Chardin's quote, human evolution could be regarded as a historical process that is not pushed from behind by the force of causal necessity but pulled towards a kind of universal attractor, a temporal vortex, in a way driving us to transcendence. And this would be a different way of understanding who we —humans— are, and what our existence means in relation to the universe. Hence why astrobiologists are interested in in technological Singularity(ies). That is because even humans might be replaced by machines —or hybrids— and that would bring us closer to reaching some kind of extraterrestrial civilization, since more advanced intelligent life forms, possibly evolved elsewhere in the Universe, could have experienced their own Singularity a long time ago (Gale et al., 2020).

The physicist Frank J. Tipler, amongst other outstanding scientists, argues that some kind of technological singularity is inevitable and our fate —as the human species— is to live forever, expanding and then finally reaching what he calls the "Omega point" (Tipler, 2003:2007) prior to the next Big Bang. Human beings then could be situated as a transitory evolutionary phenomenon connecting biological evolution to biocultural evolution and to technocultural evolution —the potential birth of technological life, in accordance with the theory of atechnogenesis; as a sort of intermediary between basic evolutionary mechanisms and the next cosmic evolution level, ultimately leading to the hypothetical arrival of an "Omega civilization" … when mankind will transcend biology (Last, 2015:2020).

By definition, physics conceives Singularity as something beyond which we cannot see; something unfathomable … something appearing on the horizon which, following some people's predictions, will happen in the near future: in 2045, 25 years from now, according to Kurzweil (Kurzweil, 2006). Other reputed scientists such as the co-founder of Microsoft, Paul Allen, are much more skeptical about it (Allen, 2011) and think it might take longer. However, nearly everyone agrees on 3 main points: 1) technological innovation is growing exponentially and having a stronger impact than ever before; 2) AI and cloud computing will be the basic drivers; and 3) reaching a certain peak may mean reaching a tipping point where the accelerating pace of increasingly smart machines will most probably outrun all human capabilities and endanger the social fabric —and even mankind itself, as the late Stephen Hawking already warned.

REFERENCES

Allen, P. G. (2020, April 2). Paul Allen: The Singularity Isn't Near. *MIT Technology Review*. https://www.technologyreview.com/2011/10/12/190773/paul-allen-the-singularity-isnt-near/

Ambrose, S. H. (2001). Paleolithic Technology and Human Evolution. *Science*, *291*(5509), 1748–1753. doi:10.1126cience.1059487 PMID:11249821

Andrew Ng: Artificial Intelligence is the New Electricity. (2017, February 2). [Video]. YouTube. https://www.youtube.com/watch?v=21EiKfQYZXc

Ashton, K. (2009). That 'Internet of Things' Thing. *RFID Journal*. https://www.rfidjournal.com/that-internet-of-things-thing?4986

Bas, E. (2015). The Future Internet: Alternative Visions (Public Administration and Information Technology (17). In R. Ono & J. Winter (Eds.), *The Liquid Self: Exploring the Ubiquitous Nature of the Future Internet and Its Pervasive Consequences on Social Life* (1st ed., pp. 193–215). Springer.

Bas, E., & Guillo, M. (2015). Participatory foresight for social innovation. FLUX-3D method (Forward Looking User Experience), a tool for evaluating innovations. *Technological Forecasting and Social Change*, *101*, 275–290. doi:10.1016/j.techfore.2015.06.016

Bauman, Z. (2006). *Liquid Times: Living in an Age of Uncertainty* (1st ed.). Polity.

Bejan, A. (2020). Human evolution is biological & technological evolution. *Bio Systems*, *195*, 104156. doi:10.1016/j.biosystems.2020.104156 PMID:32492467

Byrne, E. (1979). Technology and Human Existence. *The Southwestern Journal of Philosophy*, *10*(1), 55–69. doi:10.5840wjphil19791015

Chalmers, D. J. (2010). The singularity: A philosophical analysis. *Journal of Consciousness Studies*, *17*(9–10), 7–65.

Cisco. (2014). *Cisco Live!* [Video]. IBM Watson Media. http://www.ustream.tv/ciscolive

Cisco. (2015). *Cisco Live!* [Video]. IBM Watson Media. https://www.cisco.com/c/r/en/us/internet-of-everythingioe/index.html

Clark, A. (2004). Natural-Born Cyborgs: Minds, Technologies, and the Future of Human Intelligence (Illustrated ed.). Oxford University Press.

Clark, A., & Chalmers, D. (1998). The Extended Mind. *Analysis*, *58*(1), 7–19. doi:10.1093/analys/58.1.7

Daemmrich, A. (2017). Invention, Innovation Systems, and the Fourth Industrial Revolution. *Technology and Innovation*, *18*(4), 257–265. doi:10.21300/18.4.2017.257

Dator, J. (1996). Futures studies as applied knowledge. In R. A. Slaughter (Ed.), *New Thinking for a New Millennium* (pp. 105–115). Routledge.

Emerging trends in social media. (2014, July 3). *MIT News*. https://news.mit.edu/2014/emerging-trends-social-media

Evans, D. (2011). *The Internet of Things: How the Next Evolution of the Internet Is Changing Everything*. Academic Press.

Gale, J., Wandel, A., & Hill, H. (2019). Will recent advances in AI result in a paradigm shift in Astrobiology and SETI? *International Journal of Astrobiology*, *19*(3), 295–298. doi:10.1017/S1473550419000260

Grignolio, A., Mishto, M., Faria, A. M., Garagnani, P., Franceschi, C., & Tieri, P. (2014). Towards a liquid self: How time, geography, and life experiences reshape the biological identity. *Frontiers in Immunology*, *5*, 153. doi:10.3389/fimmu.2014.00153 PMID:24782860

Joy, B. (2018, October 2). Why the Future Doesn't Need Us. *Wired*. https://www.wired.com/2000/04/joy-2/

Koch, C. (2011) Christof Koch at Singularity Summit 2011: 'The neurobiology and mathematics of consciousness.' *Exponential Times*. http://www.exponentialtimes.net/videos/christof-koch-singularity-summit-2011-neurobiology-and-mathematics-consciousness

Kreisberg, J. C. (2018, January 17). A Globe, Clothing Itself with a Brain. *Wired*. https://www.wired.com/1995/06/teilhard/

Kurzweil, R. (1999). Spiritual Machines, the merging of man and machine. *The Futurist*, *33*(9), 16–21.

Kurzweil, R. (2001, March 7). *The Law of Accelerating Returns*. Kurzweil Accelerating Intelligence. https://www.kurzweilai.net/the-law-of-accelerating-returns

Kurzweil, R. (2005). *TED talk*. https://www.ted.com/talks/ray_Kurzweil_the_accelerating_power_of_technology?language=eu

Kurzweil, R. (2006). *The Singularity Is Near: When Humans Transcend Biology*. Penguin Books.

Lahoz-Beltra, R. (2014, May 14). *The "crisis of noosphere" as a limiting factor to achieve the point of technological singularity*. https://arxiv.org/abs/1405.3378

Last, C. (2015). Big Historical Foundations for Deep Future Speculations: Cosmic Evolution, Atechnogenesis, and Technocultural Civilization. *Foundations of Science*, *22*(1), 39–124. doi:10.100710699-015-9434-y

Last, C. (2020). Atechnogenesis and Technocultural Evolution. *World-Systems Evolution and Global Futures*, 165–188. doi:10.1007/978-3-030-46966-5_9

Logan, R. K. (2013). McLuhan extended and the Extended mind thesis (EMT). Avant: Journal of Philosophical-Interdisciplinary Vanguard, 4(2-3), 45-58. 12849/40202013.0709.0003

Lovelock, J., & Margulis, L. (1974). Atmospheric homeostasis by and for the biosphere: The gaia hypothesis. *Tellus, 26*(1–2), 2–10. doi:10.1111/j.2153-3490.1974.tb01946.x

Lyon, D. (2009, September 16). Ray Kurzweil Wants to Be a Robot. *Newsweek.* https://web.archive.org/web/20100413081045/http://www.newsweek.com/id/197812

Marshall, K. P. (1999). Has technology introduced new ethical problems? *Journal of Business Ethics, 19*(1), 81–90. doi:10.1023/A:1006154023743

McLuhan, M. (1994). *Understanding Media: The Extensions of Man*. The MIT Press.

Miraz, M. H., Ali, M., Excell, P. S., & Picking, R. (2015, September). A review on Internet of Things (IoT), Internet of Everything (IoE) and Internet of Nano Things (IoNT). *2015 Internet Technologies and Applications (ITA).* doi:10.1109/ITechA.2015.7317398

Nazaretyan, A. P. (2014). Singularity Hypotheses: A Scientific and Philosophical Assessment. Minds and Machines, 24(2), 245–248. doi:10.100711023-014-9340-z

Olson, E. T. (2011). The Extended Self. *Minds and Machines, 21*(4), 481–495. doi:10.100711023-011-9258-7

Rothenberg, D. (1993). *Hand's End. Technology and the limits of nature*. University California Press. doi:10.1525/9780520913202

Schwab, K. (2017). The Fourth Industrial Revolution (Illustrated ed.). Currency.

Staff, W. (2017, June 5). The Year 2020, Explained. *Wired.* https://www.wired.com/2000/07/the-year-2020-explained/

Stout, D., Passingham, R., Frith, C., Apel, J., & Chaminade, T. (2011). Technology, expertise and social cognition in human evolution. *The European Journal of Neuroscience, 33*(7), 1328–1338. doi:10.1111/j.1460-9568.2011.07619.x PMID:21375598

Stout, D., Toth, N., Schick, K., & Chaminade, T. (2008). Neural correlates of Early Stone Age toolmaking: Technology, language and cognition in human evolution. *Philosophical Transactions of the Royal Society of London. Series B, Biological Sciences, 363*(1499), 1939–1949. doi:10.1098/rstb.2008.0001 PMID:18292067

Talks, T. E. D. (2006, November 14). *The accelerating power of technology* [Video]. Kurzweil. https://www.ted.com/talks/ray_kurzweil_the_accelerating_power_of_technology?language=eu doi:10.1111/j.1365-2966.2007.11895.x

Taylor, T. (2010). *The Artificial Ape: How Technology Changed the Course of Human Evolution (MacSci)* (1st ed.). St. Martin's Press.

Tipler, F. J. (2003). Intelligent life in cosmology. *International Journal of Astrobiology, 2*(2), 141–148. doi:10.1017/S1473550403001526

Vance, A. (2010, June 14). In the Singularity Movement, Humans Are So Yesterday. *The New York Times*. https://www.nytimes.com/2010/06/13/business/13sing.html

Vernadsky, V. I. (1998). *The Biosphere – Complete Annotated Edition*. Copernicus / Springer Verlag.

Vinge, V. (1993). The Coming Technological Singularity: How to Survive in the Post-Human Era. In NASA (ed.), Vision-21: Interdisciplinary Science and Engineering in the Era of Cyberspace (pp. 11-22), Cleveland, OH: NASA Lewis Research Center.

Weissberger, A. (2014). TiECon 2014 Summary-Part 1: Qualcomm Keynote & IoT Track Overview. *IEEE ComSoc*. Available online: https://community.comsoc.org/blogs/alanweissberger/tiecon-2014-summarypart-1-qualcomm-keynote-iot-track-overview

Wiener, N. (1965). Cybernetics, Second Edition: or the Control and Communication in the Animal and the Machine (2nd ed.). The MIT Press.

XLSemanal. (2020, November 3). Roger Penrose, Nobel de Física: "El reto de la física es explicar cómo funciona la conciencia." *XLSemanal*. https://www.xlsemanal.com/personajes/20201006/roger-penrose-nobel-fisica-agujeros-negros.html

Section 2

Technology-Oriented Developments

Chapter 5
Breakthroughs in Information Technology and Their Implications for Education and Health

Xinwei Shi
https://orcid.org/0000-0003-2303-0870
Tsinghua University, China

Jian Lei
Beihang University, China

ABSTRACT

This chapter reviews the major developments in information technology (IT) since 2000 and the contributions these developments have made to industrial growth. Recently, big data, the internet of things, artificial intelligence, the fifth-generation mobile communications (5G), and other new network-based IT are accelerating the promotion and application of new business models, new formats, and new sectors. IT has become the main driving force of the digital economy. The chapter also discusses likely IT breakthroughs during 2021-40 and the potential opportunities these developments might bring in the education, health, and other sectors. It is expected that the new generation of IT will be driven by the 5G/6G, the internet of everything, and artificial superintelligence.

INTRODUCTION

From the perspective of social development history, mankind has experienced the agricultural revolution, the industrial revolution, and is experiencing the information technology (IT) revolution. The IT revolution has brought a qualitative leap in productivity, and has had a profound impact on the development of international politics, economy, and the society. The term IT was first named by Harold J. Leavitt and Thomas L. Whisler in 1958 in their Harvard Business Review article (Leavitt and Whisler, 1958). IT

DOI: 10.4018/978-1-7998-6772-2.ch005

generally refers to the processing and distribution of data through using computer systems, software, and networks. Currently, the global IT evolution is reshaping the original industrial chain and value chain, and completely stepping into a new stage where a group of new technologies are emerging including the 5G, artificial intelligence, Internet of Things and so forth.

Currently, the COVID-19 pandemic is changing or has already changed our collective recognition of addressing uncertainty as there is no reference case for this kind of crisis in living memory (Durodié, 2020). Importantly, the pandemic is more global in scope, more profoundly impactful and far-reaching, and more complex than any other crisis that today's decision-makers in every sector have experienced[1]. Much evidence, however, has shown that the IT has contributed a lot for people in different sectors to confront the COVID-19 pandemic (Chamola et al., 2020; Gozes et al., 2020; Vaishya et al., 2020). In future, therefore, the need for IT has been proved to become more important than any other industries. Notwithstanding, the research on a systematic review of IT developments in past two decades and accordingly a specific forecast on IT breakthroughs in future 2020 and 2040 is yet to be developed. This chapter aims to bridge such gap.

From 2000 to 2020 we have seen three major IT development trends including digitalization, networking and intelligence (Brennen et al., 2016; Haenlein et al., 2019; Winzer et al., 2018). We need to grasp the focus of the generation of IT, predicting its likely developments in the next two decades (seen from *Figure 1*). First, the Digitization is seen as the foundation of social informatization, and its development trend is the comprehensive digitization of society. Second, the Networking provides a physical carrier for information dissemination, and its development trend is the Internet of Everything. Third, the Intelligence reflects the level of information application, and its development trend is a new generation of intelligence, the so-called super-intelligence (i.e., mock-human emotion). In the future two decades, we believe that the IT breakthroughs will be further driven by Big Data, 5G/6G and Strong AI. The potential path of technological breakthroughs and the continued development trajectory of business models will be reflected through practical examples in the chapter.

Figure 1. Development Trends of IT from 2000 to 2040

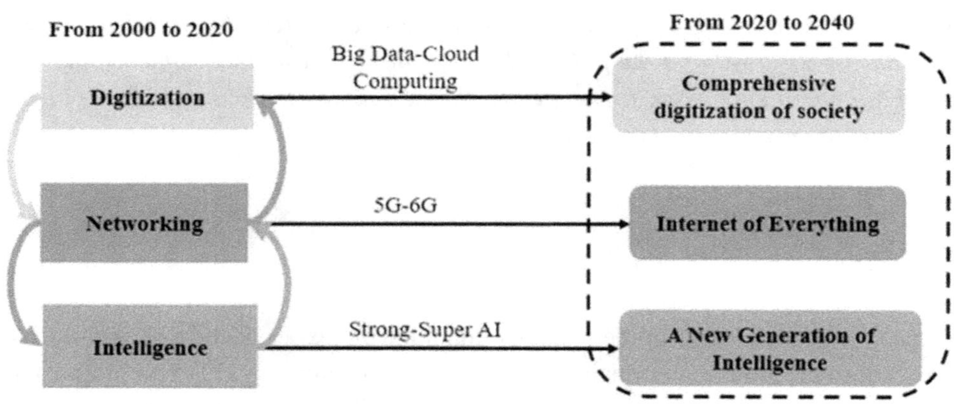

IT DEVELOPMENTS DURING 2001-2020

IT is a broad field that includes all functions and processes relating to computers or technology in an organization, which largely refer to computers and networks (Leavitt and Whisler, 1958). The IT related are usually about physical hardware, operating systems, applications, data storage, databases, servers, and so on. Specifically, the information life cycle of IT is involved with four links, including the acquisition, transmission, processing, and application of information (Matsuyama et al., 2005). Four key technologies that promote the developments of IT system include integrated circuit technology, computer technology, communication technology and software technology.

First, Robert Noyce, founder of Fairchild Semiconductor, in 1959 he invented the world's first integrated circuit (Berlin et al., 2001). In 1968, Noyce and his several other Fairchild employees established Intel. The world's first CPU was produced in 1971 (Danowitz et al., 2012). On the whole, the integrated circuit technology is seen as the foundation of IT development.

Second, in 1936, Alan Turing in 1936 proposed the Turing machine model (Van Leeuwen et al., 2001). On top of this, Von Neumann in 1945 claimed that binary systems should be applied into digital computers and executed in order. Insofar the 'Turing machine' model, as the computer architecture, has still been used. Except for the hardware system technology, the operating system technology also experienced a huge development contributing a lot to the development of computers. Three mainstream independent operating systems include Windows, Linux, and Unix. Moreover, due to the emerging rise of cloud computing, increasing attention has been paid to the cloud operating system technology.

Third, in 1948 Claude Shannon proposed information theory and a mathematical model of the information communication process, which provides the foundation for the modern communication technology (Shannon, 1948). After that, this theory develops rapidly. For example, mobile access technologies such as 3G, 4G and 5G have developed at an incredible speed. Such rapid advancement of digital communication technologies to a considerable extent are promoting the integration of telecommunication networks, television networks and the internet.

Fourth, as known, software technology and computer systems cannot be separable because they are seen as the connector between information and the physical realities. The software technologies, including Internet, and cloud computing, are increasingly evolving towards the networking, which largely facilitates the technology applications. More importantly, software technology has completely embedded in the traditional industries, which further accelerate the digitalization and intelligence in many and various industries, and lead to the era of the Internet of Everything.

As a result, from 2000 to 2020, we have found that IT development trends tend to be the digitalization, networking and intelligence. The popularity of personal computers promotes the digitization process in the early 2000s. Benefiting from the breakthroughs in storage and large-scale integration technology scopes, the computer manufacturing cost is significantly reduced (Gür, 2018). Digitization refers to the digitization of information representation and processing methods in which the data is based on the information carrier in the form of encoding. In short, all data is digitized. The core essence of digitization is the deep understanding and use of big data generated by the integration of IT revolution and economic and social activities.

Since from 2010, the breakthroughs in transmit, such as 4G and WiFi, drove the networking process development. The Internet has become the main approach for people to obtain information, exchange information, and consume information. The Internet is a system of interconnected computer networks that link devices globally using the World Wide Web. It uses internet protocol suites such as TCP/IP

(Transmission Control Protocol/Internet Protocol) to provide end-to-end data communication by specifying how data should be packetized, addressed, routed, transmitted and received. The internet consists of local and global networks for public, private, business, academic, and government use, all linked by a wide array of wireless, electronic and optical networking technology. In education industry, the distance learning became possibility, which is beyond the geographical restrictions (Schneider and Council, 2020). Moreover, the telemedicine effectively eases the contradiction of lack of medical resources. In short, transmit technology breakthrough promoted the networking in 2010s.

Nonetheless, the Internet is only concerned with the interconnection between people and the interconnection of services and services. The Internet of Things is a natural extension and expansion of the Internet (Nauman et al., 2020). In the later 2010s, accompanied by networking, data volume surge. Owing to the improvement of computing power, such as GPU, the data processing capacity has increased rapidly (Phillips et al., 2020). Intelligence reflects the quality attributes of information products. Intelligence includes perception ability, memory and thinking ability, learning and adaptation ability, behavioral decision-making ability, and so forth. Deep learning is an excellent representative of a new generation of artificial intelligence technology (Bahri et al., 2020).

Next, we will look at some major IT breakthroughs in future 2020 and 2040, and their implications on the chosen sectors.

IT'S PROSPECTIVE DEVELOPMENTS DURING 2021-2040

From the long run, IT will continually develop towards the high-speed digitalization, networking and strong intelligence. First, the-state-of-art technologies such as quantum computing will change the traditional computing industries with faster but secure computing technology capacities (Gyongyosi and Imre, 2019). As known, software's technology development will redesign future information infrastructure, innovating firms' business processes. Based on big data and machine autonomous learning, future artificial intelligence (AI) will make contributions to breakthroughs in machine perception, natural language processing (NLP), machine intelligence and so forth. Further, the intelligent robots and assisted learning will further promote the new technology application. More importantly, Gartner (2020) highlights 30 technologies in the next ten years, most of which relates to AI (Figure 2). In the following, we will discuss the future prospective developments of IT during the period of 2021 and 2040.

Digitalization: From Computerization To Digitalization

The value of big data is predominantly realized through big data technology. Big data technology is the extension and development of statistical methods, computer technology, and artificial intelligence technology (Oussous et al., 2018). While IT is penetrating into each sector, the data is increasingly playing an important role in value creations. To put it simply, the big data technology (i.e., software technology or new algorithms) is to deeply mine the value of data resources from increasing unstructured data such as pictures, online documents, videos and so forth that are available from computers and mobile sets. To process massive data, the 'cloud computing' concept was proposed by Google CEO Eric Schmidt, which belongs to one IT application mode and service mode of providing IT resources via Internet (Dingyu, 2018). With growing use of internet-based applications in people's daily lives, the cloud computing service will be full of high demand. As such, the digitalization set off a wave in various industries. For instance,

Figure 2. Gartner Hyper Cycle for Emerging Technologies (Gartner, 2020)

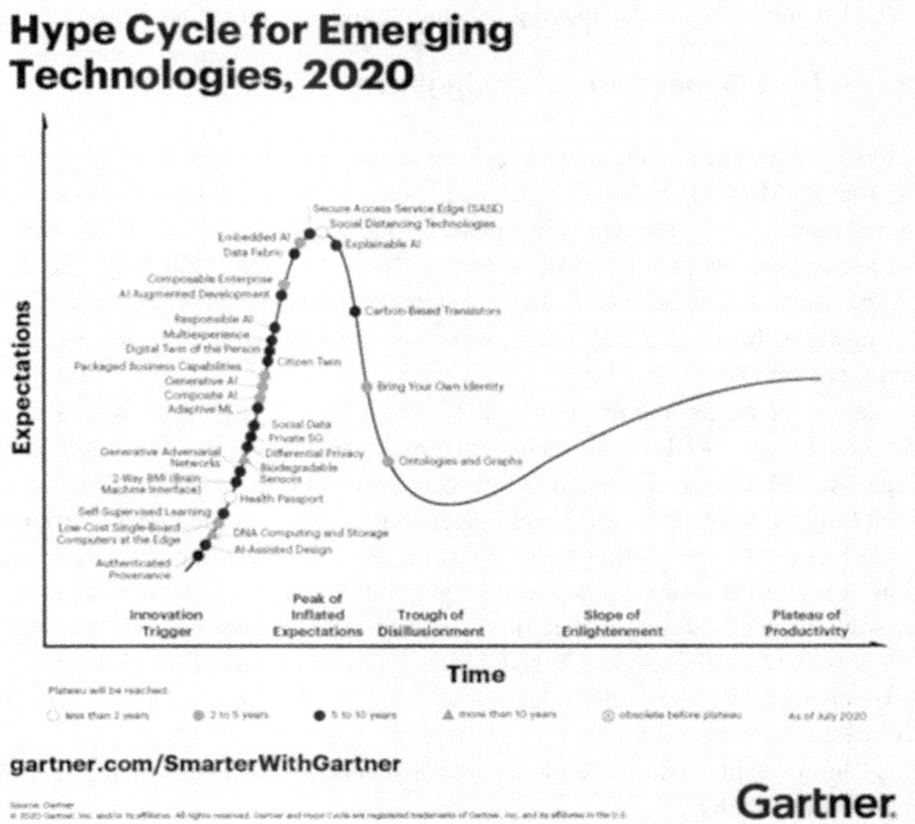

the teaching slides popularized around the world. In medical field, with the help of the development of transmit (such as 3G), hospital information system (HIS) improves hospital operating efficiency. As a consequence, the development of electronic IT and popularization of software is an essential force for digitalization in 2000s.

Networking: Internet of Everything

The Internet of Things (IoT) can be seen as one kind of intelligent network that refers to a comprehensive interconnection of between people and people, people and things, and things and things (Madakam et al., 2015). The IoT is an integrated part of future Internet. This can be achieved through a vast number of technologies, including radio frequency identification (RFID), near field communication (NFC), infrared (IR) sensors, and many more.

The IoT technology mainly relates to three links. First, it refers to the use of RFID, GPS, cameras, and other sensing, capturing subsets to acquire real-time data on objects. Second, the IoT technology relates to the reliable transmission that the information interaction and sharing can be achieved via various communication networks at anytime and anyplace. The third IoT technology relies on the intelligent processing that refers to the massive amounts of data can be analyzed via big data technology so as to extract valuable information and then make wise decisions in the real economic society. Currently, the application of IoT technology has expanded to education, medical and other sectors.

Intelligence: A New Generation of Intelligence

Importantly, the great progress of algorithm, for instance Artificial Intelligence (AI), played a key role in intelligent process (Milsson, 2014). For instance, the AI-based teaching robot assistant replaced the repetitive, monotonous and regular works in teachers' routines (Zhang et al., 2019). The AI algorithms outperformed humans in recognizing medical images (Ker et al., 2017). It helps the doctor diagnose the disease, improving medical efficiency. In space industry, the optimal parameters are computed with machine learning algorithms. All in all, computing power and AI algorithms have become the influential factors in intelligent era (Topol, 2019).

Some people speculate that the development of AI will go through three stages: one is the weak artificial intelligence stage (ANI, Artificial Narrow Intelligence); the second is the strong artificial intelligence stage (AGI, Artificial General Intelligence); the third is the super artificial intelligence stage (ASI, Artificial Super Intelligence) (Barr and Feigenbaum, 2014). In the 'weak artificial intelligence' stage, ANI only focuses on completing a specific task, such as speech recognition, image recognition and translation. It is good at a single aspect of artificial intelligence, similar to advanced bionics. They exist only for solving specific tasks of specific types, and most of them are statistical data from which models can be derived. Google's AlphaGo and AlphaGo Zero are typical 'weak artificial intelligence'. Although they can defeat the world-class champions in the field of chess, AlphaGo and AlphaGo Zero can only play chess (Wang et al., 2016).

Above all, IT has been in existence for thousands of years and has been evolving since the beginning of time, and it is still evolving.

FUTURE IT BREAKTHROUGHS

From 5G to 6G

Currently, the 5G is the fifth generation of mobile network and wireless communication. 5G aims to offer much faster speeds and ultra-reliable device-to-device connections to keep up with the ever-changing ways in which the internet is being used (Andrews et al., 2014). Based on the main characteristics of 5G, the popularization and development of intelligence will be greatly strengthened; first, the interconnection of all things enabled by 5G will realize the full digitization of the physical world. Many devices that cannot be digitized and connected will achieve networking via the use of 5G. Secondly, 5G will greatly expand the scope of smart applications. By 2030, a new network standard will emerge that provides even greater speeds than 5G, namely the 6G (seen from Figure 3). 6G will facilitate the Industry 4.0 revolution, accelerate the development of smart cities, revolutionize the health-care sector through innovations such as mobile edge computing, virtualization and AI, and so forth (Wave Networking Group, 2020).

Figure 3. the Sixth Generation of mobile network (Wave Networking Group, 2020)

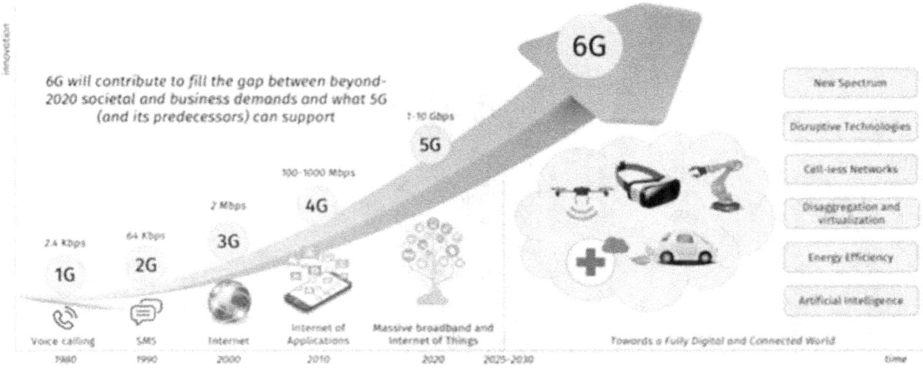

Internet of Everything

Compared with 4G, 5G has three distinct characteristics including high speed low latency, and high capacity, which stands the connecting power for the Internet of Everything (Boudreau et al., 2009). The term 'The internet of everything' is often seen as the next stage of Internet of Thing (Snyder and Byrd, 2017). It refers to consumer products and devices that connect to the internet and have expanded digital features. In short, it is a holistic ecosystem where everything is interconnected. It certainly includes the machine-to-machine communication, people-to-machine and people-to-people communication through technology (Figure 4).

Artificial Super-Intelligence (ASI)

The ASI refers to that the AI technology has the capacity of performing the tasks that are beyond human's capabilities or mind to think (Gill, 2016). In future, ASI will become firms' core production and innovation engine. In the 'artificial super-intelligence' stage, ASI belongs to human-level artificial intelligence, which is comparable to humans in all aspects. A machine with ASI is not only a tool, but also a 'thinking'. A perceptual and self-aware ASI can think, plan, solve problems, think abstractly, understand complex ideas, and learn quickly, and so forth.

Undoubtedly, above the three future major IT breakthroughs will make great contributions to the industrial developments in the sectors of education and health.

INDUSTRIAL IMPLICATIONS OF PROSPECTIVE IT DEVELOPMENTS

This part will mainly select two sectors including the education and health to depict the implications of historical and future IT breakthroughs. Artificial intelligence (AI) can be seen as a branch of computer science for producing intelligent machines (Barr and Feigenbaum, 2014). AI technology can even be applied into address hospital information connectivity, health insurance payment. In the near future, increasing magnitude accumulation in medical information and data, plus more effective algorithm

Figure 4. Internet of Everything (Miraz et al., 2015)

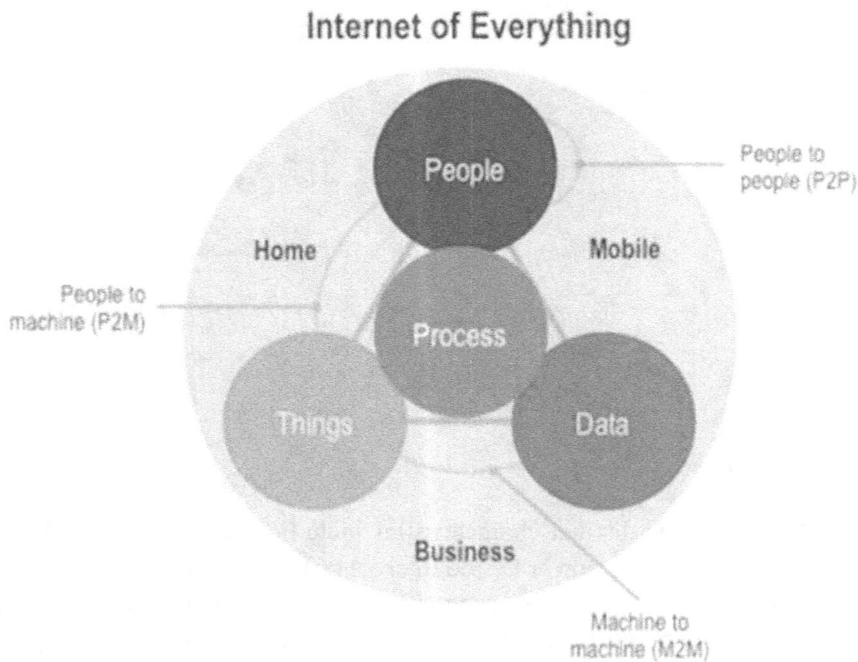

updates, there will be more new application places. In addition, these prospective IT developments will play a significant role on industrial changes.

Education

In short, the AI technology developed is to let machines to simulate human's cognitive capabilities, which is of great application opportunities in education (Beck et al., 1996). We list several examples here. The end pursuit of AI technology to change education lies in the realization of each students' individualization and autonomy. As known, ancient education styles stress the word of mouth from the teachers to students (Bloomer, 2015). The modern school system gives the chance of receiving school education to everyone. With the increasing emergence of mobile internet, more personalized education is highly required by the most people (Tekin et al., 2015). The development of modern IT technologies, especially AI technology, will assist to educate based on each student's characters and aptitudes (Koller, 2011). Abundant online teaching resources, more intelligent diagnosis and reminder, are providing students a more independent learning environment. In addition, the AI platform can automatically collect and capture data from students' autonomous learning activities at any time, eventually forming a data set. Then the platform will produce a diagnostic evaluation report for each student based on a specific analysis model.

Furthermore, teachers' teaching work in the cognitive aspect will be greatly liberated due to the further development of AI technology (Rajeshwari and Krishna Prasad, 2020). Some parts of teaching conceptual and procedural knowledge, for example, can be replaced by machines. Through the use of AI, teachers can free themselves from much repetitive mechanical labor, and students' work are assessed by more intelligent diagnosis system. Meanwhile, students are also provided by more intelligent support and services of online learning (Mayer, 2019). As a consequence, teachers will have more time to provide more individualized communication with students.

Health

In 2018, IDC predicts that the global technology and service expenditures that support the digital transformation exceed 1 trillion dollars, and the healthcare is the most important part (Rydning, 2018). There are two major approaches that have been seen as the most obvious changes in the medical and health industry, including the innovation of drugs and medical device products and changes in treatment technology (Kalis et al., 2018). And the increasing improvement of informatization and digitalization can largely accelerate the improvement of medical quality, the efficiency of medical treatment, and medical management.

Medical institutions require more intelligent data platforms to collect and analyze patient data at real time in order to improve the quality of patient care (He et al., 2019). For instance, IBM Watson Health integrates AI-powered clinical decision support to assist hospital's electronic health record system so as to help adoption and efficiency of the patient care (Smalley, 2017). Moreover, benefited from the application of AI technology, the remote diagnosis and treatment can efficiently address the challenges of limited high-quality doctors (Shah and Chircu, 2018). AI enables to analyze large data sets based on clinical trials, health records, and so forth. Through combing AI and clinical medical information, there can be intelligent diagnosis and treatment of diseases including cancer, cardiovascular and relevant diseases. The establishment of an intelligent platform in hospitals will facilitate multi-department collaboration and management, and even external cooperation.

POTENTIAL SPECULATIONS ON IT DEVELOPMENTS BEYOND 2040

Beyond 2040, the most conservative prediction will be that there are over 100 billion devices that include mobile devices, wearable devices, medical equipment, automobiles and so on being connected to the Internet (Feng et al., 2018). The data and information involved will bring a new revolution to our future daily work and lives.

Moreover, the further advancement of social technology may be predicted that it will assist people to establish own social contracts and social structures via online communities (Schneider, 2018). And it will be increasingly hard to control public opinion.

CONCLUSION

Above all, this chapter reviews the developments of IT in the past two decades from 2000 to 2020, and make some predictions of IT developments during 2021-2040. From the perspective of evolutionary economics, the popularization of any kind of technologies is inextricably linked to the local economic and social developments at that time. In this chapter, the education and health sector are selected to discuss the potential implications of future prospective IT developments. It is significantly worthwhile to continually pay more attention to the IT implications on the education and health sector.

REFERENCES

2020-2022 Emerging Technologies for Enterprises. (2020). *Gartner*. https://www.gartner.com/en/information-technology/trends/emerging-technology-roadmap

Andrews, J. G., Buzzi, S., Choi, W., Hanly, S. V., Lozano, A., Soong, A. C., & Zhang, J. C. (2014). What will 5G be? *IEEE Journal on Selected Areas in Communications*, *32*(6), 1065–1082.

Bahri, Y., Kadmon, J., Pennington, J., Schoenholz, S. S., Sohl-Dickstein, J., & Ganguli, S. (2020). Statistical mechanics of deep learning. *Annual Review of Condensed Matter Physics*.

Barr, A., & Feigenbaum, E. A. (Eds.). (2014). *The Handbook of Artificial Intelligence* (Vol. 2). Butterworth-Heinemann.

Beck, J., Stern, M., & Haugsjaa, E. (1996). *Applications of AI in Education*. XRDS.

Berlin, L. R. (2001). Robert Noyce and Fairchild Semiconductor, 1957-1968. *Business History Review*, 63–101.

Bloomer, W. M. (Ed.). (2015). *A companion to ancient education* (Vol. 120). John Wiley & Sons.

Boudreau, G., Panicker, J., Guo, N., Chang, R., Wang, N., & Vrzic, S. (2009). Interference coordination and cancellation for 4G networks. *IEEE Communications Magazine*, *47*(4), 74–81.

Brennen, J. S., & Kreiss, D. (2016). Digitalization. *The international encyclopedia of communication theory and philosophy*, 1-11.

Chamola, V., Hassija, V., Gupta, V., & Guizani, M. (2020). A comprehensive review of the COVID-19 pandemic and the role of IoT, drones, AI, blockchain, and 5G in managing its impact. *IEEE Access: Practical Innovations, Open Solutions*, *8*, 90225–90265.

Danowitz, A., Kelley, K., Mao, J., Stevenson, J. P., & Horowitz, M. (2012). CPU DB: Recording microprocessor history. *Communications of the ACM*, *55*(4), 55–63.

Dingyu, Z. (2018). Cloud computing technology and its application in robot control. *Iop Conference*, *392*(6).

Durodié, B. (2020). Handling uncertainty and ambiguity in the COVID-19 pandemic. *Psychological Trauma: Theory, Research, Practice, and Policy*, *12*(S1), S61.

Feng, P., Li, Q., Zhang, P., & Chen, Z. (2018). Logical acquisition method based on data migration for Android mobile devices. *Digital Investigation*, *26*, 55–62.

Gill, K. S. (2016). *Artificial super intelligence: beyond rhetoric*. Academic Press.

Gozes, O., Frid-Adar, M., Greenspan, H., Browning, P. D., Zhang, H., Ji, W., . . . Siegel, E. (2020). *Rapid ai development cycle for the coronavirus (covid-19) pandemic: Initial results for automated detection & patient monitoring using deep learning ct image analysis*. arXiv preprint arXiv:2003.05037.

Gür, T. M. (2018). Review of electrical energy storage technologies, materials and systems: Challenges and prospects for large-scale grid storage. *Energy & Environmental Science*, *11*(10), 2696–2767.

Gyongyosi, L., & Imre, S. (2019). A survey on quantum computing technology. *Computer Science Review*, *31*, 51–71.

Haenlein, M., & Kaplan, A. (2019). A brief history of artificial intelligence: On the past, present, and future of artificial intelligence. *California Management Review*, *61*(4), 5–14.

He, J., Baxter, S. L., Xu, J., Xu, J., Zhou, X., & Zhang, K. (2019). The practical implementation of artificial intelligence technologies in medicine. *Nature Medicine*, *25*(1), 30–36.

Kalis, B., Collier, M., & Fu, R. (2018). 10 promising AI applications in health care. *Harvard Business Review*.

Ker, J., Wang, L., Rao, J., & Lim, T. (2017). Deep learning applications in medical image analysis. *IEEE Access: Practical Innovations, Open Solutions*, *6*, 9375–9389.

Koller, D. (2011). Death knell for the lecture: Technology as a passport to personalized education. *New York Times*, 5.

Leavitt, H. J., & Whisler, T. L. (1958). Management in the 1980's. *Harvard Business Review*, *14*, 41–48.

Madakam, S., Lake, V., Lake, V., & Lake, V. (2015). Internet of Things (IoT): A literature review. *Journal of Computer and Communications*, *3*(05), 164.

Matsuyama, S., Akagiri, K., & Suginuma, K. (2005). *U.S. Patent Application No. 10/985,729*. US Patent Office.

Mayer, R. E. (2019). Thirty years of research on online learning. *Applied Cognitive Psychology*, *33*(2), 152–159.

Miraz, M. H., Ali, M., Excell, P. S., & Picking, R. (2015). *A review on Internet of Things (IoT), Internet of Everything (IoE) and Internet of Nano Things (IoNT). 2015 Internet Technologies and Applications*. ITA. doi:10.1109/itecha.2015.7317398

Nauman, A., Qadri, Y. A., Amjad, M., Zikria, Y. B., Afzal, M. K., & Kim, S. W. (2020). Multimedia Internet of Things: A comprehensive survey. *IEEE Access: Practical Innovations, Open Solutions*, *8*, 8202–8250.

Nilsson, N. J. (2014). *Principles of artificial intelligence*. Morgan Kaufmann.

Oussous, A., Benjelloun, F. Z., Lahcen, A. A., & Belfkih, S. (2018). Big Data technologies: A survey. *Journal of King Saud University-Computer and Information Sciences*, *30*(4), 431–448.

Phillips, J. C., Hardy, D. J., Maia, J. D., Stone, J. E., Ribeiro, J. V., Bernardi, R. C., ... Tajkhorshid, E. (2020). Scalable molecular dynamics on CPU and GPU architectures with NAMD. *The Journal of Chemical Physics*, *153*(4), 044130.

Rajeshwari, M., & Krishna Prasad, K. (2020). IBM Watson Industry Cognitive Education Methods. *International Journal of Case Studies in Business, IT, and Education*, *4*(1), 38–50.

Rydning, D. R. J. G. J. (2018). *The digitization of the world from edge to core*. International Data Corporation.

Schneider, N. (2018). An Internet of ownership: Democratic design for the online economy. *The Sociological Review*, *66*(2), 320–340.

Schneider, S. L., & Council, M. L. (2020). Distance learning in the era of COVID-19. *Archives of Dermatological Research*, 1–2.

Shah, R., & Chircu, A. (2018). IoT and AI in healthcare: A systematic literature review. *Issues in Information Systems*, *19*(3).

Shannon, C. E. (1948). A mathematical theory of communication. *The Bell System Technical Journal*, *27*(3), 379–423.

Smalley, E. (2017). *AI-powered drug discovery captures pharma interest*. Academic Press.

Snyder, T., & Byrd, G. (2017). The internet of everything. *IEEE Computer Architecture Letters*, *50*(06), 8–9.

Tekin, C., Braun, J., & van der Schaar, M. (2015, April). eTutor: Online learning for personalized education. In *2015 IEEE International Conference on Acoustics, Speech and Signal Processing (ICASSP)* (pp. 5545-5549). IEEE.

Tian, Y., Ma, J., Gong, Q., Sengupta, S., Chen, Z., Pinkerton, J., & Zitnick, L. (2019, May). Elf opengo: An analysis and open reimplementation of alphazero. In *International Conference on Machine Learning* (pp. 6244-6253). PMLR.

Topol, E. J. (2019). High-performance medicine: The convergence of human and artificial intelligence. *Nature Medicine*, *25*(1), 44–56.

Vaishya, R., Javaid, M., Khan, I. H., & Haleem, A. (2020). Artificial Intelligence (AI) applications for COVID-19 pandemic. *Diabetes & Metabolic Syndrome*, *14*(4), 337–339.

Van Leeuwen, J., & Wiedermann, J. (2001). The Turing machine paradigm in contemporary computing. In *Mathematics unlimited—2001 and beyond* (pp. 1139–1155). Springer.

Wang, F. Y., Zhang, J. J., Zheng, X., Wang, X., Yuan, Y., Dai, X., . . . Yang, L. (2016). Where does AlphaGo go: From church-turing thesis to AlphaGo thesis and beyond. *IEEE/CAA Journal of Automatica Sinica, 3*(2), 113-120.

Wave Networking Group. (2020). *6G*. Retrieved from http://mmwave.dei.unipd.it/research/6g/

Winzer, P. J., Neilson, D. T., & Chraplyvy, A. R. (2018). Fiber-optic transmission and networking: The previous 20 and the next 20 years. *Optics Express*, *26*(18), 24190–24239.

Zhang, B., Xia, H., & Lim, H. O. (2019, December). Development of an AI based teaching assisting system. In *Proceedings of the International Conference on Artificial Intelligence, Information Processing and Cloud Computing* (pp. 1-5). Academic Press.

ENDNOTE

[1] Source: Deloitte https://www2.deloitte.com/global/en/pages/about-deloitte/articles/covid-19/covid-19--confronting-uncertainty-through---beyond-the-crisis-.html

Chapter 6
Revisiting Our Quantum World:
Applications to Education, Health, and Security

Lester Ingber
https://orcid.org/0000-0003-1502-3910
Physical Studies Institute LLC, USA

ABSTRACT

We now know that quantum mechanics has been a fundamental structure of our world since the universe came into being. However, it has been only a century since the experimental and theoretical discoveries of quantum mechanics were made. We are becoming increasingly aware of its many implications and applications. In particular, there are implications across many disciplines that most likely will affect education, health, and security. Examples are given of the need to start education as early as possible in schools, the use of nano-robots to deliver drugs targeted to specific molecular sites, and to developing new cryptographic systems to safeguard our privacy.

1. INTRODUCTION

As seen by differences in two reviews of this chapter, in the space permitted and without introducing equations, it is likely not possible to make this presentation both easily readable and detailed for most readers. The choice of examples are those of the author, and care has been taken to offer readability and depth at level to convey the importance of the Quantum World to all readers.

While some attribute the birth of computers to Charles Babbage circa 1837 since he designed his Analytic Engine with a rather full set of instructions, that machine was never actually constructed, e.g., see https://en.wikipedia.org/wiki/Analytical_Engine .

The history of computers is relatively new, less than 90 years old, within the lifetime of some readers of this book! Clearly, humans have been quite busy during this time, developing powerful computers to replace pen/pencil and paper.

An outline of the short (in time only) history of computers includes:

DOI: 10.4018/978-1-7998-6772-2.ch006

- The mathematical foundations of computing are generally attributed to Turing in 1936 (Turing, 1937).
- The first computer is usually considered to be the ENIAC, completed in 1946.
- The first quantum algorithm that has had far-reaching implications is mostly attributed to Shor in 1994 (Shor, 1994).
- The physical-mathematical foundations of quantum computing are generally attributed to Benioff in 1980 (Benioff, 1980).
- The initial history of quantum computers *per se* is usually attributed to Feynman in 1982 (R. Feynman, 1982).
- The first commercially available quantum computer is generally considered to be made by D-WAVE in 2017, building on a small 2-qubit quantum computer built circa 1997-1998.

The basic unit of a quantum computer is a qubit (quantum bit), the quantum analog of a classical-computer bit. A classical bit can be a state 0 or 1, whereas a qubit can be in a linear combination, called a superposition, of 0 and 1. For a classical computer with 2 bits, there are 4 possible states, but only one state can be realized at any given time. For a quantum computer with 2 qubits, there are 4 possible states, all of which can be realized simultaneously, and "superposition" of all linear combinations are possible. There are multiple theories for testing, in some cases offering radically different interpretations of reality based on quantum mechanics, e.g., https://www.scientificamerican.com/article/this-twist-on-schroedingers-cat-paradox-has-major-implications-for-quantum-theory/ .

When such a quantum state is subject to decoherence, e.g., by a classical measurement process, it becomes a classical state, wherein it becomes just one classical state. This decoherence is at the heart of why it is so difficult to build a quantum computer, as any classical perturbation causes quantum states to collapse into classical states. Decoherence still is referred to by many people as the "collapse" of the wave-function; however, even some of the issues reported here are not well explained using the term collapse.

1.1. Factorization of Large Numbers Using Shor's Algorithm

Most people who read about quantum computers associate the technology with new methods of computation that will be game changers for security. That is because many articles have been written on how this new technology will change the way just about all major passwords are encrypted using factorization of large numbers.

1.2. Faster Searches Using Grover's Algorithm

The probability of a quantum state is given by the absolute square of the wave-function of that state.

Because the wave-function lives in a complex space with both real and imaginary numbers, the absolute square multiplying the wave-function times its Hermitian transpose, changing the sign of the imaginary part, gives a real number, whereas the simple square will not unless the imaginary part is zero.

In 1996, Grover developed an algorithm to statistically search spaces of size N in time on the order of the square root of N, whereas a classical algorithm requires time statistically on the order of N (Grover, 1997). Basically, Grover's algorithm uses the wave-function to perform the search, whereas classical algorithms must sample the probability of states. A description (with equations) is given in https://www.quantiki.org/wiki/grovers-search-algorithm .

1.2.1. Quantum Key Distribution

As expected, since quantum computers may crack most classical encryption that depends on factorization, there have been efforts to find methods of quantum encryption. One such method is quantum key distribution (QKD) (Fang *et al*, 2020), which recently has been used to transmit secure messages up to and back from a satellite.

1.3. Many Quantum Computers

There are many quantum computers now available from many companies/institutions. About a year ago, the author cited 17 (Ingber, 2018b).

D-WAVE (Canada)
DeepMind (Canada)
Facebook
Google
IBM
Intel
Microsoft
National Laboratory for Quantum Information Sciences (China)
Nippon Telegraph and Telephone
NOKIA Bell Labs
NSA
Post-Quantum
Rigetti
Russian Quantum Center
Toshiba
Quantum Circuits
Quantum Technologies (European Union)

As of the time of writing this chapter, there is an explosion of companies and institutions implementing quantum computers. For example, see https://eas.caltech.edu/engenious/16/chair_message. Caltech is also entering this domain using Amazon's AWS platform; see https://www.caltech.edu/about/news/preparations-begin-aws-center-quantum-computing-caltech.

The advances in the number of qubits available for computation have been expanding exponentially; see https://www.statista.com/chart/17896/quantum-computing-developments/. In October 2019, Google announced a breakthrough in computational power; see https://www.weforum.org/agenda/2019/10/quantum-computers-next-frontier-classical-google-ibm-nasa-supremacy. In August 2020 Quantum Base announced business plans to build a computer with 1000 trillion qubits; see https://www.forbes.com/sites/daveywinder/2020/08/18/goodbye-passwords-hello-unbreakable-quantum-ids-containing-1000-

trillion-atoms-quantum-base-qid-lancaster-university. Also, in August 2020 IBM announced a major breakthrough in https://www.zdnet.com/article/ibm-hits-new-quantum-computing-milestone/.

Although companies are exploring different machines, the scales of thought of quantum computation are expanding rapidly. Quantum computing is being taken quite seriously even as it struggles to compete with classical computing on large scales.

1.4. Quantum Mechanics Is Still A Mystery

It must be appreciated that quantum mechanics (QM) is still a mystery to us all. There are unresolved issues about the basic assumptions, called axioms, on which QM rests. There is a steady stream of research examining these issues; for example, https://phys.org/news/2020-08-quantum-paradox-reveals-contradiction-widely.html reports on such aspects published as this chapter is being written (Bong *et al*, 2020). The issue arises when considering Wigner's observer observing the observer of a quantum experiment, a point raised by Wigner in 1961 (Wigner, 1961). It is well established that when a macroscopic body, like a large machine or a person, interacts with a quantum event, the wave-function describing that event suffers decoherence, and the event is pinned into a probabilistic event which is what is typically measured. That is, the measurement just measures one of the possible manifestations of the wave-function, not the wave-function itself. When consideration is made for an observer observing a sensitive quantum system making a quantum measurement on a different quantum system, the analysis can fork into different possibilities as yet not well understood.

1.5. What Is At Stake?

Why is this new quantum technology important, and how can it affect our lives? Some rationale will be given in this chapter. Keep in mind that just about all of this is still considered as being speculative, in terms of what new quantum technologies can deliver on large social scales. However, in terms of a financial options calculation, where the expected gain/loss over the duration of the option is a sum of products of expected payoffs/losses times probabilities of becoming mature technology, readers will see that the expected gain is quite large for humanity.

It may "hurt" to try to understand some quantum technologies — in the sense of getting emotionally disturbed by being presented with a lot of new information. To make this transition a bit easier, we will look at two example systems, one that was with us in the womb — our brain, and the other a vital part of our social interactions — money. Then, we may also get some understanding of another system, with the key word "blockchain" that many readers may have seen in the press, and how developments in the quantum world may deeply affect us soon.

The following sections will describe quantum influences in several disciplines. Becoming familiar with the quantum world requires education, which should be acquired at as early an age as possible (in single digits), to develop intuitions that will help us to learn more if we wish, and at least prepare us to make informed decisions affecting all our lives. Neuroscience will lead to new drugs. Quantum cryptology will lead to new and secure methods of transactions, including those involving money. Quantum blockchains will lead to new secure ledgers of information that affect bank transactions, government databases, and storing and retrieving your personal information. These are much too important to leave in the hands of a few people, even if you feel you could trust them. Instead, we all must become vital participants in our futures.

2. EDUCATION

The quantum world is not "intuitive." Before 1900, no one truly imagined the nature of the quantum world. Even Democritus and Leucippus, his mentor, both circa 400-500 B.C., who are credited with postulating "atomism", had no clue as the nature of the atomic world. Indeed, it took experiments only possible in the 1900's to understand the nature of quantum mechanics. For example, the famous double-slit experiment by Davisson and Germer in 1927 (Davisson & Germer,1927), which may have been modeled on a similar experiment performed by Young in 1801, showed that truly atomic particles possess a wave-particle duality. Quantum entities (like an electron) sometimes behave as particles, e.g., in collisions, and sometimes behave as waves, e.g., in defraction like in the double-slit experiments. When they behave as waves, they can demonstrate interference patterns, which we now attribute to the wave-function describing their paths in space-time. When they behave as particles, it is the absolute square of this wave-function that describes the probability of their space-time location.

Quantum particles also exhibit the "uncertainty principle", wherein variables that are mathematically "conjugate" within the experimentally verified Planck constant $h = 1.0545718$ X 10-34 meters-squared kilogram/sec. For example, momentum and positions are conjugate variables, as are energy and time.

There are limits to the precision/resolution that can be obtained by both members of conjugate variables, the product being limited to h. This means that the resolution of momentum multiplied by the resolution of position (in the same system) can only be measured to within h; at their minimum levels of resolution, an increase of resolution of one of them incurs a corresponding loss of resolution in the other.

The primary take-home lesson is that all this is not intuitive. We only know the above because of verified experiments and verified theories since circa 1900.

So, how can we expect children to enter the new world of quantum technologies that are rapidly being created, based solely on their interactions with the world devoid of knowledge of these quantum experiments and theories? The clear answer is that they cannot, and they will not, learn any of this unless they are exposed to these experiments and theories. While the mathematics and physics theory needed to understand quantum mechanics may require some years to acquire, basic intuitions of this world can easily be imparted by exposing children to many experiments as early as pre-kindergarten, even before their formal education begins! Many people who are not exposed to the quantum world will be clueless and helpless to make informed decisions, e.g., like voting on funding new technologies or on selecting drugs based on these technologies. Therefore, we must start as early as possible, educating students and citizens about the quantum world in which they live.

There are several online sites that offer courses in quantum computing, e.g.,_https://www.edx.org/learn/quantum-computing_https://www.coursera.org/courses?query=quantum%20computing_and in quantum mechanics, e.g., https://www.edx.org/learn/quantum-physics-mechanics_https://www.coursera.org/courses?query=quantum%20physics

Most courses are free.

There also are attempts to teach quantum mechanics without equations, e.g., see https://arstechnica.com/science/2021/01/the-curious-observers-guide-to-quantum-mechanics/ .

3. NEUROSCIENCE

As mentioned above, quantum mechanics may be influential in our thought processes, e.g., as described below. It is unclear whether new advances in quantum mechanics may bring forth new ways of developing these thought processes, e.g., perhaps aided by a Brain-Computer Interface (BCI), e.g., see https://en.wikipedia.org/wiki/Brain%E2%80%93computer_interface .

It is likely possible to control brain function by targeting specific mechanisms at molecular levels, instead of using drugs that are widely systemic and have multiple contraindications. As an example, discussed in a recent paper (Ingber, 2018a), large-scale neuronal firings can influence regenerative molecular $_{Ca}{}^{2+}$ generated at tripartite neuron-astrocyte quantum-scale sites. This suggests a mechanism by which nano-robots may control background synaptic activity that in turn control attention and memory, by delivering pharmaceutical agents. This interaction, if indeed found to be present in the neocortex, would represent interactions over huge ranges, interactions between quantum molecular scales and highly synchronous neural firings among millions of neurons.

Of the 10^{11} cells in the human brain, about half are neurons, and the other half are glial cells. Astrocytes likely comprise most glial cells. There are many papers which examine the roles of astrocytes on synaptic processes (Bellinger, 2005; Innocenti *et al*, 2000; Scemes & Giaume, 2006; Agulhon *et al*, 2008; Pereira & Furlan, 2009; Reyes & Parpura, 2009; Araque & Navarrete, 2010; Banaclocha *et al*, 2010; Volterra *et al*, 2014).

Regenerative intercellular calcium waves (ICWs) can travel over 100s of astrocytes, encompassing many neuronal synapses. These ICWs are documented in the control of synaptic activity (Ross, 2012).

Ca^{2+} (concentration of Ca^{2+}) increases release probabilities at synaptic sites, likely due to $_{Ca}{}^{2+}$ triggering release of gliotransmitters. (Free waves are considered here, not intracellular astrocyte calcium waves in situ which also increase neuronal firings.)

During selective attention tasks, partially confirmed by fitting scalp electroencephalographic (EEG) $_{Ca}{}^{2+}$ recordings, free regenerative waves, arising from tripartite interactions, couple to the magnetic vector potential **A** produced by highly synchronous collective firings. **A** is derived from the neocortical dipole moment **Q** averaged over many neurons, in turn proportional to the current **I** which develops the electric potential on the scalp measured as EEG.

3.1. Reasonable Estimates

Estimates used for **Q** come from experimental data. When coherent activity among many macro columns associated with short-term memory (STM) is considered, **A** may be orders of magnitude larger. There is direct coherence between $_{Ca}{}^{2+}$ waves and the activity of **A**. A simple classical calculation shows $q\mathbf{A}$, where q is the charge of a Ca^{2+} ion, from macroscopic EEG to be on the order of 10^{-28} kg-m/s, while the momentum **p** of a Ca^{2+} ion is on the order of 10^{-30} kg-/ms.

The recent XSEDE.org project developed by the author (Ingber, 2011; Ingber, 2012b; Ingber, Pappalepore & Stesiak, 2014; Ingber, 2015; Ingber, 2016b), investigates Top-Down influences of macroscopic patterns of neuronal firings, measured by scalp EEG during attentional memory tasks, on Bottom-Up free microscopic Ca^{2+} ions created tripartite interactions, which interact with background synaptic activity that in turn influence large synchronous firings of neurons.

During the development of this project, it was recognized that the proposed mechanism for this Top-Down Bottom-Up synergy could be influenced by nano-robots, to deliver pharmaceutical agents under conscious control of a person (Ingber, 2015).

3.2. EEG Influences on CA Waves (EICW)

3.2.1. SMNI Model of Neocortical Interactions

Since 1980 a Statistical Mechanics of Neocortical Interactions (SMNI) model has been developed in a couple score of papers, scaling synaptic interactions to neuronal firings, to minicolumnar firings, to macro columnar firings, to regional firings — the scale measured by scalp EEG (Ingber, 1982; Ingber, 1983; Ingber, 1984; Ingber, 1994; Ingber, 1997; Ingber, 2012b). At the columnar level, SMNI has detailed properties of short-term memory (STM), e.g., capacity (auditory 7 ± 2 and visual 4 ± 2), duration, stability, primacy versus recency rule, and Hick's law. Only recently have neuroscientists confirmed with some experiments (Salazar *et al*, 2012; Asher, 2012), that some STM are processed by highly synchronized patterns of neuronal firings.

At the regional scale SMNI has derived the EEG wave equation (Ingber & Nunez, 2010), and has been tested by fitting EEG Training data using Adaptive Simulated Annealing (ASA) (Ingber, 1993; Ingber, 2012a), testing on out-of-sample Testing sets, including centers sponsored by the National Institute on Alcohol Abuse and Alcoholism (NIAAA) (X.L. Zhang, Begleiter, Porjesz, Wang & Litke, 1995; X.L. Zhang, Begleiter & Porjesz, 1997; X.L. Zhang, Begleiter, Porjesz & Litke, 1997). This data set was used in earlier SMNI studies (Ingber, 1997; Ingber, 1998).

3.2.2. Neurology Considered

Astrocytes influence glutamate, the main excitatory neurotransmitter in the neocortex, by taking in some glutamate released by presynaptic neurons and converting it into glutamine which can enter presynaptic neurons where it can be re-converted into glutamate. GABA is the main inhibitory neurotransmitter in the neocortex, produced by inhibitory neurons by also utilizing glutamic acid (which when stripped of a hydrogen atom is glutamate) from astrocytes (Patel *et al*, 2001; Walls *et al*, 2015). Ca^{2+} waves arise from nonlinear cooperative regenerative processes from internal stores, involving several biochemical steps (Ross, 2012; Pitta *et al*, 2012; Goldberg *et al*, 2010).

3.2.3. Classical/Quantum Vector Potential influence on Ca^{2+} Momenta

Columnar firings develop electromagnetic fields described by a magnetic vector potential, referred to as the SMNI vector potential (SMNI-VP). Early experiments included the "Smoking Gun" that implicates top-down interactions at molecular scales (Ingber, 2011; Ingber, 2012b). Some papers calculated the approach in a classical physics context (Ingber, 2012c). After these papers were published, detailed interactions were described of SMNI-VP firing states with Ca^{2+} waves, in both classical and quantum mechanics (Ingber, 2011; Ingber, 2012b), and quantum contexts (Ingber, Pappalepore & Stesiak, 2014; Ingber, 2015), using the "canonical momentum" $\mathbf{p} + q\mathbf{A}$. The theoretical construct of the canonical momentum $\mathrm{n} = \mathbf{p} + q\mathbf{A}$ is firmly entrenched in classical and quantum mechanics (R.P. Feynman, 1949; Schulman, 1981). See https://www.feynmanlectures.caltech.edu/II_15.html for a nice introduction.

Columnar firing states are reasonably modeled as a wire/neuron with current **I** along a length z observed from a perpendicular distance r from a line of thickness r_0. The integral (sum of contributions) of **I** over r gives the magnetic vector potential **A** (Jackson, 1962). This gives an insensitive log dependence on distance. The contribution to **A** includes many minicolumnar lines of current from 100's to 1000's of macro columns, contributing to large synchronous bursts of EEG (Srinivasan *et al*, 2007). **E** and **B**, are derivatives of **A** with respect to *r*, giving inverse-polynomial sensitivities. They do not possess this logarithmic insensitivity and do not linearly accumulate strength across macro columns. This study is robust against much theoretical modeling as experimental data are used wherever possible.

The author used path-integrals to derive a closed-form solution of the propagation of the Ca^{2+} wave-packet composed of many Ca^{2+} ions (Ingber, 2018a). This wave function was used to calculate the probability distribution of this wave-packet, which was used to modify synaptic interactions in the SMNI model, tested by fitting this to EEG data. The calculation is quite sensitive to h which is a test of the sensitivity to quantum interactions at the large scale of columnar firings of neurons.

3.2.4. Fitting SMNI Model Including EICW To EEG Data

The influence of **A** on the background-noise of synaptic parameters SMNI used ASA to fit 28 parameters across a circuitry underlying 6 electrode sites using NIAAA data (Ingber, Pappalepore & Stesiak, 2014; Ingber, 2015).

The momenta of Ca^{2+} ions are influenced during EEG events like N100 and P300 potentials common in selective attention tasks (Srinivasan *et al*, 2007). Codes for SMNI fits to EEG data (Ingber, 1997; Ingber, Ca^{2+} 1998) were used as templates. The influence of Ca^{2+} waves is tested by parameterizing synaptic parameters to be dependent on Ca^{2+} wave activity. To date, results of fits to EEG data only demonstrate that fits to an **A** model are reasonably better than fits to the no-**A** model.

3.3. Conscious Control of Nano-Robots Influencing Attention and Memory

It has been demonstrated that nano-robots may be used in this context (Ingber, 2015), assuming

Ca^{2+}-wave momentum-sensors acting as piezoelectric devices (Alivisatos *et al*, 2013; Wang, 2012). At the beginning of a Ca^{2+} wave (100's of ms), a change of momentum is of the order of 10^{-30} kg-m/s. A Ca^{2+} wave packet of 1000 ions with onset time of 1 ms produces a force on the order 1 Newton = 1 kg-m/s^2). Nano-robots drawn to this site could deposit chemicals/drugs that interact with the Ca^{2+}-wave process. If the receptor area of the nanosystem were 1 nm^2 (resolution of scanning confocal electron microscopy (SCEM)), this would require a pressure sensitivity of 10^{-6} Pa (1 Pa = 1 pascal = 1 N/m^2).

The nano system could be switched on/off at a regional/columnar level by sensitivity to local electric/magnetic fields.

3.4. Free Will

In addition to researching STM and multiple scales of neocortical interactions, there is interest in possible quantum influences on highly synchronous neuronal firings to understand possible connections to consciousness and "Free Will" (FW) (Ingber, 2016a; Ingber, 2016b).

If neuroscience ever establishes experimental evidence from quantum-level processes of tripartite synaptic interactions with large-scale synchronous neuronal firings, then FW may be established using the Conway-Kochen quantum no-clone "Free Will Theorem" (FWT) (Conway & Kochen, 2006; Conway & Kochen, 2009).

Thus, a $_{Ca}^{2+}$ quantum wave-packet could generate a state proven to have not previously existed, since quantum states cannot be cloned. Therefore, thoughts generated (in part) by quantum processes can indeed be unique.

3.5. Quantum Zeno Effects

The wave function of the Ca^{2+} wave packet was calculated by the author, and it was demonstrated that despite multiple collisions during their regenerative processes over long durations of hundreds of ms, typical $_{Ca}^{2+}$ waves support a Zeno or "bang-bang" effect which may promote long coherence times (Facchi, Lidar & Pascazio, 2004; Facchi & Pascazio, 2008; Wu *et al*, 2012; Giacosa & Pagliara, 2014; P. Zhang *et al*, 2014; Kozlowski *et al*, 2015; Patil *et al*, 2015; Muller *et al*, 2016).

Of course, the Zeno/"bang-bang" effect may exist only in special contexts, since decoherence among particles is known to be one of the fastest processes known (Preskill, 2015).

However, the constant collisions of Ca^{2+} ions, causing some ions to enter or leave the wave packet, may perpetuate at least part of the wave, permitting the Zeno/"bang-bang" effect. qPATHINT provides an opportunity to explore the coherence stability of the wave due to serial shocks of this process. qPATHINT was developed by the author and has been used to calculate the evolution of the quantum path-integral, an algorithm developed bBy Feynman in 1948 (R.P. Feynman, 1948) to propagate short-time probabilities over long time spans. PATHINT is the underlying classical code developed for classical systems (Wehner & Wolfer, 1983a; Wehner & Wolfer, 1983b; Wehner & Wolfer, 1987; Ingber, Fujio & Wehner, 1991; Ingber, Wehner *et al*, 1991), further developed by the author for arbitrary dimensions.

The path-integral presents a mathematically equivalent representation of both multivariate stochastic differential equations with multiplicative noise (nonlinear coefficients of Gaussian noise) and of multivariate partial differential equations, but also offering numerical and intuitive advantages such as derivations of common concepts like force, momentum (Langouche *et al*, 1979).

4. FINANCE

4.1. Quantum Money

Many countries are exploring the use of "digital currencies". E.g., see https://www.federalreserve.gov/newsevents/speech/brainard20200813a.htm. These include the use of blockchains for security but controlled by central governments. This can be viewed as just the first step to considering new currencies, which of course will compete with, if not dominate, distributed blockchains like Bitcoin. While blockchains are often associated with distributed systems, even centralized systems are looking at blockchains simply because of their efficiency in recording ledgers.

Soon, quantum computing will continue to be investigated for applications to financial products. Financial derivative products will be developed to offer hedging as well as speculation. Then, qPATHTREE and qPATHINT will be poised to calculate financial derivatives in these quantum complex spaces. This presents contexts beyond using quantum computers to calculate financial derivatives, since the space of the dependent variables themselves may live in quantum worlds (Baaquie *et al*, 2002; Piotrowski *et al*, 2005; Accardi & Boukas, 2007; Meyer, 2009; Aaronson & Christiano, 2012; Jogenfors, 2016).

4.2. Financial Options on Quantum Money

PATHINT and the quantum version qPATHINT were developed to numerically calculate the path integral, especially for serial changes in time in the presence of random shocks — not approachable with standard Monte Carlo techniques. The codes are written for arbitrary N dimensions, and have been used for several papers in both classical and quantum systems (2000; Ingber & Wilson, 2000; Ingber, 2016a; Ingber, 2017a; Ingber, 2017b).

Many systems propagate in the presence of continual "shocks", including: future dividends, changes in interest rates, and changes in asset distributions used in American options algorithms which are included in these calculations. Even such classical random shocks have been introduced into qPATHINT quantum systems, demonstrating how hybrid classical-quantum calculations can be developed.

Similarly, the development of hybrid classical-quantum systems currently is a high priority for many developers of quantum computers, e.g., see https://www.dwavesys.com/press-releases/d-wave-announces-general-availability-first-quantum-computer-built-business .

5. BLOCKCHAINS

The design of the "Blockchain" is generally attributed to Nakamoto in 2008 (https://en.wikipedia.org/wiki/Satoshi_Nakamoto). A copy of his 2008 white paper can be downloaded from https://bitcoin.org/bitcoin.pdf .

The blockchain is a ledger that is validated, entry by entry, for any item or contract. Still today, this validation via encryption pretty much hiding identities of people utilizing them is at the core of much strain in crypto-products like crypto-currency (such as Bitcoin, Ethereum and Ripple). Proof of

Work was the original design and still has the largest scale via Bitcoin, but this requires lots of energy and time for the constant verification process; Bitcoin's potential drain on energy is even considered to contribute to climate change!

While theoretically, anyone can participate in the validation process, in practice only a handful of large data banks as well as consuming towns and cities effectively do most of the validation. This has given rise to some products like Ripple that do away with the democracy of validation by validators, and these are endorsed by many large institutions, e.g., many banks, which prefer this more efficient control over resources, dismaying many ardent supporters of blockchains.

As in SMNI, the core of the quantum no-clone "Free Will Theorem" (FWT) theorem can have important applications in this domain. For example, quantum currency cannot be cloned. Such currencies are exceptional candidates for very efficient blockchains, e.g., since each "coin" has a unique identity (Meyer, 2009; Aaronson & Christiano, 2012; Bartkiewicz *et al*, 2016; Jogenfors, 2016). As in SMNI, there are issues about the decoherence time of such "coins".

ACKNOWLEDGMENT

Some of the work cited has used supercomputer resources of the Extreme Science and Engineering Discovery Environment (XSEDE.org), where the author has been Principal Investigator (PI) of several projects since February 2013.

REFERENCES

Aaronson, S., & Christiano, P. (2012). *Quantum money from hidden subspaces*. MIT.

Accardi, L., & Boukas, A. (2007). *The quantum Black-Scholes equation*. doi:10.1145/2213977.2213983

Agulhon, C., Petravicz, J., McMullen, A. B., Sweger, E. J., Minton, S. K., Taves, S. R., Casper, K. B., Fiacco, T. A., & McCarthy, K. D. (2008). What is the role of astrocyte calcium in neurophysiology? *Neuron*, *59*(6), 932–946. doi:10.1016/j.neuron.2008.09.004 PMID:18817732

Alivisatos, A. P., Andrews, A. M., Boyden, E. S., Chun, M., Church, G. M., Deisseroth, K., Donoghue, J. P., Fraser, S. E., Lippincott-Schwartz, J., Looger, L. L., Masmanidis, S., McEuen, P. L., Nurmikko, A. V., Park, H., Peterka, D. S., Reid, C., Roukes, M. L., Scherer, A., Schnitzer, M., . . . Zhuang, X. (2013). Nanotools for neuroscience and brain activity mapping. *American Chemical Society Nano, 7*(3), 1850-1866. https://dx.doi.org/10.1021/nn401288k

Araque, A., & Navarrete, M. (2010). Glial cells in neuronal network function. *Philosophical Transactions of the Royal Society of London. Series B, Biological Sciences*, 2375–2381.

Asher, J. (2012). *Brain's code for visual working memory deciphered in monkeys NIH-funded study*. NIH Press Release. https://www.nimh.nih.gov/news/science-news/2012/in-sync-brain-waves-hold-memory-of-objects-just-seen.shtml

Baaquie, B. E., Coriano, C., & Srikant, M. (2002). *Quantum mechanics, path integrals and option pricing: Reducing the complexity of finance*. National U Singapore.

Banaclocha, M. A. M., Bo'okkon, I., & Banaclocha, H. M. (2010). Long-term memory in brain magnetite. *Medical Hypotheses*, *74*(2), 254–257.

Bartkiewicz, K., Cernoch, A., Chimczak, G., Lemr, K., Miranowicz, A., & Nori, F. (2016). *Experimental quantum forgery of quantum optical money*. Adam Mickiewicz University.

Bellinger, S. (2005). Modeling calcium wave oscillations in astrocytes. *Neurocomputing*, *65*(66), 843–850.

Benioff, P. (1980). The computer as a physical system: A microscopic quantum mechanical Hamiltonian model of computers as represented by Turing machines. *Journal of Statistical Physics*, *22*(5), 563–591.

Bong, K.-W., Utreras-Alarcon, A., Ghafari, F., Liang, Y.-C., Tischler, N., Cavalcanti, E. G., Pryde, G. J., & Wiseman, H. M. (2020). A strong no-go theorem on the Wigner's friend paradox. *Nature Physics*, *17*(Aug), 1745–2481.

Conway, J., & Kochen, S. (2006). *The free will theorem*. Princeton U. J.

Davisson, C., & Germer, L. H. (1927). Diffraction of electrons by a crystal of nickel. *Physical Review*, *30*, 705–741.

De Pitta, M., Volman, V., Berry, H., Parpura, V., Volterra, A., & Ben-Jacob, E. (2012). Computational quest for understanding the role of astrocyte signaling in synaptic transmission and plasticity. *Frontiers in Computational Neuroscience*, *6*(98), 1–25.

Facchi, P., Lidar, D. A., & Pascazio, S. (2004). Unification of dynamical decoupling and the quantum Zeno effect. *Physical Review A.*, *69*(032314), 1–6.

Facchi, P., & Pascazio, S. (2008). Quantum Zeno dynamics: Mathematical and physical aspects. *Journal of Physics. A. Mathematical Nuclear and General*, *41*(493001), 1–45.

Fang, X.-T., Zeng, P., Liu, H., Zou, M., Wu, W., Tang, Y.-L., Sheng, Y.-J., Xiang, Y., Zhang, W., Li, H., Wang, Z., You, L., Li, M.-J., Chen, H., Chen, Y.-A., Zhang, Q., Peng, C.-Z., Ma, X., Chen, T.-Y., & Pan, J.-W. (2020). Implementation of quantum key distribution surpassing the linear rate-transmittance bound. *Nature Photonics*, *14*, 422–425.

Feynman, R. (1982). Simulating physics with computers. *International Journal of Theoretical Physics*, *21*(6/7), 467–488. https://dx.doi.org/10.1007/BF02650179

Feynman, R. P. (1948). Space-time approach to non-relativistic quantum mechanics. *Reviews of Modern Physics*, *20*(2).

Feynman, R. P. (1949). Space-time approach to quantum electrodynamics. *Physical Review*, *76*(6).

Giacosa, G., & Pagliara, G. (2014). Quantum Zeno effect by general measurements. *Physical Review A.*, *052107*, 1–5.

Goldberg, M., De Pitta', M., Volman, V., Berry, H., & Ben-Jacob, E. (2010). Nonlinear Gap Junctions Enable Long-Distance Propagation of Pulsating Calcium Waves in Astrocyte Networks. *Public Library of Science Computational Biology.*, *6*(8), 1–14.

Grover, L. K. (1997). *Quantum computers can search rapidly by using almost any transformation*. Bell Labs.

Ingber, L. (1982). Statistical mechanics of neocortical interactions. I. Basic formulation. *Physica D. Nonlinear Phenomena*, *5*, 83–107.

Ingber, L. (1983). Statistical mechanics of neocortical interactions. Dynamics of synaptic modification. *Physical Review A.*, *28*, 395–416.

Ingber, L. (1984). Statistical mechanics of neocortical interactions. Derivation of short-term-memory capacity. *Physical Review A.*, *29*, 3346–3358.

Ingber, L. (1993). *Adaptive Simulated Annealing (ASA). Global optimization C-code*. Caltech Alumni Association.

Ingber, L. (1994). Statistical mechanics of neocortical interactions: Path-integral evolution of short-term memory. *Physical Review. E*, *49*(5B), 4652–4664.

Ingber, L. (1997). Statistical mechanics of neocortical interactions: Applications of canonical momenta indicators to electroencephalography. *Physical Review. E*, *55*(4), 4578–4593. https://www.ingber.com/smni97_cmi.pdf

Ingber, L. (1998). Statistical mechanics of neocortical interactions: Training and testing canonical momenta indicators of EEG. *Mathematical and Computer Modelling*, *27*(3), 33–64.

Ingber, L. (2000). High-resolution path-integral development of financial options. *Physica A*, *283*(3-4), 529–558.

Ingber, L. (2011). Computational algorithms derived from multiple scales of neocortical processing, In *Pointing at Boundaries: Integrating Computation and Cognition on Biological Grounds*. Springer. doi:10.100712559-011-9105-4

Ingber, L. (2012a). Adaptive Simulated Annealing, In *Stochastic global optimization and its applications with fuzzy adaptive simulated annealing*. Springer. https://www.ingber.com/asa11_options.pdf

Ingber, L. (2012b). Columnar EEG magnetic influences on molecular development of short-term memory, In *Short-Term Memory: New Research.* Nova. https://www.ingber.com/smni11_stm_scales.pdf

Ingber, L. (2012c). Influence of macrocolumnar EEG on Ca waves. *Current Progress Journal, 1*(1), 4-8. https://www.ingber.com/smni12_vectpot.pdf

Ingber, L. (2015). Calculating consciousness correlates at multiple scales of neocortical interactions. In *Horizons in Neuroscience Research.* https://www.ingber.com/smni15_calc_conscious.pdf

Ingber, L. (2016a). Path-integral quantum PATHTREE and PATHINT algorithms. *International Journal of Innovative Research in Information Security.*, *3*(5), 1–15.

Ingber, L. (2016b). Statistical mechanics of neocortical interactions: Large-scale EEG influences on Molecular processes. *Journal of Theoretical Biology*, *395*, 144–152.

Ingber, L. (2017a). Evolution of regenerative Ca-ion wave-packet in neuronal-firing fields: Quantum path- integral with serial shocks. *International Journal of Innovative Research in Information Security.*, *4*(2), 14–22.

Ingber, L. (2017b). Options on quantum money: Quantum path-integral with serial shocks. *International Journal of Innovative Research in Information Security*, *4*(2), 7–13.

Ingber, L. (2018a). Quantum calcium-ion interactions with EEG.Sci. 1(7), 1-21.

Ingber, L. (2018b). Quantum Variables in Finance and Neuroscience. In *Lecture Plates 2018: QVFN*. Physical Studies Institute. https://www.ingber.com/path18_qpathint_lecture.pdf

Ingber, L., Fujio, H., & Wehner, M. F. (1991). Mathematical comparison of combat computer models to Exercise data. *Mathematical and Computer Modelling*, *15*(1), 65–90.

Ingber, L., & Nunez, P. L. (2010). Neocortical Dynamics at Multiple Scales: EEG Standing Waves, Statistical Mechanics, and Physical Analogs. *Mathematical Biosciences*, *229*, 160–173.

Ingber, L., Pappalepore, M., & Stesiak, R. R. (2014). Electroencephalographic field influence on calcium Momentum waves. *Journal of Theoretical Biology*, *343*, 138–153.

Ingber, L., Wehner, M. F., Jabbour, G. M., & Barnhill, T. M. (1991). Application of statistical mechanics methodology to term-structure bond-pricing models. *Mathematical and Computer Modelling*, *15*(11), 77–98.

Ingber, L., & Wilson, J. K. (2000). Statistical mechanics of financial markets: Exponential modifications to Black-Scholes. *Mathematical and Computer Modelling*, *31*(8/9), 167–192.

Innocenti, B., Parpura, V., & Haydon, P. G. (2000). Imaging extracellular waves of glutamate during calcium signaling in cultured astrocytes. *The Journal of Neuroscience: The Official Journal of the Society for Neuroscience*, *20*(5), 1800–1808.

Jackson, J. D. (1962). *Classical Electrodynamics*. Wiley & Sons.

Jogenfors, J. (2016). *Quantum bitcoin: An anonymous and distributed currency secured by the no-cloning theorem of quantum mechanics*. Linkoping U.

Kozlowski, W., Caballero-Benitez, S. F., & Mekhov, I. B. (2015). *Non-Hermitian dynamics in the quantum Zeno limit*. U Oxford.

Langouche, F., Roekaerts, D., & Tirapegui, E. (1979). Functional integral methods for stochastic fields. *Physica A*, *95*, 252–274.

Meyer, K. (2009). *Extending and simulating the quantum binomial options pricing model* (Thesis). U Manitoba. http://hdl.handle.net/1993/3154

Muller, M. M., Gherardini, S., & Caruso, F. (2016). *Quantum Zeno Dynamics through stochastic protocols*. U Florence.

Patel, A. B., Rothman, D. L., Cline, G. W., & Behard, K. L. (2001). Glutamine is the major precursor for GABA synthesis in rat neocortex in vivo following acute GABA-transaminase inhibition. *Brain Research*, *919*(2), 207–220.

Patil, Y. S., Chakram, S., & Vengalattore, M. (2015). Measurement-induced localization of an ultracold lattice gas. *Physical Review Letters*, *115*(140402), 1–5.

Pereira, A. Jr, & Furlan, F. A. (2009). On the role of synchrony for neuron-astrocyte interactions and perceptual conscious processing. *Journal of Biological Physics*, *35*(4), 465–480.

Piotrowski, E. W., Schroeder, M., & Zambrzycka, A. (2005). Quantum extension of European option pricing based on the Ornstein-Uhlenbeck process. *Physica A*, *368*(1), 176–182.

Preskill, J. (2015). *Quantum Mechanics. Lecture Notes*. Caltech. http://www.theory.caltech.edu/people/preskill/ph219/

Reyes, R. C., & Parpura, V. (2009). The trinity of Ca2+ sources for the exocytotic glutamate release from astrocytes. *Neurochemistry International*, *55*(3), 1–14.

Ross, W. N. (2012). Understanding calcium waves and sparks in central neurons. *Nature Reviews. Neuroscience, 13*, 157–168.

Salazar, R. F., Dotson, N. M., Bressler, S. L., & Gray, C. M. (2012). Content-specific fronto-parietal synchronization during visual working memory. *Science, 338*(6110), 1097–1100.

Scemes, E., & Giaume, C. (2006). Astrocyte calcium waves: What they are and what they do. *Glia, 54*(7), 716–725.

Schulman, L. S. (1981). *Techniques and Applications of Path Integration*. J. Wiley & Sons.

Shor, P. W. (1994) Algorithms for quantum computation: discrete logarithms and factoring. *Proceedings 35th Annual Symposium on Foundations of Computer Science, IEEE Xplore*, 124-134. https://ieeexplore.ieee.org/document/365700 and https://dx.doi.org/10.1109/SFCS.1994.365700

Srinivasan, R., Winter, W. R., Ding, J., & Nunez, P. L. (2007). EEG and MEG coherence: Measures of functional connectivity at distinct spatial scales of neocortical dynamics. *Journal of Neuroscience Methods, 166*(1), 41–52.

Turing, A. M. (1937). On computable numbers, with an application to the entscheidungsproblem. *Proceedings of the London Mathematical Society, 42*(1), 230–265.

Volterra, A., Liaudet, N., & Savtchouk, I. (2014). Astrocyte Ca2+ signalling: An unexpected complexity. *Nature Reviews. Neuroscience, 15*, 327–335.

Walls, A. B., Waagepetersen, H. S., Bak, L. K., Schousboe, A., & Sonnewald, U. (2015). The glutamine-glutamate/GABA cycle: Function, regional differences in glutamate and GABA production and effects of interference with GABA metabolism. *Neurochemical Research, 40*(2), 402–409.

Wang, Z. L. (2012). *Piezoelectric Nanogenerators for Self-Powered Nanosensors and Nanosystems*. John Wiley Sons. https://dx.doi.org/10.1002/047134608X.W8136

Wehner, M. F., & Wolfer, W. G. (1983a). Numerical evaluation of path-integral solutions to Fokker-Planck equations. I. *Physical Review A., 27*, 2663–2670.

Wehner, M. F., & Wolfer, W. G. (1983b). Numerical evaluation of path-integral solutions to Fokker-Planck equations. II. Restricted stochastic processes. *Physical Review A., 28*, 3003–3011.

Wehner, M. F., & Wolfer, W. G. (1987). Numerical evaluation of path integral solutions to Fokker-Planck equations. III. Time and functionally dependent coefficients. *Physical Review A., 35*, 1795–1801.

Wigner, E. P. (1961). Remarks on the mind-body problem. In I. J. Good (Ed.), *The Scientist Speculates* (pp. 171–184). Heineman.

Wu, S. L., Wang, L. C., & Yi, X. X. (2012). Time-dependent decoherence-free subspace. *Journal of Physics. A. Mathematical Nuclear and General, 405305*, 1–11.

Zhang, P., Ai, Q., Li, Y., Xu, D., & Sun, C. (2014). Dynamics of quantum Zeno and anti-Zeno effects in an open system. *Science China. Physics, Mechanics & Astronomy, 57*(2), 194–207.

Zhang, X. L., Begleiter, H., & Porjesz, B. (1997). Do chronic alcoholics have intact implicit memory? An ERP study. *Electroencephalography and Clinical Neurophysiology*, *103*, 457–473.

Zhang, X. L., Begleiter, H., Porjesz, B., & Litke, A. (1997). Electrophysiological evidence of memory impairment in alcoholic patients. *Biological Psychiatry*, *42*, 1157–1171.

Zhang, X. L., Begleiter, H., Porjesz, B., Wang, W., & Litke, A. (1995). Event related potentials during object recognition tasks. *Brain Research Bulletin*, *38*(6), 531–538.

Chapter 7
Biotechnology:
Recent Developments, Emerging Trends, and Implications for Business

Qing-Ping Ma
https://orcid.org/0000-0002-4379-0378
The University of Nottingham, Ningbo, China

ABSTRACT

Biotechnology utilizes biological systems or living organisms to create, develop, or make products. This chapter reviews the current state of biotechnology and examines its future trends. Currently, biotechnology plays key roles in medicine, agriculture, and industry. In medicine, vaccines which still rely on biological systems for their production, are the best tools to prevent infectious diseases; antibodies and RNA/DNA probes have been crucial in detecting and treating diseases; and genetic editing and gene therapy is making it possible to treat hereditary diseases. In agriculture, biotechnology is generating crops that produce high yields and need fewer inputs, crops that need fewer applications of pesticides, and crops with enhanced nutrition profiles. In industry, biotechnology is being utilized in food processing, metal ore processing, the production of chemicals, and reducing energy consumption and pollution.

INTRODUCTION

Biotechnology is a technology that utilizes biological systems and living organisms or their parts to develop, create or make various products. It is based on biology and can be considered a branch of biology. Although biotechnology is a contemporary buzz word, it has been around for more than 6,000 years since mankind began to use the biological processes of microorganisms to make useful food products, such as bread, beer and cheese, and to preserve dairy products (Arranz-Otaegui, Carretero, Ramsey, Fuller, & Richter, 2018; Dietrich, Heun, Notroff, Schmidt, & Zarnkow, 2012; Salque et al., 2013). In modern times, biotechnology has been playing an even greater role in improving our life and the environment.

Biotechnology uses living organisms and their parts to produce desired products. Thus, brewing and baking bread (using yeast, a living organism) fall within the concept of biotechnology, while direct utiliza-

DOI: 10.4018/978-1-7998-6772-2.ch007

tion of macroscopic biomaterials such as harvesting crops for grains and cutting trees for timbers is not conventionally considered as biotechnology. An important development in the history of the discipline was the use of selective breeding to increase the production of crops and livestock for use as food, and to enhance their appearance, making them more attractive as ornamental plants and pets (Freedman & Wayne, 2017; Piperno, Ranere, Holst, Iriarte, & Dickau, 2009; Purugganan & Fuller, 2009). Darwin (1968) discussed how selective breeding produced desired changes over time in his 1859 book, *On the Origin of Species*. With the discovery of microbes as pathogens and causes of fermentation and the emergence of genetics in the nineteenth century (Bordenave, 2003; Lane, 2015; Weiling, 1991), biotechnology experienced enormous growth and played an increasingly important role not only in agriculture, but also in medicine and industry.

The advent of genetic engineering in the 1970s opened new avenues for the development and application of biotechnology (Jackson, Symons, & Berg, 1972; Jaenisch & Mintz, 1974). Previously, traditional processes usually utilized living organisms in their natural form or modified them by selective breeding. Modern biotechnology involves more advanced modification of biological systems or organisms, because genetic engineering provides new possibilities for making changes in an organisms' genetic material (DNA) to produce desired products unimaginable with traditional breeding. The development of biotechnology is closely related to progress in many different scientific disciplines including biochemistry, cell biology, molecular biology, embryology, genetics, immunology, and microbiology. Many techniques and methods widely today used in biotechnology were invented by researchers in these related disciplines. Today, biotechnology is experiencing exponential growth, with new technologies and products being developed every year in medicine, agriculture, and industry.

The aim of this chapter is to introduce basic concepts of biotechnology, review the current state of the discipline, and examine future trends. The rest of the chapter is organized as follows: the next section introduces biotechnology in medicine; the following section looks at biotechnology in agriculture; the fourth section examines industrial biotechnology; the fifth section introduces bioinformatics; the sixth section looks at the impact of biotechnology on the environment and considers how biotechnology can be used to tackle environmental problems; and the final section summarizes the chapter and its major conclusions.

BIOTECHNOLOGY IN MEDICINE

The two main functions of medicine are the prevention of disease and the treatment of disease and injuries. For both functions, biotechnology has been playing an important role and will play an even greater role in the future. Biotechnology contributes to the progress and success of medicine in the following areas: 1) providing new reagents, new methods and new drugs for the prevention, diagnosis and treatment of disease; 2) providing new materials for medical research that advance our understanding of physiology, immunology, pathology of diseases and pharmacology of drugs, which in turn help in the development of new drugs, new vaccines and new treatments; 3) making it possible to personalize medicine and to cure by gene therapy those diseases that are caused by defective or mutated genes; and 4) making it possible to produce new biomaterials for treatment and transplantation that replace dysfunctional tissues and organs in the body.

New Reagents and New Drugs For Prevention, Diagnosis and Treatment

Infectious diseases caused by pathogenic microbes and protozoans have been a serious threat to human health and even life. The earliest applications of biotechnology were vaccines and antiserums used to prevent and treat infectious diseases. Immunoassay and DNA/RNA detection are indispensable tools in the diagnosis and detection of pathogens. Since developing new drugs from small molecules is becoming more difficult lately, more resources have been put into research on antibodies and DNA or RNA as therapeutic drugs by the pharmaceutical industry.

Prevention of Infectious Diseases By Vaccination

Infection by a pathogen usually induces an immunoreaction in human body. This generates an immunity against future infection of that pathogen, which is the physiological mechanism for inoculation with vaccines, or vaccination. Around 1000AD, traditional Chinese doctors began using materials collected from smallpox patients with the aim of inducing a mild infection that would prevent more severe forms of infection (Thein, Goh, & Phua, 1988). Jenner's use of the much less virulent cowpox to immunize people against smallpox, a viral disease with high mortality, is the most spectacular prophylactic success so far, eventually leading to the elimination of smallpox on Earth (Riedel, 2005; Willis, 1997). Jenner's cowpox vaccine succeeded because cowpox and smallpox share the antigen for immunoreaction so that immunity to cowpox also protects people against smallpox. Other vaccines usually use inactivated (i.e., killed) or attenuated pathogens.

With the progress in microbiology and cell biology to isolate pathogens, grow them in medium or cultured cells and tissues, medical and pharmaceutical professionals have more instruments in their toolkits for producing effective vaccines against a variety of diseases. In vitro cultivation of pathogens with medium or cell and tissue culture makes vaccine production safer, more efficient, and permits the use of live pathogens which are attenuated or made non-virulent. Pasteur found that culturing bacteria or virus under some conditions might reduce their virulence, so that vaccines can be made against a disease when there is no natural weak form of the disease like cowpox. Molecular cloning technology can produce a pathogen's proteins (antigens) that induce immune reaction in vivo against that pathogen and use such proteins conjugated with other compounds as vaccine. With this approach, Novavax, a biotechnology company has bioengineered the coronavirus's spike proteins to form knucklebone-shaped nanoparticles as experimental vaccine for clinical trial, injected together with its proprietary Matrix-M adjuvant to induce immunity against SARS-CoV-2 (Keech et al., 2020).

Another approach is DNA vaccination, i.e., the introduction of a piece of DNA, which codes (a part of) the pathogen protein, into a less virulent virus (used as the vector) and uses this transformed vector virus as a vaccine. Johnson & Johnson is developing an adenovector vaccine, which introduces a piece of DNA from SARS-CoV-2 into a common cold-causing adenovirus that has been genetically changed so that it cannot replicate in the body (Mercado et al., 2020). The University of Oxford in collaboration with the biopharmaceutical company AstraZeneca has also transferred the SARS-CoV-2 spike protein DNA into a weakened version of an adenovirus to produce a DNA vaccine against COVID-19 (Folegatti et al., 2020). This vaccine is now being used clinically.

RNA vaccination uses an RNA containing vector, such as lipid nanoparticles to transfer the RNA sequence that codes for antigens into host cells in the body so that transfected host cells produce the antigens that induce immunoreaction. One of its advantages over DNA vaccination is that mRNA is translated in the cytosol without the need to enter the cell nucleus, averting the risk of being integrated to the host genome (Verbeke, Lentacker, Smedt, & Dewitte, 2019). The pharmaceutical company Pfizer in collaboration with the German biotech company BioNTech developed an mRNA vaccine (Pious & Ingole, 2020), which was the first COVID-19 vaccine approved for clinical use. Moderna Therapeutics, a Massachusetts-based biotech company, in collaboration with the National Institutes of Health has also developed an mRNA vaccine (Philippidis, 2020), which has also been approved for clinical use against COVID-19. Governments in many countries have signed large contracts with those pharmaceutical and biotech companies to purchase COVID-19 vaccines to immunize their residents against the disease.

Detection of Pathogens and Diagnosis of Diseases

An important step in diagnosing and treating infectious diseases is to detect and identify those microbes, bacteria, fungi, mycoplasma, chlamydia, virus and protozoans. Protozoology as well as sub-disciplines of microbiology such as bacteriology, mycology, and virology have been developed to study those microorganisms and protozoans. Early studies on bacteria usually involved growing bacteria on medium (bacterial culture) and dyeing them for observation and initial classification under microscope. Robert Koch made essential contribution in establishing the criteria for identifying which bacterium is responsible for causing a particular disease, Koch's postulates (A. S. Evans, 1976; Walker, LeVine, & Jucker, 2006). Growing and detecting virus are more difficult because viruses are submicroscopic infectious agents that replicate only inside the living cells of an organism. Later biochemical reactions and immune reactions are used for identification of pathogens. Immunoreaction began to be widely used because of its specificity and sensitivity, firstly with antiserum, i.e. serological reaction, and then with purified antibodies. Radioimmunoassay, enzyme link immunosorbent assay (ELISA), immunofluorescence techniques, immunocytochemistry, and immunohistochemistry, etc. have greatly increased the detection sensitivity of antigens which indicate the existence of pathogens in vitro or in situ. The monoclonal antibody technology enables production of antibodies against one specific antigenic determinant or epitope (Waldmann, 1991). Immunoassay is also used to measure many important substances in the blood to check up health conditions or help diagnose various diseases. Following each invention of these technologies, new biotech startups mushroomed because of the opportunities to commercialize test toolkits as well as reagents and instruments.

The development of molecular biology has further enhanced the toolkits for detecting pathogens. DNA sequencing has long been automated, which enables fast identification of virus that causes a new disease. It took only about a month from the first identified COVID-19 case to the sequencing of SARS-CoV-2 RNA (Zhu et al., 2020). The invention and subsequent improvement with heat-resistant enzymes of the polymerase chain reaction (PCR) method provides an indispensable tool for molecular biology and rapid detection of pathogens. PCR exponentially amplifies nucleic acid fragments, such that in theory one molecule of viral DNA (reverse transcribed from viral RNA) is sufficient to be detected. DNA or RNA probes for hybridization detection can be synthesized in little time, whereas antibodies needed for immunoassay can take much longer to produce by injecting antigens to animals or preparing hybridoma cells for monoclonal antibody. Therefore, the sequencing of SARS-CoV-2 RNA immediately created business opportunities for biotech firms to produce test kits and reagents for diagnosis. Currently detection of SARS-CoV-2 relies mainly on detecting viral nucleic acid. DNA microarray is a technology that miniaturizes DNA detection samples together with automation such that some DNA microarray systems can do as many as a million tests at once, thus greatly increasing the DNA detection throughput (Heller, 2002).

Genetic testing can identify defective genes which underlie high morbidity of certain diseases in some families, thus allows the genetic diagnosis of vulnerabilities to inherited diseases (Katsanis & Katsanis, 2013). In a broader sense, it can test the possible presence of genetic diseases, or mutant forms of genes associated with increased risk of developing genetic disorders. The results of a genetic test can confirm or rule out a suspected genetic condition or help determine a person's chance of developing or passing on a genetic disorder. Genetic testing is also used to determine a child's parentage or blood relations between people. In the future, genetic testing may be used to forecast a person's health conditions under different circumstances in terms of lifestyle, diet, work and environment, and she/he can adjust her/his circumstance to have the health condition she/he desires.

Immunoassays and DNA/RNA reactions need tools to measure them for detection and diagnosis of pathogens or defective genes. Biosensors provide convenient tools for monitoring physiological parameters of the body as well as biological processes. A biosensor is an analytical device that combines a biological component with a physicochemical detector to detect a chemical substance. Leland C. Clark and Champ Lyons invented the first biosensor in 1962 (Clark Jr & Lyons, 1962). Today biosensors are usually based on some forms of the metal-oxide-semiconductor field-effect transistor (MOSFET), which was invented by Mohamed M. Atalla and Dawon Kahng in 1959 (Bassett, 2002). The first bio-field-effect transistor (BioFET), which was an ion-sensitive field-effect transistor (ISFET) and invented by Piet Bergveld in 1970 (Toumazou & Georgiou, 2011), is a special type of MOSFET, where the metal gate is replaced by an ion-sensitive membrane, electrolyte solution and reference electrode. It is widely used in biomedical applications such as detection of DNA hybridization, biomarker detection from blood, antibody detection, glucose measurement, and pH sensing (Schöning & Poghossian, 2002). More BioFETs, including gas sensor FET (GASFET), pressure sensor FET (PRESSFET), chemical FET (ChemFET), reference ISFET (REFFET), enzyme-modified FET (ENFET) and immunologically modified FET (IMFET), were developed during the 1970s and 1980s. They have been widely used to measure physical, chemical, biological and environmental parameters (Bergveld, 1985). DNA field-effect transistor (DNAFET), gene-modified FET (GenFET) and cell-potential FET (CPFET) had been developed by the early 2000s (Schöning & Poghossian, 2002), and they can be used for detecting specific DNA sequences, gene modifications and cell potentials respectively. It can be expected that biosensors and detection systems will continue to improve and provide faster and more convenient detection and diagnosis.

Treatment With Biotechnological Products

Pathogens can induce immunoreaction which is the production in the body of antibodies that neutralize the future invading pathogens. The antiserum of animals and humans that have been infected by a pathogen may be used to treat diseases caused by that pathogen. The first disease successfully treated with antiserum is diphtheria by Emil Adolf von Bering in 1890, who also developed the anti-tetanus serum to treat tetanus (Behring, 1890). Antiserum treatments were attempted for many infectious diseases and snake bites before more effective treatments are found. Anti-rabies serum and antivenin are still the essential preventive treatment for dog and snake bites respectively. Antiserum and antibodies have been tried for treating COVID-19; US President Donald Trump has been treated with the still experimental monoclonal antibody cocktail from the biotech company Regeneron. With better understanding of pathophysiological processes of many diseases such as cancer, monoclonal antibodies against specific cytokines, receptors or enzymes can be used. Some antibodies have been successfully used in treating cancers (Weiner, Surana, & Wang, 2010).

Specially designed DNA or RNA fragments can also be used as therapeutic medicine (Hurley, 2002; Pai et al., 2006). Unlike antibodies that neutralize cytokines, receptors and enzymes essential for the pathogenesis of a disease after they have been produced, DNA and RNA drugs can block the translation of those protein from their mRNA templates such that they would not be produced. One important advantage of using DNA or RNA therapeutics is that designing and synthesizing them could be much faster than making monoclonal antibodies. Making a monoclonal antibody against a certain target is usually faster than developing a small molecular drug nowadays. That is why big pharmaceutical companies have increased their efforts in developing biological therapeutics, which previously attract mainly biotech startups.

Bioactive substances extracted from fungi, plants and animals or their metabolic products have been important sources of pharmaceuticals. In 1928, Alexander Fleming discovered the mold *Penicillium*, which led to the purification of the antibiotic compound, penicillin, formed by the mold by Howard Florey, Ernst Boris Chain and Norman Heatley (Ligon, 2004). Since the discovery of Penicillin, antibiotics derived from fungi have been indispensable for treating bacterial diseases and saved innumerable lives. Hormones extracted from animal tissues have been used in treating diseases which are symptoms of deficiency of certain hormones in the body such as diabetes. Insulin used for treating diabetes was previously extracted from the pancreas of abattoir animals (cattle or pigs) and the porcine and bovine insulins are slightly different from human insulin such that they could induce immunoreaction which reduces their effectiveness in long-term use. There are many hormones that exist in too small quantities such that it is not feasible to use extracts from animal tissues for clinical treatment. In 1978 Genentech inserted a plasmid vector that carries humanized insulin gene into the bacterium *Escherichia coli* (Johnson, 1983). The genetically engineered bacteria are able to produce large quantities of synthetic human insulin at relatively low cost. With genetic engineering technologies becoming more accessible and convenient to use, many hormone and cytokines that were previously unavailable for clinical use are widely available at relatively low costs. Therapeutic and diagnostic proteins and peptides can now be produced by genetically engineered bacteria, fungi and plants (Daniell, Streatfield, & Wycoff, 2001; Hiatt, Caffferkey, & Bowdish, 1989; Kiefer et al., 1992). Natural or modified forms of human hormones, cytokines and antibodies will be more widely used in treating various diseases in the future, which present enormous business opportunities.

New Materials and Methods For Medical and Pharmaceutical Research

Biomedical and pharmaceutical researches play an essential role in preventing and treating diseases and improving people's health. Advances in biomedical research have promoted development of biotechnology, which in turn provides new materials and methods that facilitate biomedical and pharmaceutical research. For example, inbred animals have facilitated biomedical research by reducing variability or providing certain desired features (Festing, 2010). Selective breeding has produced strains of animals with propensity for pathological conditions as models for some human diseases, which can be used for investigating pathogenesis as well as testing potential treatments. Tissue culture and cell lines enable more controlled environment and test conditions than in vivo experiments. Cell-based tests are also more similar to the in vivo interactions between compounds and cells than those using subcellular fractions.

Cell-based tests together with miniaturization and automation provide high throughput screening capacity for identifying drug candidates.

While traditional selective breeding is time consuming and limited in choices to develop animal models of human diseases, transgenic technology is much faster and versatile in producing animal models deficient of certain genes (by gene knock out) or expressing certain genes (by gene knock in). Gene knock-out and knock-in can be used to investigate the function of a receptor, ion channel, cytokine, enzyme, and hormone, etc., and the capacity of the body to compensate the deficiency of a certain protein. Transgenic animals can be used to produce animal pathological models for investigating pathological changes as well as for testing potential treatments (Ferrante et al., 2000; Guy, Cardiff, & Muller, 1992; Kato et al., 1998; Li et al., 2000; Refolo et al., 2000). Genetically engineered cell lines which express certain receptors, ion channels and enzymes facilitate biochemical, physiological and pharmaceutical researches. Expression of cloned receptors or ion channels in Xenopus oocytes have been widely used in electrophysiological research to characterize those receptors and ion channels.

Genetic engineering has produced cells and animals that express fluorescent proteins. Since their fluorescence change indicates variations in physiological and biochemical conditions, they provide powerful and convenient tools for investigating physiological functions and cellular responses to bioactive agents. Various fluorescent proteins or dyes that are sensitive to voltage, calcium or other ions are used to investigate physiological function and pharmacological effects and to screen compounds for drug candidates. High throughput and high content screenings using fluorescent dyes in cell lines with miniaturization and automation have become indispensable tools in developing new small molecule drugs.

Personalized Medicine and Gene Therapy

Pharmacogenomics is the technology that analyses how genetic makeup affects an individual's response to drugs (W. E. Evans & Relling, 1999). It has long been known that people respond to the same drug differently in terms of efficacy and adverse effects. A drug, which cures many patients without noticeable adverse effects, may cause severe adverse effects in some patients. Pharmacogenomics investigates the influence of genetic variation on drug responses in patients by correlating gene expression or single-nucleotide polymorphisms with a drug's efficacy or toxicity. Once the correlation is well established, clinicians can use this knowledge to optimize drug therapy for individual patients, i.e. "personalized medicine". Its objective is to ensure maximum efficacy with minimal adverse effects with respect to the patients' genotype (Becquemont, 2009), such that drugs and drug combinations are optimized for each individual's unique genetic makeup (Squassina et al., 2010).

Genetic testing helps identify defective or mutated genes before they cause severe problems. Many hereditary diseases are caused by such defective or mutant genes, and they were incurable in the past. Gene therapy promises to treat such diseases and other disorders by inserting a gene into a patient's cells instead of using drugs or surgery (Verma et al., 2000). Several approaches are being tried for gene therapy: 1) replacing a mutated gene with a healthy copy of the gene; 2) inactivating, or "knocking out," a mutated gene that is functioning improperly; and 3) introducing a new gene into the body to help fight a disease. Certain viruses are often used as vectors to deliver the new gene by infecting the cell, and the viruses are modified so they cannot cause disease when used in people. When used as vectors, retroviruses integrate their genetic material including the new gene into a chromosome in host cells; adenoviruses introduce their DNA into the nucleus of the cell, but the DNA is not integrated into a chromosome. There are many gene therapy clinical trials on various diseases. The U.S. Food and Drug

Administration (FDA) approved the first gene therapy, Kymriah (tisagenlecleucel) for certain pediatric and young adult patients with a form of acute lymphoblastic leukemia (ALL), on 30 August 2017. Several others have been approved since then (Food and Drug Administration, 2020), and more can be expected in the future. On January 21, 2021, the website MedicalStartups (https://www.medicalstartups.org/top/genetherapy/) displayed a list of the top 58 gene therapy startups in the world.

The discovery of a molecular tool known as CRISPR-Cas9 for gene-editing technologies makes gene therapy and other genetic engineering procedures more feasible (Ran et al., 2013). CRISPR-Cas9 allows researchers to remove and insert DNA in the desired locations with precision. Before the advent of CRISPR-Cas9, two approaches based on zinc finger nucleases (ZFNs) and transcription activator-like effector nucleases (TALENs) respectively were used to make site-specific double-stranded breaks in DNA. They are inconvenient because they rely on protein-DNA binding and engineering fusion proteins is more difficult than designing an RNA sequence. Designing an RNA sequence that binds specifically to a unique target DNA sequence requires only knowledge of the Watson-Crick base-pairing rules. CRISPR-Cas9 uses RNA-DNA binding to guide nuclease activity, which simplifies the design and enables application to a broad range of target sequences. Thus, the CRISPR-Cas9 system is preferable to the fusion protein designs required for using ZFNs or TALENs, and it will facilitate future development of genetic engineering and gene therapy. Researchers may find even more powerful gene-editing tools in the future.

New Biomaterials For Treatment and Transplantation

Lymphocyte therapies have been used to treat cancers. It was found that peripheral blood mononuclear cells (PBMC) cultured in the presence of interleukin-2 (IL-2) can result in a type of non-specific killer cell that can kill different tumor cells, which is called lymphokine-activated killer (LAK) cell. Then it is used together with IL-2 in adoptive immunotherapy to treat some types of cancer (Grimm, Mazumder, Zhang, & Rosenberg, 1982). Another adoptive cell therapy used to treat cancer is the tumor infiltrating lymphocyte (TIL) therapy, which uses lymphocytes isolated from tumor and cultured in the presence of IL-2 for one to two weeks (Dudley, Wunderlich, Shelton, Even, & Rosenberg, 2003). TIL therapy appears to be more effective than LAK therapy.

Stem cells are characterized by their ability to differentiate and mature into other types of cells with specialized functions as well as to self-generate, a phenomenon where they divide and produce more stem cells. Therefore, they might be used as a repair tool to replenish other cells that are worn out or damaged (Strauer & Kornowski, 2003). One of the most common types of stem cell therapy is the adult stem cell transplant, which is used to treat a variety of blood cancers and disorders, including leukemia, lymphoma, and multiple myeloma. This procedure can be done using bone marrow or peripheral blood stem cells, and the U.S. FDA has approved several products of hematopoietic progenitor cells from cord blood (Food and Drug Administration, 2020). Stem cells are expected to play a large role in treating various diseases in the near future.

Organ transplantation has played an important role in prolonging life and improving the quality of life for many patients. With the success and growth of transplantation, there is a severe shortage of organ supply. Genetically modified pigs have been suggested as organ donors for xenotransplantation (Cooper, Gollackner, & Sachs, 2002). If genetic engineering can solve the issue of human anti-pig cytotoxicity, it might alleviate the severe shortage of organ supply especially for patients with terminal illness. Genetically engineered cells may become a tool for tissue engineering and regenerative medicine. Future biotechnology might be able to use cultured cells to construct by 3D printing artificial organs with the same function and structure as natural organs. Many startups are set up to develop such artificial organs. On January 6, 2021, the website MedicalStartups (https://www.medicalstartups.org/top/organs/) listed Carmat, SynCardia, Organovo, Tissium, BiVACOR, Cyfuse Biomedical, BioAesthetics, Regenovo Biotechnology, Panderum Technologies, Beta-O2, Nanofiber Solutions, and Cornea Biosciences as the top 12 startups creating artificial organs in the world. It is possible to generate tissue committed stem cells for regenerative therapies and use them to regenerate damaged tissues and organs.

BIOTECHNOLOGY IN AGRICULTURE

Biotechnology has played an important role in agricultural development by generating crops with high yields and fewer inputs, developing crops with better flavor or enhanced nutrition profiles, improving crop insect resistance, enhancing crop herbicide tolerance, lowering volumes of agricultural chemicals required by crops and facilitates the use of more environmentally sustainable farming practices. The main aims of using biotechnology in agriculture are firstly, to increase output of crops; secondly, to improve the flavor and nutritional profiles; and thirdly, to reduce the use of pesticides and other chemicals as well as labor and energy in agriculture.

The domestication of crops and livestock by early humans is actually a process of selective breeding. For thousands of years, humans have used selective breeding to improve the production of crops and livestock. In selective breeding, organisms with desirable characteristics are mated to produce offspring with the same characteristics which is strengthened each generation. This technique has produced sweet corns, fragrant rice, and chicken for laying more eggs as well as those growing fast to provide more meat, etc. Traditional breeding through hybridization has resulted in many crops with desirable characteristics such as high yield or resistance to diseases, pests, drought and saline soil. Hybrid rice strains play an important role in the fast growth of grain outputs in the past four decades in China.

Genetic engineering can not only provide far greater control over genetic structure of plants and animals than previously afforded by methods such as selective breeding and mutation breeding, but also produce new strains with new traits that do not occur naturally in the species (Mittler & Blumwald, 2010). For example, a plant can be engineered to express a pesticide, thereby ending the need of external application of pesticides. Bt corn is such a case (Saxena, Flores, & Stotzky, 1999). Commercial sale of genetically modified foods began in 1994, when Calgene first marketed its Flavr Savr delayed ripening tomato (Gershon & Lehrman, 1994). Genetically modified (GM) crops have primarily focused on cash crops in high demand by farmers such as soybean, corn, canola, and cotton, which have been widely adopted by farmers. Those GM crops usually have new traits such as resistance to certain pests, diseases, stressful environmental conditions, resistance to chemical treatments (e.g. resistance to an herbicide), reduction of spoilage, or improving the nutrient profile of the crop. Plants have been genetically engineered to produce pharmaceutical agents, biofuels, and other industrially useful goods, as well as for bioremediation. GM livestock have also been developed and the U.S. FDA approved the first GM salmon for commercial production and consumption in 2015 (Waltz, 2016).

The wide adoption of GM crops by farmers has caused concerns in some consumers and environmentalists. Environmentalists are concerned with invasion by GM crops to the natural flora. Consumers worry mainly about the safety of consuming GM food and some researchers have reported adverse effects in experimental animals fed on GM food. However, there is a scientific consensus that currently available food derived from GM crops poses no greater risk to human health than conventional food, and those reported adverse effects all come from researches that lack scientific rigor (Domingo & Bordonaba, 2011; Nicolia, Manzo, Veronesi, & Rosellini, 2014). To ensure safety, each GM food needs to be tested on a case-by-case basis before introduction (Haslberger, 2003). The legal and regulatory status of GM foods varies by country, with some nations banning them and others permitting them with various degrees of regulatory restriction.

Biotechnology is also applied to agricultural processes. Micropropagation is the propagation of plants from tissue cultures, which can be used for the selection and domestication of plants with increased efficiency and productivity. Fertile farmland is a scarce resource and most land areas in the world are not good for farming. It might be possible in future that transgenic plants can be purposely designed to grow under specific environments such that hitherto unfarmable land will be used to produce food and other agricultural products. For example, the creation of enhanced seeds that resist extreme environmental conditions of arid regions might transform deserts into farmland (Hu & Xiong, 2014). It is also hoped that biotechnology might produce more environmentally friendly solutions than the current agriculture.

BIOTECHNOLOGY IN INDUSTRY

Application of biotechnology for preparing certain types of food and drink has a very long history. The making of wine, beer, cheese and many sauces for cooking involves fermentation processes, which were introduced in early Mesopotamia, Egypt, China and India. Modern industrial fermentation processes still use the same basic biological mechanisms. In brewing, starch (carbohydrates) from grains is converted into sugar by enzymes in malted grains and then into alcohol by added specific yeasts to produce beer. The process of lactic acid fermentation has produced preserved foods such as soy sauce. The production of leavened bread also uses fermentation. These processes of fermentation are the first use of biotechnology to convert a food source into another form, although the underlying mechanism was not fully understood until Louis Pasteur's work in 1857 (Pasteur, 1857, 1858).

With better understanding of microbiology and the mechanism underlying industrial fermentation, scientists and engineers explored ways to improve the existing production processes and to develop new processes for manufacturing products other than food and drink. In 1917, Chaim Weizmann first used a pure microbiological culture in an industrial process, to produce acetone form corn starch using *Clostridium acetobutylicum*. It was during World War I when the United Kingdom desperately needed acetone to manufacture explosives. More non-food products have been manufactured by industrial fermentation in the following years, and production of many antibiotics such as penicillin involves a fermentation process. Naturally present bacteria have been used by the mining industry in bioleaching, which is the extraction of metals from their ores through the use of living organisms, one of several applications within biohydrometallurgy. Bioleaching is much cleaner than the traditional heap leaching with cyanide and several methods are used to recover copper, zinc, lead, arsenic, antimony, nickel, molybdenum, gold, silver and cobalt. Biotechnology tends to consume less in resources than traditional processes used to produce industrial goods.

The emergence of molecular cloning and genetic engineering spawned the modern biotech industry. In 1971, Paul Berg made progress in his gene splicing experiments. Herbert W. Boyer and Stanley N. Cohen successfully transferred genetic material into a bacterium in 1972, such that the imported material would be reproduced (S. N. Cohen, Chang, Boyer, & Helling, 1973). On June 16, 1980, the United

States Supreme Court ruled that a genetically modified microorganism could be patented in the case of *Diamond v. Chakrabarty* (Robinson & Medlock, 2005), which significantly increased the commercial viability of a biotechnology industry. The intellectual property rights granted to genetically modified organisms (GMOs) stimulated the growth of biotech industry. Significant progress has been made in creating GMOs that enhance the diversity of applications, which in turn enhances economic viability of the biotechnology industry.

Industrial fermentation has been used to generate products in sectors such as chemicals, food and feed, detergents, paper and pulp, textiles and biofuels. Using renewable raw materials to produce a variety of chemicals and fuels with biotechnology help advance towards lowering greenhouse gas emissions and move away from a petrochemical-based economy. Bio-refineries are being built across North America to test and refine technologies to produce biofuels and chemicals from renewable biomass, which can help reduce greenhouse gas emissions. Rising demand for biofuels is good news for the biotechnology sector, which has allowed the U.S. farming industry to rapidly increase its supply of corn and soybeans by developing genetically modified seeds that resist pests and drought. Corn and soybeans are the main inputs into biofuels, and the U.S. Department of Energy estimates that ethanol usage could reduce U.S. petroleum-derived fuel consumption by up to 30% by 2030 (Anderson et al., 2012; Hoekman, 2009). By increasing farm productivity, biotechnology boosts biofuel production. Microorganisms involved in industrial fermentation can be genetically modified to have more desired properties. Enzymes can be used as industrial catalysts to either produce valuable chemicals or destroy hazardous or polluting chemicals.

A large number of biotech companies are developing pharmaceutical products, vaccines, antibodies, cytokines and hormones etc., as aforementioned. They often focus on the early stage of the research and development, and collaborate with big pharmaceutical companies during the later stage development such as clinical trials, because big pharmaceutical companies have more expertise and resources for such large scale operations. Many biotech companies supply reagents and toolkits for researchers in various institutions to conduct experiments with biotechnology as their main business. Today biotechnology is so widely used by researchers in biomedical laboratories that demand for reagents and equipment is high. Cell and tissue culture, various immunoassays, molecular cloning, PCR, DNA and RNA detection all need high quality reagents as well as equipment. Production and supply of high quality reagents and equipment are an important part of the biotechnology industry.

Exploration of natural resources especially biological resources has always been an important maneuver of mankind. Domestication of crops and livestock is an important milestone in the history of mankind. Compared with land resources, marine resources are relatively under-explored, although marine aquaculture has become an important sector in some economies and provided more abundant aquatic food at cheaper prices than previous reliance on wild resources. Marine technology for aquaculture needs knowledge of those marine species and technical know-how on how to farm them. It also belongs to biotechnology. Marine biopolymers such as alginate and chitin/chitosan are highly abundant but underexploited renewable biomasses (Kurita, 2006; Skaugrud, Hagen, Borgersen, & Dornish, 1999). They may be explored to provide novel materials, enzymes and other bioactive materials. Besides their natural biological and structural functions, they might be engineered to produce new biomaterials with novel functionalities. Biofuels have been produced with photosynthetic micro-algae (Maeda, Yoshino, Matsunaga, Matsumoto, & Tanaka, 2018). There are more marine resources in the oceans to be explored by biotechnology, which implies plenty of business opportunities.

BIOINFORMATICS

Bioinformatics is an interdisciplinary field that addresses biological problems using computational techniques to understand biological data. It may also be referred to as computational biology. Bioinformatics combines biology, computer science, information engineering, mathematics and statistics to analyze and interpret the biological data obtained from researches on structures and functions of nucleic acids, proteins, subcellular organelles, cellular organizations, and networks and systems as well as spatial structure of macromolecules. Literature analysis and development of software for bioinformatics are also classified in the field of bioinformatics.

Bioinformatics has established techniques that allow extraction of useful results from large amounts of raw data. It aids in sequencing and annotating genomes in genetic studies as well as in the analysis of gene and protein expression and regulation. It plays an essential role in structural biology by assisting the simulation and modeling of DNA, RNA, proteins as well as bio-molecular interactions (Dawson, Maciejczyk, Jankowska, & Bujnicki, 2016; Kmiecik et al., 2016; Sim, Minary, & Levitt, 2012). Functional genomics, structural genomics, and proteomics rely heavily on bioinformatics, which has become a key component in the biotechnology and pharmaceutical sector. Protein structure prediction is an important application of bioinformatics. Text mining of biological literature may provide novel cues for further researches which had previously not been recognized by people (A. M. Cohen & Hersh, 2005). Bioinformatics helps analyze and catalogue the biological pathways and networks that are an important part of systems biology.

Several approaches have been developed to analyze the location of organelles, genes, proteins, and other components within cells. Microscopic pictures allow us to locate both organelles as well as molecules. The localization of proteins often contains information about the role of a protein. A protein found in the nucleus is likely to be involved in gene regulation or splicing. By contrast, a protein found in mitochondria is more likely to be involved in respiration or other metabolic processes. Therefore, protein localization is an important factor in protein function prediction. Bioinformatics has established protein subcellular localization prediction resources including protein subcellular location databases, and prediction tools, and experimental scientists can use these resources to develop their research ideas and then test their ideas by experiments. With the rapid accumulation of biological data from all scientific disciplines, bioinformatics will play an even greater role and provide ample business opportunities for entrepreneurs. The website BioPharmGuy (https://biopharmguy.com/links/company-by-name-bioinformatics.php) currently (February 7, 2021) lists 122 bioinformatics companies worldwide.

BIOTECHNOLOGY, ENVIRONMENT AND LAW

Environmentalists are concerned with the negative consequences of human exploration of natural resources, which have resulted in extinction of some species and deterioration of natural environment. Human activities also produce large quantities of waste which may exceed nature's capacity to degrade and absorb them, worsening the environmental issue. However, biotechnology may also provide solutions to environmental problems. Environmental biotechnology is the application of biotechnology for solving environmental problems. Many different techniques and methods can be used to tackle environmental problems such as treatment of waste using naturally occurring or GM microorganisms, development of more environment-friendly processes to replace existing ones, and adoption of more environment-friendly products such as biodegradable plastics.

Microorganisms play an essential role in degrading and cleaning wastes of both naturally occurring and man-made. Thus, understanding the distribution and functions of microorganisms is important for environmental protection. Microbial ecology studies the interactions and basic principles that determine the distribution and abundance of microorganisms. It tries to understand microbial ecosystems and therefore provides a basic platform for analyses of ecological problems as well as for development of new technologies that tackle environmental problems. Progress in microbial ecology will facilitate the development of environmental biotechnology that tries to manipulate microorganisms in the ecosystems to remedy damaged environment by cleaning and reducing waste (Thavasi, Jayalakshmi, & Banat, 2011). Ananda Chakrabarty had modified a bacterium of the genus *Pseudomonas* capable of breaking down crude oil, which involves the transfer of entire organelles between strains of the *Pseudomonas* bacterium rather than gene manipulation. He proposed to use the modified bacterium in treating oil spills (Ezezika & Singer, 2010). There may be plenty of business opportunities in developing microorganisms that can efficiently treat wastes especially those difficult to degrade.

While biotechnology can be applied to recycle used materials, treat waste, and cleanup sites contaminated by industrial activities (bioremediation), it might also be used to produce biological weapons. It should not be overlooked that bioterrorism might use virulent microorganisms and toxins to cause diseases and death in humans, livestock and crops. Accidents in research laboratories causing leaks of virulent pathogens can also pose a serious threat to public health. Flow of genetic material from transgenic organisms into wild strains may cause loss of biodiversity. The potential negative impact of biotechnology on the environment and society requires adequate regulation of biotechnological research and materials that could pose threats to the environment or people.

The current regulations are mainly focused on two aspects: 1) the use of genetic engineering technology, to establish the legal basis for assessing and managing the risks associated with its use; and 2) the release of GMOs, including genetically modified crops and genetically modified fish. There are differences in the regulation of GMOs between countries, especially between the US and Europe (Gaskell, Bauer, Durant, & Allum, 1999). Regulation also varies with the intended use of the products of the genetic engineering, with crops not intended for food use being generally not reviewed by authorities responsible for food safety. The European Union (EU) is stricter on approval for cultivation within the EU than for import and processing. The cultivation of GMOs has caused concerns about flow of genetic material from GM crops into non-GM ones. The regulations on coexistence of GM and non-GM crops also differ with country, leading to different incentives for cultivation of GM crops.

CONCLUSION

This chapter has reviewed the current state of biotechnology and its contribution to human health, economy and environment. From domestication of crops and livestock and production of beer, wine, cheese and leavened bread to vaccines protecting people from infectious diseases and antibiotics for treating bacterial diseases, human society depends on biotechnology. Almost every important discovery or invention in biotechnology leads immediately to business opportunities and new startups mushroom in that field. The advent of molecular cloning and genetic engineering have greatly enhanced mankind's ability to modify the genome of bacteria, fungi, cell lines, plants and animals for the benefits of mankind.

CRISPR-Cas9 as a tool for gene editing makes genetic engineering and gene therapy more facile and convenient. To conclude this chapter, we can look forward to the future of biotechnology and project what biotechnology will achieve in the near future.

What mankind wants to achieve with biotechnology is no more than getting methods and products to improve health and improve quality of life, which depends on better ability to biological processes and efficient application of such ability. With rapid progress in genetic engineering, we can expect that technologies to correct genetic defects and introduce preferred traits into the genomes will become conventional in treating human hereditary diseases and providing medicinal biologics for preventing and treating non-hereditary diseases by 2030. DNA, RNA and protein vaccines will become the standard prophylactics, and antibodies and modified cytokines play major roles in treating various diseases. Developing gene therapies and producing medicinal biologics continue to provide ample business opportunities. Artificial organs created from cultured cells may start to be used as treatment for some diseases.

By 2035, technologies to improve crops and livestock for more nutritious and more delicious foods may blur the boundary between foods from plants and those from animals. Genes that give beef, lamb, pork, chicken and seafood their distinct flavors and nutrition can be introduced into the genome of plants and fungi to produce meat-like foods in farmland or culture tanks, which will substantially increase energy efficiency on food production. Genetically modified microorganisms will be more widely used in many industrial processes. On the medical front, artificial organs will be more widely used.

By 2040, we may expect that regeneration mechanism is well understood and disabilities and injuries will be treated by regenerative technologies. New organs can be grown in situ or transplanted to replace the damaged or lost ones. Mechanisms for longevity and health are sufficiently understood and gene therapy can be used to transfer the health and longevity traits to people who do not have those traits. People's life expectancy will be greatly increased.

Progresses in biotechnology depend largely on our understanding of basic mechanisms of life. Some serendipitous discoveries may accelerate progress of biotechnology and make what we expect to come much earlier. There may also be barriers that are difficult to surmount, then progress would be slower. No matter whether the path for progress is a smooth one or not, biotechnology will play a more important role in medicine, agriculture and industry in the future.

REFERENCES

Anderson, J., DiCicco, D., Ginder, J., Kramer, U., Leone, T., Raney-Pablo, H., & Wallington, T. (2012). High octane number ethanol–gasoline blends: Quantifying the potential benefits in the United States. *Fuel*, *97*, 585–594. doi:10.1016/j.fuel.2012.03.017

Arranz-Otaegui, A., Carretero, L. G., Ramsey, M. N., Fuller, D. Q., & Richter, T. (2018). Archaeobotanical evidence reveals the origins of bread 14,400 years ago in northeastern Jordan. *Proceedings of the National Academy of Sciences of the United States of America*, *115*(31), 7925–7930. doi:10.1073/pnas.1801071115 PMID:30012614

Bassett, R. K. (2002). *To the digital age: Research labs, start-up companies, and the rise of MOS technology*. Johns Hopkins University Press.

Becquemont, L. (2009). Pharmacogenomics of adverse drug reactions: Practical applications and perspectives. *Pharmacogenomics, 10*(6), 961–969. doi:10.2217/pgs.09.37 PMID:19530963

Behring, E. v. (1890). Ueber das zustandekommen der diphtherie-immunität und der tetanus-immunität bei thieren. *Deutsche medizinsche Wochenschrift, 16*, 1113–1114.

Bergveld, P. (1985). The impact of MOSFET-based sensors. *Sensors and Actuators, 8*(2), 109–127. doi:10.1016/0250-6874(85)87009-8

Bordenave, G. (2003). Louis Pasteur (1822–1895). *Microbes and Infection, 5*(6), 553–560. doi:10.1016/S1286-4579(03)00075-3 PMID:12758285

Clark, L. C. Jr, & Lyons, C. (1962). Electrode systems for continuous monitoring in cardiovascular surgery. *Annals of the New York Academy of Sciences, 102*(1), 29–45. doi:10.1111/j.1749-6632.1962.tb13623.x PMID:14021529

Cohen, A. M., & Hersh, W. R. (2005). A survey of current work in biomedical text mining. *Briefings in Bioinformatics, 6*(1), 57–71. doi:10.1093/bib/6.1.57 PMID:15826357

Cohen, S. N., Chang, A. C., Boyer, H. W., & Helling, R. B. (1973). Construction of biologically functional bacterial plasmids in vitro. *Proceedings of the National Academy of Sciences of the United States of America, 70*(11), 3240–3244. doi:10.1073/pnas.70.11.3240 PMID:4594039

Cooper, D. K., Gollackner, B., & Sachs, D. H. (2002). Will the pig solve the transplantation backlog? *Annual Review of Medicine, 53*(1), 133–147. doi:10.1146/annurev.med.53.082901.103900 PMID:11818467

Daniell, H., Streatfield, S. J., & Wycoff, K. (2001). Medical molecular farming: Production of antibodies, biopharmaceuticals and edible vaccines in plants. *Trends in Plant Science, 6*(5), 219–226. doi:10.1016/S1360-1385(01)01922-7 PMID:11335175

Darwin, C. (1968). *On the origin of species by means of natural selection*. John Murray.

Dawson, W. K., Maciejczyk, M., Jankowska, E. J., & Bujnicki, J. M. (2016). Coarse-grained modeling of RNA 3D structure. *Methods (San Diego, Calif.), 103*, 138–156. doi:10.1016/j.ymeth.2016.04.026 PMID:27125734

Dietrich, O., Heun, M., Notroff, J., Schmidt, K., & Zarnkow, M. (2012). The role of cult and feasting in the emergence of Neolithic communities. New evidence from Göbekli Tepe, south-eastern Turkey. *Antiquity, 86*(333), 674–695. doi:10.1017/S0003598X00047840

Domingo, J. L., & Bordonaba, J. G. (2011). A literature review on the safety assessment of genetically modified plants. *Environment International, 37*(4), 734–742. doi:10.1016/j.envint.2011.01.003 PMID:21296423

Dudley, M. E., Wunderlich, J. R., Shelton, T. E., Even, J., & Rosenberg, S. A. (2003). Generation of tumor-infiltrating lymphocyte cultures for use in adoptive transfer therapy for melanoma patients. *Journal of Immunotherapy, 26*(4), 332-342.

Evans, A. S. (1976). Causation and disease: The Henle-Koch postulates revisited. *The Yale Journal of Biology and Medicine, 49*(2), 175–195. PMID:782050

Evans, W. E., & Relling, M. V. (1999). Pharmacogenomics: Translating functional genomics into rational therapeutics. *Science*, *286*(5439), 487–491. doi:10.1126cience.286.5439.487 PMID:10521338

Ezezika, O. C., & Singer, P. A. (2010). Genetically engineered oil-eating microbes for bioremediation: Prospects and regulatory challenges. *Technology in Society*, *32*(4), 331–335. doi:10.1016/j.techsoc.2010.10.010

Ferrante, R. J., Andreassen, O. A., Jenkins, B. G., Dedeoglu, A., Kuemmerle, S., Kubilus, J. K., Kaddurah-Daouk, R., Hersch, S. M., & Beal, M. F. (2000). Neuroprotective effects of creatine in a transgenic mouse model of Huntington's disease. *The Journal of Neuroscience: The Official Journal of the Society for Neuroscience*, *20*(12), 4389–4397. doi:10.1523/JNEUROSCI.20-12-04389.2000 PMID:10844007

Festing, M. F. (2010). Inbred strains should replace outbred stocks in toxicology, safety testing, and drug development. *Toxicologic Pathology*, *38*(5), 681–690. doi:10.1177/0192623310373776 PMID:20562325

Folegatti, P. M., Ewer, K. J., Aley, P. K., Angus, B., Becker, S., Belij-Rammerstorfer, S., ... Clutterbuck, E. A. (2020). Safety and immunogenicity of the ChAdOx1 nCoV-19 vaccine against SARS-CoV-2: A preliminary report of a phase 1/2, single-blind, randomised controlled trial. *Lancet*, *396*(10249), 467–478. doi:10.1016/S0140-6736(20)31604-4 PMID:32702298

Food and Drug Administration. (2020). *Approved Cellular and Gene Therapy Products*. Retrieved from https://www.fda.gov/vaccines-blood-biologics/cellular-gene-therapy-products/approved-cellular-and-gene-therapy-products

Freedman, A. H., & Wayne, R. K. (2017). Deciphering the origin of dogs: From fossils to genomes. *Annual Review of Animal Biosciences*, *5*(1), 281–307. doi:10.1146/annurev-animal-022114-110937 PMID:27912242

Gaskell, G., Bauer, M. W., Durant, J., & Allum, N. C. (1999). Worlds apart? The reception of genetically modified foods in Europe and the US. *Science*, *285*(5426), 384–387. doi:10.1126cience.285.5426.384 PMID:10411496

Gershon, D., & Lehrman, S. (1994). Genetically altered tomato set to get the green light. *Nature*, *368*(6472), 574–574. doi:10.1038/368574a0 PMID:8145830

Grimm, E. A., Mazumder, A., Zhang, H., & Rosenberg, S. (1982). Lymphokine-activated killer cell phenomenon. Lysis of natural killer-resistant fresh solid tumor cells by interleukin 2-activated autologous human peripheral blood lymphocytes. *The Journal of Experimental Medicine*, *155*(6), 1823–1841. doi:10.1084/jem.155.6.1823 PMID:6176669

Guy, C. T., Cardiff, R., & Muller, W. J. (1992). Induction of mammary tumors by expression of polyomavirus middle T oncogene: A transgenic mouse model for metastatic disease. *Molecular and Cellular Biology*, *12*(3), 954–961. doi:10.1128/MCB.12.3.954 PMID:1312220

Haslberger, A. G. (2003). Codex guidelines for GM foods include the analysis of unintended effects. *Nature Biotechnology*, *21*(7), 739–741. doi:10.1038/nbt0703-739 PMID:12833088

Heller, M. J. (2002). DNA microarray technology: Devices, systems, and applications. *Annual Review of Biomedical Engineering*, *4*(1), 129–153. doi:10.1146/annurev.bioeng.4.020702.153438 PMID:12117754

Hiatt, A., Caffferkey, R., & Bowdish, K. (1989). Production of antibodies in transgenic plants. *Nature, 342*(6245), 76–78. doi:10.1038/342076a0 PMID:2509938

Hoekman, S. K. (2009). Biofuels in the US–challenges and opportunities. *Renewable Energy, 34*(1), 14–22. doi:10.1016/j.renene.2008.04.030

Hu, H., & Xiong, L. (2014). Genetic engineering and breeding of drought-resistant crops. *Annual Review of Plant Biology, 65*(1), 715–741. doi:10.1146/annurev-arplant-050213-040000 PMID:24313844

Hurley, L. H. (2002). DNA and its associated processes as targets for cancer therapy. *Nature Reviews. Cancer, 2*(3), 188–200. doi:10.1038/nrc749 PMID:11990855

Jackson, D. A., Symons, R. H., & Berg, P. (1972). Biochemical method for inserting new genetic information into DNA of Simian Virus 40: Circular SV40 DNA molecules containing lambda phage genes and the galactose operon of Escherichia coli. *Proceedings of the National Academy of Sciences of the United States of America, 69*(10), 2904–2909. doi:10.1073/pnas.69.10.2904 PMID:4342968

Jaenisch, R., & Mintz, B. (1974). Simian virus 40 DNA sequences in DNA of healthy adult mice derived from preimplantation blastocysts injected with viral DNA. *Proceedings of the National Academy of Sciences of the United States of America, 71*(4), 1250–1254. doi:10.1073/pnas.71.4.1250 PMID:4364530

Johnson, I. S. (1983). Human insulin from recombinant DNA technology. *Science, 219*(4585), 632–637. doi:10.1126cience.6337396 PMID:6337396

Kato, M., Takahashi, M., Akhand, A. A., Liu, W., Dai, Y., Shimizu, S., Iwamoto, T., Suzuki, H., & Nakashima, I. (1998). Transgenic mouse model for skin malignant melanoma. *Oncogene, 17*(14), 1885–1888. doi:10.1038j.onc.1202077 PMID:9778055

Katsanis, S. H., & Katsanis, N. (2013). Molecular genetic testing and the future of clinical genomics. *Nature Reviews. Genetics, 14*(6), 415–426. doi:10.1038/nrg3493 PMID:23681062

Keech, C., Albert, G., Cho, I., Robertson, A., Reed, P., Neal, S., ... Zhou, H. (2020). Phase 1–2 trial of a SARS-CoV-2 recombinant spike protein nanoparticle vaccine. *The New England Journal of Medicine, 383*(24), 2320–2332. doi:10.1056/NEJMoa2026920 PMID:32877576

Kiefer, M. C., Schmid, C., Waldvogel, M., Schläpfer, I., Futo, E., Masiarz, F., Green, K., Barr, P. J., & Zapf, J. (1992). Characterization of recombinant human insulin-like growth factor binding proteins 4, 5, and 6 produced in yeast. *The Journal of Biological Chemistry, 267*(18), 12692–12699. doi:10.1016/S0021-9258(18)42332-0 PMID:1377672

Kmiecik, S., Gront, D., Kolinski, M., Wieteska, L., Dawid, A. E., & Kolinski, A. (2016). Coarse-grained protein models and their applications. *Chemical Reviews, 116*(14), 7898–7936. doi:10.1021/acs.chemrev.6b00163 PMID:27333362

Kurita, K. (2006). Chitin and chitosan: Functional biopolymers from marine crustaceans. *Marine Biotechnology (New York, N.Y.), 8*(3), 203–226. doi:10.100710126-005-0097-5 PMID:16532368

Lane, N. (2015). The unseen world: reflections on Leeuwenhoek (1677) 'Concerning little animals'. *Philosophical Transactions of the Royal Society B: Biological Sciences, 370*(1666), 20140344.

Li, M., Ona, V. O., Guégan, C., Chen, M., Jackson-Lewis, V., Andrews, L. J., ... Przedborski, S. (2000). Functional role of caspase-1 and caspase-3 in an ALS transgenic mouse model. *Science*, *288*(5464), 335–339. doi:10.1126cience.288.5464.335 PMID:10764647

Ligon, B. L. (2004). Penicillin: Its discovery and early development. *Seminars in Pediatric Infectious Diseases*, *15*(1), 52–57. doi:10.1053/j.spid.2004.02.001 PMID:15175995

Maeda, Y., Yoshino, T., Matsunaga, T., Matsumoto, M., & Tanaka, T. (2018). Marine microalgae for production of biofuels and chemicals. *Current Opinion in Biotechnology*, *50*, 111–120. doi:10.1016/j.copbio.2017.11.018 PMID:29232601

Mercado, N. B., Zahn, R., Wegmann, F., Loos, C., Chandrashekar, A., Yu, J., ... Tostanoski, L. H. (2020). Single-shot Ad26 vaccine protects against SARS-CoV-2 in rhesus macaques. *Nature*, *586*(7830), 583–588. doi:10.103841586-020-2607-z PMID:32731257

Mittler, R., & Blumwald, E. (2010). Genetic engineering for modern agriculture: Challenges and perspectives. *Annual Review of Plant Biology*, *61*(1), 443–462. doi:10.1146/annurev-arplant-042809-112116 PMID:20192746

Nicolia, A., Manzo, A., Veronesi, F., & Rosellini, D. (2014). An overview of the last 10 years of genetically engineered crop safety research. *Critical Reviews in Biotechnology*, *34*(1), 77–88. doi:10.3109/07388551.2013.823595 PMID:24041244

Pai, S., Lin, Y., Macaes, B., Meneshian, A., Hung, C.-F., & Wu, T. C. (2006). Prospects of RNA interference therapy for cancer. *Gene Therapy*, *13*(6), 464–477. doi:10.1038j.gt.3302694 PMID:16341059

Pasteur, L. (1857). Mémoire sur la fermentation alcoolique. *Comptes Rendus. Chimie*, *45*(6), 1032–1036.

Pasteur, L. (1858). Nouveaux faits concernant l'histoire de la fermentation alcoolique. *Comptes Rendus. Chimie*, *47*, 1011–1013.

Philippidis, A. (2020). Moderna's COVID-19 Vaccine Speeds to Phase III Trial After More Positive Data. *Clinical OMICs*, *7*(4), 10–11. doi:10.1089/clinomi.07.04.17

Pious, N., & Ingole, S. (2020). Race for COVID-19 Vaccine. *Trends in Biomaterials & Artificial Organs*, *34*(S3), 62–65.

Piperno, D. R., Ranere, A. J., Holst, I., Iriarte, J., & Dickau, R. (2009). Starch grain and phytolith evidence for early ninth millennium BP maize from the Central Balsas River Valley, Mexico. *Proceedings of the National Academy of Sciences of the United States of America*, *106*(13), 5019–5024. doi:10.1073/pnas.0812525106 PMID:19307570

Purugganan, M. D., & Fuller, D. Q. (2009). The nature of selection during plant domestication. *Nature*, *457*(7231), 843–848. doi:10.1038/nature07895 PMID:19212403

Ran, F. A., Hsu, P. D., Wright, J., Agarwala, V., Scott, D. A., & Zhang, F. (2013). Genome engineering using the CRISPR-Cas9 system. *Nature Protocols*, *8*(11), 2281–2308. doi:10.1038/nprot.2013.143 PMID:24157548

Refolo, L. M., Pappolla, M. A., Malester, B., LaFrancois, J., Bryant-Thomas, T., Wang, R., Tint, G. S., Sambamurti, K., & Duff, K. (2000). Hypercholesterolemia accelerates the Alzheimer's amyloid pathology in a transgenic mouse model. *Neurobiology of Disease*, *7*(4), 321–331. doi:10.1006/nbdi.2000.0304 PMID:10964604

Riedel, S. (2005). Edward Jenner and the history of smallpox and vaccination. *Baylor University Medical Center Proceedings*, *18*(1), 21–25. doi:10.1080/08998280.2005.11928028 PMID:16200144

Robinson, D., & Medlock, N. (2005). Diamond v. Chakrabarty: a retrospective on 25 years of biotech patents. *Intellectual Property & Technology Law Journal, 17*(10), 12-15.

Salque, M., Bogucki, P. I., Pyzel, J., Sobkowiak-Tabaka, I., Grygiel, R., Szmyt, M., & Evershed, R. P. (2013). Earliest evidence for cheese making in the sixth millennium BC in northern Europe. *Nature*, *493*(7433), 522–525. doi:10.1038/nature11698 PMID:23235824

Saxena, D., Flores, S., & Stotzky, G. (1999). Insecticidal toxin in root exudates from Bt corn. *Nature*, *402*(6761), 480–480. doi:10.1038/44997 PMID:10591205

Schöning, M. J., & Poghossian, A. (2002). Recent advances in biologically sensitive field-effect transistors (BioFETs). *Analyst (London)*, *127*(9), 1137–1151. doi:10.1039/B204444G PMID:12375833

Sim, A. Y., Minary, P., & Levitt, M. (2012). Modeling nucleic acids. *Current Opinion in Structural Biology*, *22*(3), 273–278. doi:10.1016/j.sbi.2012.03.012 PMID:22538125

Skaugrud, Ø., Hagen, A., Borgersen, B., & Dornish, M. (1999). Biomedical and pharmaceutical applications of alginate and chitosan. *Biotechnology & Genetic Engineering Reviews*, *16*(1), 23–40. doi:10.1080/02648725.1999.10647970 PMID:10819076

Squassina, A., Manchia, M., Manolopoulos, V. G., Artac, M., Lappa-Manakou, C., Karkabouna, S., Mitropoulos, K., Zompo, M. D., & Patrinos, G. P. (2010). Realities and expectations of pharmacogenomics and personalized medicine: Impact of translating genetic knowledge into clinical practice. *Pharmacogenomics*, *11*(8), 1149–1167. doi:10.2217/pgs.10.97 PMID:20712531

Strauer, B. E., & Kornowski, R. (2003). Stem cell therapy in perspective. *Circulation*, *107*(7), 929–934. doi:10.1161/01.CIR.0000057525.13182.24 PMID:12600901

Thavasi, R., Jayalakshmi, S., & Banat, I. M. (2011). Effect of biosurfactant and fertilizer on biodegradation of crude oil by marine isolates of Bacillus megaterium, Corynebacterium kutscheri and Pseudomonas aeruginosa. *Bioresource Technology*, *102*(2), 772–778. doi:10.1016/j.biortech.2010.08.099 PMID:20863694

Thein, M., Goh, L., & Phua, K. (1988). The smallpox story: From variolation to victory. *Asia-Pacific Journal of Public Health*, *2*(3), 203–210. doi:10.1177/101053958800200313 PMID:3052542

Toumazou, C., & Georgiou, P. (2011). Piet Bergveld-40 years of ISFET technology: From neuronal sensing to DNA sequencing. *Electronics Letters*, *47*(26), S7–S12. doi:10.1049/el.2011.3231

Verbeke, R., Lentacker, I., Smedt, S. C. D., & Dewitte, H. (2019). Three decades of messenger RNA vaccine development. *Nano Today*, *28*, 100766. doi:10.1016/j.nantod.2019.100766

Verma, I. M., Naldini, L., Kafri, T., Miyoshi, H., Takahashi, M., Blömer, U., & Gage, F. (2000). Gene therapy: promises, problems and prospects. In *Genes and Resistance to Disease* (pp. 147–157). Springer. doi:10.1007/978-3-642-56947-0_13

Waldmann, T. A. (1991). Monoclonal antibodies in diagnosis and therapy. *Science*, *252*(5013), 1657–1662. doi:10.1126cience.2047874 PMID:2047874

Walker, L., LeVine, H., & Jucker, M. (2006). Koch's postulates and infectious proteins. *Acta Neuropathologica*, *112*(1), 1–4. doi:10.100700401-006-0072-x PMID:16703338

Waltz, E. (2016). GM salmon declared fit for dinner plates. *Nature Biotechnology*, *34*(1), 7–8. doi:10.1038/nbt0116-7a PMID:26744960

Weiling, F. (1991). Historical study: Johann Gregor Mendel 1822–1884. *American Journal of Medical Genetics*, *40*(1), 1–25. doi:10.1002/ajmg.1320400103 PMID:1887835

Weiner, L. M., Surana, R., & Wang, S. (2010). Monoclonal antibodies: Versatile platforms for cancer immunotherapy. *Nature Reviews. Immunology*, *10*(5), 317–327. doi:10.1038/nri2744 PMID:20414205

Willis, N. (1997). Edward Jenner and the eradication of smallpox. *Scottish Medical Journal*, *42*(4), 118–121. doi:10.1177/003693309704200407 PMID:9507590

Zhu, N., Zhang, D., Wang, W., Li, X., Yang, B., Song, J., . . . Lu, R. (2020). A novel coronavirus from patients with pneumonia in China, 2019. *New England Journal of Medicine*.

KEY TERMS AND DEFINITIONS

Adenovector: A common cold-causing adenovirus that has been genetically changed for carrying DNA sequences into host cells.

Biohydrometalluigy: A subfield within hydrometallurgy which includes aspects of biotechnology.

Bioinformatics: An interdisciplinary field that addresses biological problems using computational techniques to understand biological data.

Bioleaching: The extraction of metals from their ores through the use of living organisms.

Biopolymers: Natural polymers produced by the cells of living organisms.

Biosensor: An analytical device that combines a biological component with a physicochemical detector to detect a chemical substance.

DNA Microarray: A collection of microscopic DNA spots attached to a solid surface, which can be used to measure the expression levels of large numbers of genes simultaneously or to genotype multiple regions of a genome.

DNA Vaccination: Vaccination by injection with genetically engineered plasmid containing the DNA sequence encoding the antigen(s) against which an immune response is sought, so cells directly produce the antigen, causing a protective immunological response.

Gene Therapy: A medical field which focuses on the utilization of the therapeutic delivery of nucleic acids into a patient's cells as a drug to treat disease.

Genetic Engineering: The direct manipulation of an organism's genes using biotechnology.

Genetic Testing: Using tests to identify changes in DNA sequence or chromosome structure.

Genomics: An interdisciplinary field of biology focusing on the structure, function, evolution, mapping, and editing of genomes. A genome is an organism's complete set of DNA, including all of its genes.

Molecular Cloning: A set of experimental methods in molecular biology that are used to assemble recombinant DNA molecules and to direct their replication within host organisms.

Monoclonal Antibody: An antibody made by cloning a unique antibody producing cell and all subsequent antibodies derived this way trace back to a unique parent cell and bind only to the same epitope (the part of an antigen that is recognized by the antibody).

MOSFET: The metal-oxide-semiconductor field-effect transistor is a type of insulated-gate field-effect transistor that is fabricated by the controlled oxidation of a semiconductor, typically silicon.

Personalized Medicine: A medical model that separates people into different groups—with medical decisions, practices, interventions and/or products being tailored to the individual patient based on their predicted response or risk of disease.

Polymerase Chain Reaction (PCR): A method widely used to rapidly make millions to billions of copies of a specific DNA sample through thermal cycling with primers and DNA polymerase.

Proteomics: The large-scale study of proteins to understand the proteome, which is the entire set of proteins that is produced or modified by an organism or system.

RNA Vaccination: Vaccination by a vector containing an RNA sequence that codes the antigen(s) against which an immune response is sought.

SARS-CoV-2: The coronavirus that causes COVID-19.

Selective Breeding: The process by which humans use animal breeding and plant breeding to selectively develop particular phenotypic traits (characteristics) by choosing which typically animal or plant males and females will sexually reproduce and have offspring together.

Single-Nucleotide Polymorphism: A substitution of a single nucleotide at a specific position in the genome, that is present in a sufficiently large fraction of the population.

Chapter 8
Cognitive Technologies: Applications to Cognition Itself

Fei Luo
University of Chinese Academy of Sciences, China

ABSTRACT

Cognitive science is the most important emerging discipline at the turn of the century. Since its birth, cognitive science has integrated many important branches of science, including but not limited to neuroscience, psychology, computer science, systems science, information science, and management. With the development of these related fields, cognitive science has achieved recognition and respect. However, up to date, the understanding of cognition itself remains limited. The science of meditation and mindfulness is adding new light to this field. It has been revealed that through the integration of subjective cognitive training, it may be possible to gain further understanding of cognition per se, therefore promoting new developments in cognitive technology.

INTRODUCTION

Cognition has been a fantasy for centuries. We obviously need it to observe and understand the world, but how does our cognition work? Does it work well, enabling accurate understanding? These questions spurred the creation of the discipline of cognitive science in the 1970s. The main objective of this discipline is to provide a framework for the cohesive, integrated, multidisciplinary study of the mind. Scientists in this field come from psychology, linguistics, philosophy, neuroscience, computer science, artificial intelligence, and anthropology, among other fields, seeking to illuminate how the human mind works, and why it works the way it does (for comprehensive knowledges on cognitive science, see Thagard, 2020).

Great progress has been made in the 50 years of cognitive science's existence. We have not only a better understanding of how our brains and cognition work, but also a better foundation for the understanding of cognition itself. As the historian Yuval Noah Harari stated, human beings have recently come into possession of the power to manufacture products that change not only our outer environments, but also our bodies, brains, and minds (for historical understanding of human problems, see Harari, 2018). This represents a fundamental revolution in human history, and its key is hidden in the field of cognitive science.

DOI: 10.4018/978-1-7998-6772-2.ch008

Before cognitive science can be applied in the context of this new revolution, however, we need to gain a thorough understanding of how cognition works. The purpose of this chapter is to explore this topic. First, the major developments in cognitive science and related technologies are reviewed. Next, the short-term and long-term development of this field is explored. Finally, possible impacts on education, health, and space exploration are discussed.

MAJOR MILESTONES IN COGNITIVE TECHNOLOGY DEVELOPMENT

A Brief History

Cognitive science is the interdisciplinary study of psychology and intelligence. It began to be conceptualized in the mid-1950s, with researchers in several fields developing theories of mind based on complex representation and computation. In the mid-1970s, the Cognitive Science Society was established and the journal *Cognitive Science* began to be published. Since then, more than 100 universities in North America, Europe, Asia, and Australia have established cognitive science programs, and many additional universities have offered cognitive science courses.

Attempts to understand the mind and its operation can be traced back to at least the ancient Greeks, with the efforts of philosophers such as Plato and Aristotle to explain the nature of human knowledge. Until the development of experimental psychology in the 19th century, the study of the mind was a philosophical undertaking. Wilhelm Wundt and his students pioneered a more systematic laboratory-based approach to the study of mental activity. For decades, however (through the 1950s in North America), experimental psychology was dominated by behaviorism, which virtually denies the existence of the mind. According to behaviorists such as J.B. Watson, psychology should be limited to the study of relationships between observable stimuli and behavioral responses. This approach conveniently excludes formal scientific discussion of consciousness and mental representation.

Around 1956, great changes began to take place in the intellectual world. George Miller summed up the findings of numerous studies regarding the limitations of human cognitive ability. For example, he noted that our short-term memory is limited to about seven items. He proposed that the limitations of memory could be overcome with the recoding of information in "blocks," which are psychological representations that need to be encoded and decoded. Although primitive computers had existed for only a few years at this time, pioneers such as John McCarthy, Marvin Minsky, Alan Newell, and Herbert Simon were exploring the field of artificial intelligence. In addition, Noam Chomsky countered the behaviorist hypothesis that language is a habit of acquisition, proposing a psychological grammar composed of rules to explain language understanding. These six thinkers can be regarded as the founders of cognitive science.

Multiple Methodologies

The theoretical concepts of cognitive science are unifying. Researchers in different fields have brought diverse views and methods to the study of the mind and intelligence. Although cognitive psychologists now often engage in theoretical and computational modeling, their main approach is experimental, involving human participants. Various aspects of the cognition of individuals, usually undergraduate students fulfilling course requirements, are examined in the laboratory under controlled conditions. For example, psychologists have used experiments to study errors in deductive reasoning, concept formation and use, the speed of thinking with mental imagery, and performance in solving problems by analogy. Conclusions about how the brain works must not be based solely on "common sense" and introspection, which provide a misleading picture, many of which cannot be realized. Thus, careful psychological research using experiments derived from diverse approaches is very important for the scientificity of cognitive science.

In psychology, as in any well-developed science, the solving of key problems requires the embedding of experimentation within a theoretical framework, in this case of hypothetical mental representation and processing. One of the best ways to develop a theoretical framework is through the development and testing of computational models similar to mental manipulation. To complement psychological experimentation, researchers have developed computational models that simulate all thinkable aspects of human behavior. Model design, construction, and experimental calculation form the core methodology in the field of artificial intelligence, a branch of computer science. Ideally, computational models and psychological experiments in cognitive science complement each other; in artificial intelligence, however, the power of different knowledge representation methods has been tested relatively independently of experimental psychology.

Although some linguists conduct psychological experiments or develop computational models, most currently use different methods. One main theoretical approach is to identify grammatical principles that provide the basic structures of languages. Grammar can explain why some utterances are acceptable, whereas others are not. Cognitive linguistics, however, focuses less on syntax and more on semantics and concepts.

Like cognitive psychologists, neuroscientists often conduct controlled experiments, but their observations are quite different because they are concerned directly with the nature of the brain. Researchers can insert electrodes into non-human subjects to record the discharge of a single neuron. This approach is obviously intrusive for humans; thus, magnetic and positron-based scanning devices are now used widely to observe activity in different parts of the brain while people perform various mental activities. Evidence of brain function is also gathered by examining the performance of people with recognizable brain damage, such as stroke-induced defects in a part of the brain devoted to language that result in the inability to speak sentences. Like cognitive psychology, neuroscience is usually theoretical and experimental, with theory development typically aided by the development of computational models of neuronal group behavior.

Cognitive anthropology expands the study of human thinking, considering how it works in different cultural and linguistic contexts. Cognitive scientists are increasingly aware of the need to observe brain functioning in specific physical and social environments. For cultural anthropologists, the main approach to such observation is ethnography, which requires residence among and interaction with members of a culture to a sufficient extent to elucidate their social and cognitive systems. For example, cognitive anthropologists have studied similarities and differences in words used for directions, timing, and color among cultures.

Traditionally, philosophers do not make systematic empirical observations or build computational models, although work in experimental philosophy has increased recently. Philosophy remains important to cognitive science because it deals with basic problems using experimental and computational methods. Abstract problems such as those regarding the nature of representation and computation do not need to be solved in the daily practice of psychology or artificial intelligence research, but they inevitably arise when researchers think deeply about what they are doing. Philosophy also deals with general issues, such as the relationship between mind and body, and methodological issues, such as the nature of interpretation, which are also addressed in cognitive science, as well as normative issues (e.g., how people should think) and descriptive issues (e.g., how people think). In addition to theoretical goals regarding the understanding of human cognition, cognitive science can focus on practical goals to improve cognition, which requires normative reflection on the thinking we want. No unique method has been developed for work on the philosophy of mind, but such work should focus on the acquisition of empirical results with underpinning from the best theoretical research in other fields.

Interdisciplinary work becomes more interesting as theoretically and experimentally derived conclusions about the nature of the mind converge. For example, psychological and artificial intelligence research can be combined through computational modeling of people's behavior during experiments. The best way to master the complexity of human cognition is to use a variety of methods, especially psychological and neurological experimentation and computational modeling. Theoretically, the most abundant method is to seek to understand the mind from the perspectives of representation and computation.

Representation and Computation

Throughout history, humans have tried to understand life via achievements in science and technology. Before the modern scientific era, people defined life using various philosophical and religious concepts. Since the establishment of Newtonian physics, they have begun to understand life as a precise automated machine operating according to the laws of physics. With the development of computer science, people have used hardware and software structures to understand cognition, namely representation and computation.

The current central assumption of cognitive science is that cognition is best understood via the examination of representative structures in the brain and the computational programs that operate on them. Although the representations and computations that constitute thinking are very diverse, this assumption is thought to be sufficiently universal to cover the current scope of cognitive science, including the connectionist theory of artificial neural networks developed to simulate thinking.

Most work in cognitive science is conducted under the assumption that the brain has mental representations (e.g., logical propositions, rules, concepts, images, and analogies) similar to computer data structures and computational processes (e.g., reasoning, searching, matching, rotation, and retrieval) similar to computer algorithms. This mind-as-computer analogy has been dominant in cognitive science, but a turning point has been reached with the use of another, brain-based, analogy.

Connectivists have newly conceptualized representation and computation, taking neurons and their connections as the data structure and neuronal excitation and diffusion activation as the algorithms. A complex tripartite mind–brain–computer analogy has been established in cognitive science; each element can be used to generate new ideas about the other two. No single computational model represents the mind; the use of different computer types and programming methods has generated different propositions about ways in which the mind works. Currently, most scientists use computers with serial processors that execute one instruction at a time; in contrast, the brain and some recently developed computers perform parallel processing, performing many operations simultaneously.

A main trend in cognitive science is the combination of neuroscience research with research in many areas of psychology, including cognition, society, development, and clinic. This integration is, to some extent, related to experimentation, due to the explosive development of new instruments for the study of the brain, such as functional magnetic resonance imaging, transcranial magnetic stimulation, and optogenetics. It is also related to theoretical work, as progress has been made in understanding how large numbers of neurons perform tasks that are usually explained by cognitive theories of rules and concepts.

Theoretical Considerations

Formal logic provides powerful tools for the study of the nature of representation and computation. Propositional calculus and predicate calculus are used to express large amounts of complex knowledge. Many forms of inference, such as hypothetical inference, are logical. Logical interpretation of why people make certain inferences proceeds as follows: 1) people's mental representations are similar to the sentences generated in predicate logic, 2) people have deductive and inductive procedures for generating these sentences, and 3) inferences are produced by applying these sentence-generating procedures. However, whether logic provides the core concepts of representation and computation required for the application of cognitive science is uncertain; more effective and psychologically natural computational methods may be needed to explain human thinking.

A great deal of human knowledge can be described naturally using formal rules, such as if/then statements. Many forms of thinking, such as planning, can be modeled using rule-based systems. The interpretative model used to explain why people perform a certain type of intelligent behavior is constructed as follows: 1) people have psychological rules; 2) people have programs that use these rules to search for possible solution spaces, as well as programs that generate new rules; and 3) procedures that use and form rules generate behavior. Rule-based computational models enable detailed simulation of many psychological experiments, ranging from the problem solving involved in cryptographic algorithm construction to the language use involved in skill learning. Rule-based systems are also of practical importance, as they provide suggestions for ways in which to improve learning and develop intelligent machine systems.

A concept is an important psychological representation, corresponding partly to the words in a spoken or written language. Computational and psychological factors justify the abandonment of the classical view that concepts are strictly defined. Instead, a concept can be regarded as a set of typical characteristics, applied to obtain an approximate match with the world. Patterns and scripts are more complex than concepts corresponding to words, but they are similar in that they consist of feature bundles that can be matched and applied to new situations. The use of an interpretative model within the framework of a concept-based system to explain why people perform a certain intelligent behavior proceeds as follows: 1) people have sets of concepts that are organized by creating categories, partial hierarchies, and other associated slots; 2) people have sets of conceptual applications, including propagation activation, matching, and inheritance; and 3) the program generation behavior applied to the concept. Concepts can be transformed into rules, but the bundling of information, and thus the computing processes possible, differ between concept and rule sets.

Analogy plays an important role in human thinking; it is involved in problem solving, decision making, interpretation, and language-based communication. Computational models can be used to simulate how people retrieve and map source simulation for application to a target situation. Analogy-based interpretative modeling used to explain why people perform a certain type of intelligent behavior proceeds as follows: 1) people have linguistic and visual representations of situations, which can be used as cases or analogies; 2) people retrieve, map and adapt to these analogs; and 3) the application of analogical processes to representation produces behavior. Limitation according to similarity, structure, and purpose overcomes the problem of how to find and use previous experience to help solve new problems. Not all cognition, however, is analogical. The improper use of analogies hinders thinking, but analogies can be used very effectively in education and design applications.

Vision and other types of images also play important roles in human thinking. Image representation can capture visual and spatial information better than lengthy oral description. It is very suitable for the calculative processes involved in visual representation, including inspection, searching, scaling, rotation, and transformation. Such operations are very useful for the generation of plans and explanations for applied graphical representation. The interpretative modeling of visual expression to explain why people perform certain types of intelligent behavior proceeds as follows: 1) people have intuitive impressions of various situations, 2) people scan and rotate these images, and 3) image building and manipulation produce intelligent behavior. Images aid learning and may be the sources of some metaphorical aspects of language. Psychological experiments have shown that imagery is used in visual processes such as scanning and rotation. Neurophysiological research has confirmed the existence of close physical relationships between reasoning and mental imagery and perception. Imagery is not only visual; it also involves other sensory experiences.

Networks composed of simple nodes and links can be applied very usefully to understand psychological processes involving parallel constraint satisfaction, including visualization, decision making, interpretation selection, and meaning generation for language comprehension. Connectionist models involving Hebbian learning and back propagation can be used to simulate human learning. Connectivism-based interpretative modeling of why people perform certain types of intelligent behavior proceeds as follows: 1) people's characterization involves simple processing units that are interconnected by excitatory and inhibitory connections, 2) people propagate activation through connections between units and the modification of connections, and 3) the application of communication activation and learning to units produces behavior. Various simulations of psychological experiments have shown the psychological relevance of connectionist models; however, these models can only very roughly approximate actual neural networks.

Theoretical neuroscientists attempt to develop mathematical and computational theories and models of human and other animal brain structures and processes. Unlike connectionists, they attempt to achieve more precise biological significance by simulating the behavior of large numbers of real neurons in brain regions with important functions. Computational models of the brain have become more abundant in biology; they involves not only the use of more realistic neurons, such as those with spikes and chemical pathways, but also the simulation of interactions between brain regions, such as the hippocampus and cortex. From the perspective of logic, rules, concepts, analogies, images, and connections, these models are not strictly substitutes for computational accounts, but should be matched with them to show how psychological functions are performed at the neural level. Theoretical neuroscience modeling to explain how the brain performs cognitive tasks proceeds as follows: 1) neurons in the brain are organized into groups and brain regions through synaptic connections; 2) a neural population has a spike pattern that is transformed by sensory input and other neural populations' spike patterns; and 3) the interaction of neural groups generates the performance of functions including cognitive tasks. From the perspective of theoretical neuroscience, psychological representation is a mode of neural activity, and reasoning is the transformation of this mode.

Bayesian modeling is important in cognitive science; it is applied to understand learning, vision, motion control, language, and social cognition. It also has effective applications in robotics. The Bayesian method involves the assumption that cognition is approximately optimal, in line with probability theory and especially Bayes' theorem (that the probability of given evidence is equal to the result of a priori probability of hypothesis multiplied by the conditional probability of given evidence, divided by the probability of evidence). Bayesian interpretative modeling of how the brain performs reasoning and other functions proceeds as follows: 1) thinking is characterized by statistical correlation and conditional probability calculation; 2) thinking is the ability to calculate probabilities, such as by applying Bayes' theorem; and 3) the application of probability calculation to statistical representation enables the completion of psychological tasks, such as reasoning. Although Bayesian methods have been applied successfully to describe many phenomena, the rationality of their psychological application is controversial due to the assumption of optimality and the basis of calculation in probability theory.

Since the 1950s, artificial intelligence research has been a core part of cognitive science. The most remarkable recent development in artificial intelligence has been the deep learning method, which has led to major breakthroughs in games, object recognition, and translation. Deep learning is based on connectionism and theoretical neuroscience, but involves the use of neural networks with more layers and improved algorithms and takes advantage of faster computers and a large number of instance databases. Another important innovation is AlphaGo, the world's leading Go player, which by 2016 had combined learning from examples with reinforcement learning. The idea of deep learning is spreading back to neuroscience and is beginning to influence research in cognitive psychology. Deep learning–based modeling to explain how the brain performs cognitive tasks proceeds as follows: 1) the brain has a large number of neurons organized into 6–20 layers; 2) the brain has powerful mechanisms to learn from examples and to act through successful reinforcement learning; and 3) the learning mechanism is applied to hierarchical neural networks to realize human, and sometimes even superhuman, behavior. Although deep learning has been improved greatly in some artificial intelligence systems, how it can be applied to all aspects of human cognition, including those related to imagery, emotion, and analogy, remains unclear.

Cognitive Technologies

In cognitive research, especially in neuroscience, many new technologies have been developed. These include magnetic and positron-based brain imaging, brain imaging with electroencephalography and evoked potentials, and invasive recording of the activity of multiple single neurons with microelectrode arrays. In addition, many algorithms have been developed for analysis of the data generated with the use of such technologies, which can be classified as temporal and spatial, continuous and discrete, and connectivistic and holistic. Moreover, structural and functional brain and neuron patterns have inspired the development of new computational methods for artificial intelligence, further promoting the development of cognitive modeling.

Most neuroscientific evidence generated to date has been correlational, which means that it does not answer questions about causality. Even recent connectivistic findings cannot address causality on a larger scale. For example, if recording were to generate robust evidence that a spike in neuron alpha always leads to a spike in neuron beta when I type this sentence, we could not conclude that this particular firing pattern was the cause or the result of my typing. However, this lack of theoretical or philosophical significance does not prevent scientists from applying the data generated to the development of new technologies.

Since the beginning of this century, scientists have made predictions based on correlations revealed by research in neuroscience. These include predictions about the contents of various sensory modalities and about movement tendencies during various behaviors (for more interesting works see Nicolelis, 2020). The former enables detection of what happens during sensory cognition via brain activity recording.

The latter is a step toward the prediction of what a person is going to do, immediately and in the near future, which has applications in remote or dangerous situations.

PROSPECTIVE DEVELOPMENTS IN COGNITIVE SCIENCE, 2021–2040

Identification of The Content of Cognition

For centuries, we thought we knew what was going on in our heads, until Sigmond Freud revealed that we have a subconscious mind, which may contain many folds more things than does our conscious mind (for a comprehensive understanding of modern psychological theories and therapies, see Bankart, 2007). Thus, how can we understand ourselves and others? We cannot rely on conscious reporting or reflection, as by definition we cannot understand our subconscious. The identification of correlations between brain or neural activity and sensory cognitive content will enable the identification of the entire content of our cognition, even the parts hidden from ourselves, as long as it is related to the activity we are recording.

Although much research has been performed in this direction, we have a long way to go to achieve this goal. The major problem, which often goes unnoticed, is the inability to reflect on or report subconscious activity, as we can for conscious cognition. Thus, subconscious cognitive content can blur the results of correlation studies, making prediction more difficult and less precise. In other words, unless the subjects in an experiment can focus sharply on the object of interest, the results will likely be inaccurate.

Behavior Prediction

The purpose of research involving brain–machine interfaces is to predict what the brain will do and realize that will. With the achievement of this ability, we can easily develop exoskeletons for disabled individuals and soldiers. It will also become possible to use brain activity for the remote control of starships around Mars, or robots in dangerous areas. Preliminary results of work in this direction have been reported.

Currently, the main obstacle to the achievement of such objectives is the seemingly random activity of the brain, which makes operations imprecise. For example, in the attempt to pilot an airplane using brain signals, the chance of error at present is at least one in five. For a slow robot seeking to lift cargo, such a degree of inaccuracy might be acceptable; in dangerous situations, it is not. Only highly focused brains may generate sufficiently accurate data for this type of prediction.

FEEDBACK FROM TECHNOLOGY DEVELOPMENT

Inspiration From Virtual Reality

The development of new technologies, such as virtual reality and cloud computing, has also inspired cognitive science. Since the advent of computers, display and presentation technologies have become more advanced. We now have virtual reality, augmented reality, and real 3D, designed to make us perceive that we are experiencing a part of the real world, or the real world itself.

Inspired by virtual reality technology, and with the development of sensory cognitive neuroscience, philosophers have put forward the hypothesis of "the brain in a vat." It suggests that if we could be brains cultured in nutrient tanks and fed carefully selected information that could be decoded and recognized as the real world, then we would not be able to distinguish this existence from that of intact people living in the real world. Considered in the reverse, we are actually brains in the jars of our skulls nourished by our cerebrospinal fluid. Information is constantly fed into our brains via the nervous system, and the brain codes this information to construct the world we know. Thus, we have no means of determining whether the world we know is real; because the world presented to us is the result of cognitive computation, we lack a method for the testing of its validity. If, as we generally believe to be true, we can cognize the world only through our brains, this puzzle can never be solved.

Inspiration From Cloud Computing

For centuries, we have considered ourselves to be located in our own brains, visualizing ourselves as homunculi perched behind and between our eyes. However, recent studies have revealed that perception is attributable to our reliance on input from our eyes, ears, and skin. With the help of virtual reality, we can easily become accustomed to the perception that we are in front of, behind, or beside ourselves. In this sense, theoretically, cognition has no location and does not necessarily require a fixed location. Although this finding is liberating, the development of cloud computing has further demonstrated that even computation itself does not necessarily occur in the brain. When we use our cell phones to generate search results, the search function does not occur in our phones; rather, it occurs on one or more web servers whose locations are unknown to us. Thus, is it possible that part or all of the computation generated when we cognize something occurs outside of our brains? Although this proposition initially seems to be very unlikely, a second line of thought reveals that it actually may be very likely. As I am typing this line, the idea arising in my mind is actually based on many scientific experiments and contemplations, the result of the cognitive computing of many people in recent history. Although the logic occurs in my brain, the latter is at most a node in a cloud computing network, the activity of which has led to the current idea. All I can say is that my brain is involved in this computing, which leads me to type the line as an output of representation.

Even if I am standing in front of something no other human has seen, the tools I use to observe, name, and deduct the observables are those I have learned in my lifetime. Moreover, the eyes, ears, nose, hands, and brain are all products of evolution, the results of ancient cloud computing. Hence, what at a glance may be my brain's "own" computing is actually cloud computing across a time scale of thousands or millions of years, if not even longer. In this sense, we can safely say that cognition is always cloud computing in nature. Only with the development of cloud computing technology could we begin to appreciate this fact.

Mindfulness

As mentioned previously, the successful application of future cognitive techniques will rely on strong mental focus, and our cognition is not as we have thought, seated in our brains and reflecting reality. However, this knowledge does not seem to help us very much. Our minds still wander, unable to focus. We still feel that we are facing the real world, perched in our brains and separated from anyone and anything outside of our bodies, or beyond the current moment. Why? The simple answer is that we are using cloud computing, derived from vast ranges of time and space. That which we have newly learned and agreed on consciously propagates to only a small part of the cloud; a major part of the cloud retains the old belief; in other words, it uses the old algorithm. For this reason, the resulting representation is not easy to change; the computation must be reset so that the new algorithm predominates. In the last half century, people have rediscovered that such resetting of the cognition algorithm can be achieved through meditation, and specifically mindfulness.

Humans have meditated for millennia, to enhance health and/or change how the mind works; in other words, to reshape reality. From a cognitive science perspective, this process consists of changing the computational algorithm to obtain a better representation. The methodology usually involves training in proper focus, followed by observation of the mind. In this way, ancient people learned that the mind, and hence reality, can be reshaped; they gained knowledge about relationships between humans and their surrounding environments, about what the mind looks like and how cognition works.

Modern mindfulness involves the focusing of intentional, nonjudgmental attention on what is happening in the current moment, and the maintenance of such attention. Jon Kabat-Zinn, the founder of the first mindfulness treatment center, suggested that this technique leads to the automatic arising of awareness, which he called mindfulness (to learn more about mindfulness see Kabat-Zinn, 2018). Sam Harris, author of *Waking Up*, views mindfulness as learning what consciousness really looks like; he wrote that those who practice mindfulness will find that consciousness does not have a center, or "does not look like a self" (for details about this cognitive view of mindfulness and self, see Harris, 2014). This profound understanding of consciousness, or cognition itself, will lead us to appreciate the virtual and cloud-based nature of cognition.

IMPLICATIONS FOR EDUCATION, HEALTH, AND SPACE EXPLORATION

Cognitive technologies inevitably have implications for education, health, and space exploration in the future. Education is the training of cognition; health is part of the reality represented by cognition; and in outer space, cognition is the only tool we can use to access reality (whatever it may be).

Shaping Cognition: The Nature of Education

We usually consider education to be a way to pass on knowledge. However, considering societal differences in the prioritization of knowledge to be passed on, education is actually a way to pass on culture. From a cognitive science perspective, culture is a set of processing algorithms applied to create representations of reality. Thus, education is the shaping of cognition; it changes the way in which we view the world, and the things that we can perceive and achieve.

We all have had the experience that, having learned a lot about something, we remain unable to apply that knowledge in real practice, especially when the application requires operation or performance. Such experiences demonstrate that language-based knowledge acquisition is not equivalent to skill acquisition; the former involves the memorization of sentences or concepts, whereas the latter is inscribed directly into our output systems, namely our neural connections and muscles. Practice is the only way to transform knowledge into skill (e.g., swimming, riding a bicycle, driving a car). Some skills require physical behavior (i.e., practice of the body), whereas others require mental behavior (i.e., practice of the mind).

Meditation is exactly such a practice, and it was a part of education in ancient China and India. Confucius trained students in courtesy, music, archery, war wagon driving, history literature, and mathematics. These arts train body, mind, and close cooperation between them. In India, the yogic tradition also involves training in the coordination of the body's movements with the mind (for more understandings on the yoga tradition, see Rao, 2017). These traditions later gave rise to the concept of the unity of knowledge and action, which by definition requires the transformation of ideas into behaviors; otherwise, it remains merely a concept residing. Education can be considered to be successful only after one's mind and behavior have been transformed.

The Mind and Health

With little debate, we can say that the 21st century is a century of mental disorders (to learn more about mental disorders and related topics, see Yalom, 2017). With medical advances, many traditional diseases have apparently been conquered. In addition to the COVID-19 pandemic and the still-elusive target of cancer, however, mental diseases are clearly intractable for modern medicine. This difficulty is due in part to our centuries-long ignorance of the mind. Given the close link between mind and body, many bodily diseases entail problems of the mind. In the next few decades, a fuller understanding and better mastering of the mind will probably contribute to the curing of existing and newly arising diseases, be they mental and/or physical.

Mental training was commonly used to treat disease is almost every ancient tradition. Usually, these practices were combined with the application of natural or carefully alchemically mixed doses. Individuals with well-trained minds have sophisticated focusing and observational skills, allowing them to appreciate what is occurring in the body and mind, which facilitates decision making about what medicine to apply, and when and how to do so. Thus, mental training does not oppose modern medicine; on the contrary, it will facilitate the successful application of modern medical and even pharmaceutical discoveries.

Humans are multilevel beings, operating via multiple system layers, and mind–body training and medicine may work on any one or some of these levels. We should not limit our thinking to any particular level; overcoming such a mindset will facilitate the development of new practical skills and medicines. Our theories sometimes lag far behind our practices and discoveries. Thus, sometimes we have to let the facts speak, instead of trying to limit them with our thoughts.

The Mind In Outer Space

Space exploration is a challenging human endeavor, as is the exploration of the deep mind. Many ancient seekers, such as the Buddha, realized themselves by looking up into deep space. In *Zen-Brain Horizons*, James Austin suggested that our upper brains are in charge of bodily and self-related information, whereas our lower brains govern heavenly and space-related information. Thus, the dramatic change in world view (or algorithm, from a cognitive science perspective) that occurs when one suddenly looks up after a long period of self-scrutiny can lead to enlightenment (for more about Zen and enlightenment, see Austin, 2014).

This is not the only reason that mind training may be important for the exploration of outer space. The obstacles to this endeavor, if we really have the technology to send people that far, will be loneliness and the incompatibility of the lengthy voyages with the natural human lifespan. Although modern science has not yet led to the discovery of how to safely induce human hibernation, the Chinese and Indian traditions have long involved the induction of *dhyana* for very long periods without the subjects feeling that way. Deep *dhyana* has also been argued to markedly prolong life, and is very likely to solve the life-long journey problem of space travel. Moreover, for a well-trained, focused, and mindful mind, loneliness is not a problem, but a gift, a chance to explore one's mind more deeply (more details about dhyana and other Eastern trainings, see Rao, Paranjpe, & Dalal, 2008). Hence, we may expect that future astronauts with sufficient *dhyana* training will be ready to travel across light-years of space to other solar systems, and even other galaxies.

Mind training will also help these *dhyana*-trained astronauts maintain good mental and physical health, with a little help from medicine, during the century-long space journey. Otherwise, they would need to carry along numerous and diverse medicines and medical instruments, together with many medical doctors from every discipline. This issue is commonly ignored in science fiction films; observation of the operation of a modern hospital makes clear that a team of astronauts cannot be sent into deep space with only limited medical knowledge and instruments.

Astronauts trained in mindfulness could also perform difficult tasks, such as the "feeling" of situations or the remote control of complex drones. In deep space, many things will be unfamiliar, and communication with the base on Earth, if possible at all, may be unreliable and greatly delayed. Thus, astronauts will not be able to rely solely on conscious logical deduction and analysis; they will need to mobilize their subconscious algorithms. Astronauts well-trained in mindfulness will be able to sense the results of their subconscious computing, and thus make more appropriate and prompt decisions. Their highly-focused minds will enable them to connect better with existing machines, so that they can remotely drive robots or drones in dangerous situations. All of these skills, gained through deep cognitive training, will be essential in deep space.

SPECULATIONS ON DEVELOPMENTS IN THE FIELD BEYOND 2040

The prediction of what will happen in the far future, especially in a rapidly developing field such as cognitive science, is difficult. The development of cognitive technologies will depend on the route we choose to take in this field. If we proceed on the current track, taking the ego as the only center and the reality presented on our mental screens as the only reality, we will probably end up with very strong, logical artificial intelligence technology together with weak personal minds. However, if we choose an alternate route, seriously training our minds while smartly using the help of machines and other intelligence, we may produce new generations of empowered humans.

Lost In Reality: The Weakened Future

The 21st century is widely accepted to be dominated by mental disorders. Yuval Noah Harari, author of *21 Lessons for the 21st Century*, suggested several reasons for this mental disease pandemic. First, humans are urbanizing very rapidly. As our ancestors lived and hunted in jungles for millions of years, we are not evolutionally fit to live in the concreate cities that we have only recently started to build. This inability to adapt to city life literally drives us crazy. Another reason, probably more important but less recognized, is that we are relying too heavily on algorithms and artificial intelligence, which were invented in recent decades and to which our minds have not adapted.

Pursuit of the virtual reality world view might lead to a future such as that elaborated in *The Matrix* or *The Terminator* film series. In the former, humans are weakened and enslaved by artificial intelligence; in the latter, they are drawn into a ceaseless nightmare of mutual termination. If the reality we know is created by our minds, with increasing aid from the artificial algorithms of modern science, it will be morally easy to shift to a reality created by artificial algorithms, with decreasing aid from our minds. The choice between the two is quantitative and a matter of individual preference, and will inevitably become a source of future conflict. The problem is that we will each surely consider our own choice to be the only true reality, readily denying all others.

It has been widely accepted that living in a virtual world may harm our mental health, through overexposure to elements such as online novels, video games, and virtual societies. However, we do not yet have a science-based answer to why these experiences are more harmful than, for example, reading printed novels, playing on playgrounds, and interacting with others in person. Certainly, overexposure to such real-world activities also changes the brain, via neural plasticity, and hence the mind and behavior, easily making people out of grid. However, such effects are in many cases not as serious as those of virtual reality. We may think that this difference is attributable to the ease of things in the virtual world, which is characterized by less difficulty and fewer obstacles, and thus as escapist. However, making virtual reality as difficult as real life would likely not be an improvement. Indeed, it would likely be damaging due to the reduced difference between the real and virtual worlds, which might lead some to become immersed even more deeply in virtual reality. Consider if we were to establish the rule for a virtual world that nothing could be redone, that people who died would have to start again as newborns with almost no memory from their former existences. Such a rule would likely not frustrate virtual participants, given its approximation of the real world. It would also likely not lead to boredom and a reluctance to engage with the real world. The few people responding in such a manner would be viewed by their peers as insane, and requiring psychiatric or psychological treatment.

Several factors may explain why the machine algorithm–dominated virtual reality is more harmful than the brain algorithm–dominated 'real' reality. First, as discussed above, we are not evolutionarily adapted to machine algorithms, whose underlying mechanisms are far simpler than the brain, a highly sophisticated system that runs algorithms on multiple levels together, of which we currently have a very limited understanding. Given the inferiority of machine algorithms to brain algorithms, machine-generated reality will inevitably have many defects detected by the brain. Interfacing with this defect-ridden reality may change the rules governing brain function, throwing it out of the grid. However, one can become accustomed to such changes, and future machine algorithms may resemble brain algorithms more closely, leading to the generation of a reality that cannot be differentiated from the brain-dominated reality.

Second, as discussed above, cognition is cloud like, but this nature is difficult to notice without strong mental focus and careful observation in daily life. This nature is even more hidden in machine-generated virtual reality, where links to anything outside the machine are cut off. The acclimation of the brain to virtual reality likely makes it many-fold more difficult to awaken from the phantom egos that we usually consider to be our centers. This aspect is probably the only decisive reason for our assessment of the virtual world as truly harmful, in terms of the evolution of the mind (rather than the brain). We would like our minds to function at the greatest capacity possible, namely, as part of the cloud computing system of the universe.

Cloud-Based Mind Evolution: The Augmented Future

When we think without paying attention, we have the impression that we are the thinkers of all thoughts. Closer attention, however, will reveal that there is no thinker separate from the process of thinking, or cognition. The impression that there is a thinker is created by the ego to prove its own existence. Awareness of this situation is achieved via the form of mindfulness described by Sam Harris to lead to the realization of the true self. Similar teachings have been passed down for thousands of years in China and India. When we are not limited to the ego, we will be in a better position to appreciate the cloud-like nature of our cognition.

When we think, the computing does not occur only in our brains. Many other brains, the environment, and even the universe itself and existences light years and millennia away are involved in cloud computing. The atoms that make up our bodies are the products of ancient stars that underwent supernovas. Thus, we are the results of cloud processing. The numerous ideas that arise in our minds without planning or prediction are the results of subconscious processing, which involves the receipt of information from other cloud computing nodes without our knowledge.

With focused and open minds, we should be able to integrate more information into our processing, which should lead to the generation of more precise and detailed answers to the questions in our mind. This is the true benefit of mindfulness training. Proceeding along this path, in the future we will easily merge with the environment and each other, leading us to become more environmentally friendly, compassionate, and peaceful. In short, we will have a more sustainable future.

REFERENCES

Austin, J. H. (2014). *Zen-Brain Horizons: Toward a Living Zen*. The MIT Press. doi:10.7551/mitpress/10060.001.0001

Bankart, P. C. (2007). *Talking Cures: A History of Western and Eastern Psychotherapies*. Cengage Learning.

Harari, Y. N. (2018). *Twenty-One Lessons for the Twenty-first Century*. Spiegel & Grau.

Harris, S. (2014). *Waking Up: A Guide to Spirituality Without Religion*. Simon & Schuster.

Kabat-Zinn, J. (2018). *Meditation Is Not What You Think: Mindfulness and Why It Is So Important*. Hachette Book Group.

Nicolelis, M. (2020). *The True Creator of Everything: How the Human Brain Shaped the Universe as We Know It*. Yale University Press. doi:10.2307/j.ctvt1sggf

Rao, K. R. (2017). *Foundations of Yoga Psychology*. Springer. doi:10.1007/978-981-10-5409-9

Rao, K. R., Paranjpe, A. C., & Dalal, A. K. (2008). *Handbook of Indian Psychology*. Cambridge University Press India Pvt. Ltd. doi:10.1017/UPO9788175968448

Thagard, P. (2020). Cognitive Science. In *The Stanford Encyclopedia of Philosophy*. https://plato.stanford.edu/archives/win2020/entries/cognitive-science/

Yalom, I. D. (2017). *Becoming Myself: A Psychiatrist's Memoir*. Basic Books.

Chapter 9
Autonomous Robots:
Transformative and Disruptive

Robert Finkelstein
Robotic Technology Inc., USA

ABSTRACT

Primitive autonomous robotic vehicles were first developed during World War I and deployed during wars throughout the 20th century. More recently, autonomous vehicles with cognition were enabled by new technologies, especially in sensors, processors, and software, along with advances in humanoid and other legged robots. The autonomous car, a transformative and disruptive technology, will lead over the coming decades to the development of ubiquitous autonomous robots with increasing levels of cognition for many different applications. These robots will fill nearly every economic sector and occupational niche. The potential impact of autonomous intelligent robots on society, by the end of the century, will lead to ethical and moral dilemmas, as well as impacting jobs across the globe. There is a need to provide alternative sources of income, or alternative employment, for unemployed humans. There is also a need to consider the consequence of the possible emergence of robot self-awareness, consciousness, and free will.

DISRUPTIVE AND TRANSFORMATIVE TECHNOLOGY

The term "disruptive technology" was originally coined by Clayton M. Christensen, a Harvard Business School professor (Christensen, 1997). It was used to describe a new technology that unexpectedly displaces an already established technology (Bower and Christensen, Jan.–Feb. 1995). A disruptive technology is a revolutionary change compared with an existing technology. It disrupts and redefines the competition and industry by generating new products that initially may have serious, even dangerous, flaws, or may be inferior in some ways to the products they intend to replace. But the disruptive technology is sufficiently superior to gain a foothold in the marketplace and sufficiently different from the old technology that it forces change in the companies and industry producing the old technology – and encourages the emergence of new competitors. However, a disruptive technology may need to be improved or refined because of initial disadvantages compared with the old technology. At first it may have a limited appeal

DOI: 10.4018/978-1-7998-6772-2.ch009

to the target customer base and need to prove its overall superiority to the old technology. It may serve only niche segments of the market until it is technologically or economically enhanced or widely accepted by customers. Eventually the perceived efficacy of the disruptive technology surpasses that of the old technology. Some disruptive technologies may gain massive market acceptance and penetration almost immediately, rapidly replacing the old technology, while others may take years or decades. A disruptive technology can cause an industry leader, as well as other successful competitors, to fail and be replaced by new companies better able to exploit the new technology. The invasive new and different technology induces new competitors to emerge, often from outside the old-line product industry. The old dominant companies may flail about and cannot adapt to the disruptive technology and the new marketplace.

There are many examples of disruptive technologies over the last 150 years, a few of which include: the telephone versus the telegraph; the automobile versus the horse and buggy; television versus radio; color television versus black and white; digital high definition television versus analog; cell phones versus land lines; and "smart" cell phones versus "dumb" cell phones.

We make a distinction, which not everyone does, between disruptive and transformative technologies. Some technologies are not only disruptive, but also transformative, affecting not only companies and industries, but society as a whole. In our view, a transformative technology is one that has a profound effect on the social order, including changing societal behavior, values, culture, and relationships among individuals. It often has a major impact on the economy and national security. For example, the first useful steam engine in 1776 was a disruptive and transformative technology, as was the electric light (and electricity in general), the horseless carriage, the digital computer, the cell phone and then smart phone, the Internet, and many others.

Autonomous robots, evolving with increasing cognition and physical ability over the coming decades, will be disruptive and transformative (Finkelstein, 2020). For example, new companies in the unmanned air vehicle (UAV) and autonomous ground vehicle industry have already emerged, with civil UAVs already comprising an entirely new industry. UAV technology is beginning to evolve toward autonomous, not remotely operated, UAVs. All commercial and military aircraft may be autonomous by mid-century. By 2025 autonomous cars, trucks, and buses will be commercially operational, eventually eliminating millions of jobs for professional drivers and changing the economics and methods of transporting goods and people. The societal changes due to the introduction of autonomous vehicles will be almost as profound as those which accompanied the introduction of the horseless carriage (Finkelstein, 2016).

By the end of the century, autonomous robots will be ubiquitous throughout all economic sectors, requiring new economic models to provide the goods and services for human needs and wants in the context of pervasive unemployment.

WHAT ARE ROBOTS?

Civilization advances by extending the number of important operations which we can perform without thinking. --- Alfred North Whitehead, English philosopher and mathematician (1861-1947), (Brainy Quote (1), 2020).

A man is known by the company his mind keeps. --- Thomas Baily Aldrich (Brainy Quote (2), 2020).

From the beginning, robots were determined to overthrow their human masters. That was the theme of the Karel Capek's 1920 play, RUR (Rossum's Universal Robots), in which the word "robot," derived from the Czech noun "robota" meaning "labor" or "compulsory labor" and coined by Josef Capek, Karel's older brother, was popularized along with the concept of intelligent machines (Finkelstein and Shaker, 1994). The robotic revolt of nearly a hundred years ago took place on a stage, followed by many more similar robot uprisings over the decades, in countless stories and movies. This fictional narrative remains a cautionary lesson for now as it did then. But now we are experiencing the advent of real intelligent machines which will have a real impact on humanity, for our benefit or detriment – or, as with much technology, both. Narrative fiction becomes prescient forecasting.

The notion of robot-like entities has existed in human imagination for many centuries, in tales about non-human animated entities like the Golem, or the invention of gear-driven and water-powered automatons centuries ago (Finkelstein and Shaker, 1994). RUR, the first robot story, envisioned an intelligent machine of quite advanced technology, a humanoid robot we might now call an android. Androids with human intelligence do not yet exist, but humanoid robots, albeit primitive, have been successfully demonstrated performing useful functions. And there are now many kinds of robots in laboratories, a zoo-like taxonomy of robot species – rolling, walking, creeping, crawling, flying, and swimming – which continue to proliferate, and with some legged dog-like security robots on the verge of becoming commercially available (Mixon, 2020).

The historical species of robots can be divided into various categories for convenience. They might be categorized by operational medium (space, air, ground, water) or the operational sector (such as the military, industrial, household, commercial, etc.). They might be categorized by size or endurance or whether they are military or civil. Or they might be divided according to how they are controlled, such as pre-programmed automatons, remotely controlled telerobots, supervised autonomous robots, or fully autonomous robots (Finkelstein and Shaker, 1994). In the past military robots were primarily pre-programmed; now they are primarily telerobotic (remotely controlled) and supervised autonomous (remotely monitored and occasionally remotely controlled).

Despite the proliferation of robotic air vehicles (i.e., Unmanned Air Vehicles (UAV) or drones), and the growing array of underwater and sea surface robots (i.e., Unmanned Underwater Vehicles (UUV) and Unmanned (sea) Surface Vehicles (USV), which are increasingly becoming autonomous, in this discussion we focus on autonomous ground robots, which include Unmanned Ground Vehicles (UGV).

The core notion of an advanced robot is a machine with sensors (like an animal's senses), processors (like an animal's brain), and effectors (like an animal's ability to move with arms, hands, legs, wheels, wings, fins, and propellers) (Finkelstein and Shaker, 1994). The intelligent robot is able to perceive the environment, make appropriate decisions, and act upon the environment (Albus and Meystel, 2001). Pre-programmed automatons, without the ability to sense their environment, are generally no longer considered to be robots. Many current household appliances and vehicles can be labeled robots, by definition, even if they are not fully autonomous or as intelligent as humans. A clothes dryer, for example, able to sense the moisture and weight of a load of clothes and thereby decide on the proper length of the drying cycle, and move about even in a rather limited way (round and round), can technically be considered a robot. A car with automated cruise control, able to sense the car in front and maintain speed consistent with a minimum distance while braking and accelerating as needed, is a robot. Vacuum cleaners, autonomously able to clean floors and recharge themselves, are among the first fully autonomous household robots (alas, the humanoid robot maid and butler remain elusively in the future). The key distinction between a robot and an intelligent computer, such as IBM's Watson supercomputer which won TV's Jeopardy program against the best human contestants, is the ability of the robot to sense and interact physically with the environment, such as moving from the dining room to the living room without stepping on the cat. However, the definition of robot is likely to evolve along with robot technology.

Robot autonomy may seem like simple concept, but there has been a decades-long and arduous effort by various U.S. government agencies, defense contractors, and other organizations to define what it really means and determine exactly how to measure it. For example, the National Institute of Standards and Technology (NIST) defined variables and metrics to characterize the levels of autonomy in a program called ALFUS (Autonomy Levels for Unmanned Systems) (Huang, 2008). The three ALFUS variables that determine the measure of robot autonomy are: (1) human independence, (2) mission complexity, and (3) environmental complexity. The ALFUS definition was never completed. The Society of Automotive Engineers (SAE) and the automotive industry defined levels of autonomy for cars, trucks, and buses, starting with Level 0 for no autonomy (i.e., a manually-driven vehicle) to Level 5 (i.e., a fully autonomous vehicle able to operate without human presence or control from inside or outside the vehicle) (Shuttleworth, 2019).

In 2012 DOD Directive No. 3000.09 defined an autonomous weapon system as a weapon system that, once activated, can select and engage targets without further intervention by a human operator (Congressional Research Service, 2019). In 2016 the Defense Science Board (DSB) Summer Study on Autonomy, accomplished for the Office of the Secretary of Defense, stated that an autonomous system has the ability to independently compose and select among different courses of action to accomplish goals based on its knowledge and understanding of the world, self, and situation (Finkelstein, 2016). Simply, a fully autonomous robot is one that is able to perform its mission or function without human intervention. But does the need for a human to order the robot to perform its mission or function detract from its autonomy? Short of full autonomy, a robot may function with various levels of human intervention in a mode that is called supervised autonomous or semi-autonomous. Autonomy does not require a robot to possess human-level intelligence – just sufficient intelligence to accomplish its function, just as an ant or an amoeba has sufficient intelligence to perform its functions, like finding food and reproducing (Finkelstein, 2006).

Organisms have senses to monitor their internal and external states and help control their behavior while they pursue their prime directive (survive sufficiently to reproduce). Likewise, robots have sensors with which to monitor their internal and external states and facilitate their behavior to achieve their intended purpose (which, as yet, does not include reproduction). The robot's processors (with suitable software) transform the sensory information into perception, the ability to understand what is being sensed (Meystel and Albus, 2002).

The robot's control system connects its sensors with its processors and its effectors, like the human brain connects the senses with the arms, legs, and other body parts. The control system has an architecture, which is the way its components and their relationships are organized and interact. Depending on the control system architecture, the robot may be reactive, where it responds directly to what it senses as would non-cognitive organisms, such as cockroaches. It may be deliberative, where the robot, like cognitive creatures, has a mental model of the world (a world model) and can plan its behavior, such as humans do. Or it can be a hybrid, like humans. The robot generally plans what it will do, except in emergencies, where it reacts like a person with her autonomic (subconscious) nervous system, ducking her head without conscious thought or planning, to avoid an incoming rock. Actually, most human activity, like digestion, blood circulation, balance, eye movement, etc., is without conscious thought. It remains a philosophical issue as to whether we are conscious at all (Finkelstein, 2020).

Organisms have all sorts of effectors, like arms, hands, fingers, legs, feet, toes, tentacles, pseudopods, cilia, flagella, wings, fins, and tails. Likewise, robot effectors can take many forms, such as propellers, wheels, tracks, legs, arms, hands, tentacles, and pseudopods (Finkelstein, 2006).

Most organisms, even the most primitive, are able to communicate with other organisms in the environment. Communication among organisms typically involves sight, sound, smell (biochemicals such as pheromones), or touch. Communication between robots and human supervisors, other robots, or other entities in the environment, can also involve sight, sound, and smell. But unlike organisms, robot communication often employs wireless radio frequency (RF) transmission, electric signals over wires, light signals over fiber optic cables, and light signals transmitted through the air from lasers.

Robots already have the ability for a super power not possessed by people: the machine equivalent of extrasensory perception (ESP) in the form of mental telepathy (the ability to communicate thoughts directly between minds) and psychokinesis (the ability to move or control objects with thought). Using ordinary means of communication at or near the speed of light (such as RF or Internet connectivity), robots could directly experience what other robots are sensing and share what they are "thinking." Swarms of robots (Schranz, et. al., 2020) could join their processors into a collective "mind," a kind of distributed hive mentality. Humans are increasingly connected globally by social media. But this form of communication, whether voice or text, is excruciating slow compared with direct mind-to-mind communication. Our brains remain isolated. Robot processors can be connected into an overarching framework. In addition to sharing sensory information and decision-making, robots could directly command other computer-controlled machines anywhere in the world. Robot "ESP" is still in research and development, but robotic swarms for the military and, for civilians, electronically connected vehicles to ease automotive gridlock, will soon become available.

Intelligence is a concept that is difficult to define, beyond a score on a questionable test. Our pragmatic definition of intelligence: an intelligent system is a system with the ability to act appropriately (or make an appropriate choice or decision) in an uncertain environment. An appropriate action (or choice) is that which maximizes the probability of successfully achieving the organism's or system's function, mission, or purpose. Intelligence need not be at the human level, e.g., all organisms have intelligence appropriate to their survival needs, including amoebas, ants, aardvarks, autonomous automobiles, and androids.

Learning, in our preferred definition, is the acquisition of knowledge, skill, ability, or understanding by study, instruction, or experience, as evidenced by achieving growing success (improved behavior), with respect to suitable metrics, in a fixed environment. Learning takes place when the system's behavior increases the efficiency with which data, information, and knowledge is processed so that desirable states are reached, errors avoided, or a portion of the system's environment is controlled. On the other hand, adaptation is a change in behavior (or physical structure) in response to a changed environment. It is the ability to maintain critical or essential variables within physical (or physiological) limits (e.g., homeostasis). The changed behavior (or structure) increases the probability that the system can achieve its function or purpose (e.g., maintain homeostasis) by adjusting to the new or changed environment. So, in our view, learning occurs in a fixed environment, while adaptation occurs in a changed environment. Autonomous robots are being given the ability to learn and adapt (Finkelstein, 2020).

While there are many projects to develop robot learning and intelligence, there are none yet to develop robot wisdom. It is a perception of the relativity and relationships among things. It is the ability to filter the inessential from the essential, to recognize that which is significant amongst the detail. It is the ability to see the forest as well as the trees (Finkelstein, 2020).

No robot has yet achieved human-level cognition or consciousness. But many researchers believe humanlike machine cognition – and perhaps consciousness – will occur well before the end of the century, perhaps as soon as 2030 or 2040. Machines are becoming sufficiently intelligent to perform many jobs autonomously, with little or no human supervision. The pragmatic goal is not to make machines as smart as possible, but to make them as stupid as possible and still perform their intended functions. An autonomous car need only know enough to be a good driver; it need not know astrophysics, or how to write poetry, or how to bake an apple pie.

MOBILE ROBOT HISTORY IN A NUTSHELL

Mobile robotics had their birth a century ago, during World War I, when primitive remotely operated and autonomous vehicles flew, crawled, and swam, but were too late or too few to have had an impact on the outcome of the war (Mattingly and Finkelstein, 1992; Shaker and Wise, 1988). While air and water mobile robots are also disruptive and transformative technology, space is limited so we are focusing on mobile ground robots.

The great inventor, Nikola Tesla, pioneered early radio technology in the late 19th century and, as a consequence, experimented with the remote control of vehicles using radio waves. Tesla pursued a vision of what he called "teleautomatics." In the role of a scientist-entertainer, in 1898 he demonstrated the remote control of vehicles using wireless communication before an audience in Madison Square Garden in New York City. Foreshadowing electronic warfare, he invented countermeasures to prevent jamming the control signals for the vehicles: receivers were tuned to respond only to a combination of different frequencies (Finkelstein and Shaker, 1994).

In 1900 Tesla predicted the advent of autonomous, intelligent robotic vehicles, writing that such a vehicle "will be able to follow a course laid out or to obey orders given far in advance; it will be capable of distinguishing between what it ought to do and of making experiences … of recording impressions which will definitely affect its subsequent actions." He offered his patented invention of remote control technology to the government, and received the same response as the Wright Brothers did when they first offered their flying machine invention to the government: refusal (Finkelstein and Shaker, 1994).

In 1915, a beam of light was used to control a robot. Two U.S. engineers, John Hammond and B.F. Meisser, invented the Electric Dog, which was a three-wheeled cart constructed from a child's tricycle. It could be remotely driven by aiming a searchlight on selenium cells that were part of the control system. Also in 1915, Philadelphia inventor Felix Sabah developed a prototype robot that was a modified car which could carry about 1,000 pounds of explosives. It could be driven manually to the vicinity of enemy lines where the crew would dismount from the vehicle, set the explosive warhead, then adjust and lock the steering and clutch mechanisms so that the vehicle could attack the target autonomously at speeds up to 60 mph. There were several other robotic "land torpedoes" developed during World War I, and none saw combat during World War I (Finkelstein and Shaker, 1994).

These, and other, technological advances in robotic vehicles were ignored after World War I, except, unfortunately, by the Germans. From 1939 to the end of World War II, the Germans developed and deployed in combat thousands of robotic vehicles of various sizes. Many were tracked, remotely controlled vehicles, with as much as several hundred pounds of explosive payload, deployed to attack hardened targets like pillboxes and tanks, or to clear mine fields (Finkelstein and Shaker, 1994).

The British made a minor attempt to develop robotic land vehicles in the 1940s, including a four-legged, 100 ton walking tank that, supposedly, could operate on uneven terrain. A prototype was built and tested, but the project was cancelled (Finkelstein and Shaker, 1994).

After decades of dormancy in the U.S. military, there was a surge of development in robotic vehicles during the mid-1980s – a new generation of Unmanned Ground Vehicles (UGV), as they were called by the military. Robot Defense Systems of Thornton, Colorado, built its Programmable Robot with Logical Enemy Response (PROWLER), a six-wheeled, semi-autonomous, telerobotic, all-terrain vehicle which could carry a 2,000-pound payload, such as tandem-mounted 9mm machine guns and a rocket launcher (Finkelstein and Shaker, 1994). There was a burst of artificial intelligence (AI) and robotic innovation when the U.S. Marine Corps, Army, and the Defense Advanced Research Projects Agency (DARPA) funded the development of a variety of telerobotic and autonomous vehicle prototypes. Some of the robots were large, like modified tanks and armored personnel carriers, while other were small, like the Teleoperated Mobile Anti-armor Platform (TMAP), which was remotely controlled through a fiber optic cable and designed to attack enemy tanks, and some were mid-sized, like the Surrogate Teleoperated Vehicle (STV). The Marine Corps experimented with its Tele-Operated Vehicle (TOV). The TOV was designed to create a sense of telepresence, a kind of virtual reality, giving the operator the perception of being immersed in the remote environment in which the vehicle operated (Finkelstein and Shaker, 1994). It was successfully demonstrated decades before Hollywood exploited telepresence in science fiction movies like "Avatar" and "Surrogate."

By the 1990s, there were dozens of remotely operated robots deployed for explosive ordnance disposal (EOD). The Army Research Laboratory (ARL) demonstrated modified High Mobility Multipurpose Wheeled Vehicles (HMMWVs or Humvees) as autonomous vehicles. They were able to drive autonomously on-road at highway speeds, as well as off-road through heavily wooded areas (Finkelstein and Shaker, 1994).

From the early to the late 1990s, robotic ground vehicle development accelerated. The DARPA Tactical Mobile Robot program developed small robotic vehicles especially suited for Military Operations in Urban Terrain (MOUT), such as Throwbots and Packbots. These robots could be carried and operated by a single soldier; thrown, for example, through the window of a house, or into the entrance of a cave, to provide video of who or what might be lurking inside (Finkelstein and Shaker, 1994).

In 1999 the most ambitious program in history to transform the U.S. Army was launched: the Future Combat System (FCS). FCS attempted, essentially all at once, to replace conventional combat systems, including tanks and artillery, with robotic vehicles and manned vehicles that could also be operated robotically. Sadly, there was excessive advanced technology to be developed and fielded in too short a time, with an expected cost ballooning to several hundred billion dollars. The FCS was scaled back by Congress, then terminated (U.S. General Accounting Office, 2003).

During the early 2000's DARPA sponsored many projects to develop and demonstrate legged robots (e.g., robotic dogs, mules, and cheetahs). The DARPA Challenge programs to develop and demonstrate driverless cars (2003-2007) were, surprisingly quickly, quite successful. In 2004 the first Grand Challenge, which specified a course through the desert between Barstow, California and Primm, Nevada, offered a $1 million prize for the first autonomous vehicle to complete the course without human intervention. While the satellite-based Global Positioning System (GPS) was important for navigation, it was not sufficient. In addition to GPS, onboard machine vision and perception were important for the vehicles to be able to negotiate various obstacles during the route. The prize went unclaimed because the best distance achieved was less than seven miles out of more than a hundred in the course (Finkelstein, 2016).

The second Grand Challenge, with a doubled $2 million prize, was held a year later in October 2005. It was expected to demonstrate some improvement in autonomous vehicle performance, but a winner was not expected until sometime in the future, after several more Challenges. Surprisingly, the Stanford University Racing Team won the $2 million prize with the winning time of 6 hours, 53 minutes. And even more surprisingly, five runner-up teams successfully completed the 132-mile desert terrain course, four within the time constraint of 10 hours (Finkelstein, 2016).

The Grand Challenges were followed in 2007 by the DARPA Urban Challenge with a $2 million first prize. Instead of open desert terrain, which may have been difficult to traverse but did not represent the usual environment and dynamic complexities of automotive travel, the Urban Challenge featured a mock city (or at least suburban) environment. The competing autonomous ground vehicles had to demonstrate various maneuvers, such as merging into moving traffic, navigating traffic circles, negotiating busy intersections with four-way stop signs, passing other vehicles, parking, avoiding obstacles and manually driven vehicles and pedestrians moving about. The autonomous vehicles were scored on their ability to demonstrate a variety of traffic maneuvers, obeying all California traffic laws without human intervention. Again the results were surprising, with three winners and six successful finishers (Buehler, et. al., Eds., 2009).

The DARPA Humanoid Robot Challenge, 2013 – 2015, involved 25 international teams from companies, academic institutions and government agencies, with a $2 million first prize, a $1 million second prize, and a $500,000 third prize. The robots did better than many expected, with one major winner, but their performance was still rudimentary. In the final round, each humanoid robot had 60 minutes to complete 8 tasks useful in a rescue or emergency mission, such as after an earthquake. The robots were powered by batteries, with no recharging allowed. They were tested for: perception, mounted mobility, dexterity, decision-making, dismounted mobility, and strength. Their tasks consisted of: driving a vehicle (as a human would); exiting the vehicle and opening a door; locating and closing a leaking valve; cutting through a wall; climbing stairs; and either clearing debris or walking on rough terrain (Davenport, 2015; Finkelstein, 19 Oct. 2016).

By 2014, the global race to develop and commercialize autonomous cars was on, with competitors from the automotive industry, such as General Motors and Tesla, from the advanced technology industry, such as Google and Apple, and from the service industry, such as Amazon and Uber (Burns and Shulgan, 2018). Simultaneously, smaller autonomous robots were being developed for localized delivery of food and other retail goods, typically driving on sidewalks at slow speed rather than in the street. State and local jurisdictions began passing laws and regulations to allow autonomous robots, vehicular and otherwise, to operate, usually within various constraints, such as requiring a human driver aboard to intervene in case of emergencies, or for small delivery robots to operate on sidewalks. These autonomous delivery robots have begun to proliferate in urban areas and on college campuses (Holley, 2019).

The key technologies allowing the rapid development of autonomous robots were sensing with light detection and ranging (lidar) and deep learning. As with humans, vision – and consequent perception – is a key sensor for autonomous vehicles. Passive vision, such as using stereo eyes or cameras for 3D vision, is improving in performance; but active vision, i.e., lidar, is still superior in providing high resolution knowledge of the location, shape, and movement of entities in the environment (Finkelstein, 2016).

The first generation of lidar sensors were bulky and mechanically cumbersome, vulnerable to damage, and very expensive (e.g., more than $100,000). The electro-mechanical lidars would spin to mechanically scan the scene. They would have as many as 64 lasers inside the unit. There are now rugged, solid state, flash lidars that scan the scene electronically, without the need for mechanically spinning. They are small and inexpensive (e.g., a few hundred dollars), suitable for mounting on future commercially available autonomous cars. They can directly image a scene without scanning, providing range data to objects, as well as their colors, nearly instantly. The lidar (or advanced passive 3D vision) is a crucial addition to vehicle sensors that have been used for several years to provide driver assistance features, such as automated cruise control, automated lane keeping and lane change warning, proximity warning, cross traffic and backup warning. These sensors include multiple video cameras, microwave radars, and sonars, with cars having as many as 20 sensors arrayed in various locations. All are now commonplace and inexpensively integrated with new cars to provide driver assistance, and they will soon be joined by 360-degree lidar or advanced passive vision for autonomous driving (Finkelstein, 2016).

Deep learning employs artificial neural networks (ANN), which has been a subfield of AI for many decades. Over the years ANN demonstrated some useful applications, but it largely disappointed AI researchers who had higher expectations. Moravec's paradox notes that, contrary to traditional assumptions, high-level reasoning requires very little computation; but low-level sensorimotor skills (as in perception) require enormous computational resources. It was the advent of deep learning that mitigated Moravec's paradox, leading to better perception – and autonomy. The new technologies that enabled deep learning were (1) the ability to extend ANN from a few layers of artificial neurons to more than 150 layers, interconnected with about a billion connections; (2) new fast and powerful computing chips (e.g., Nvidia's graphics processing unit intended originally for gaming computers); and (3) millions of images, now available on the Internet, for high-speed ANN processing and training (Lipson and Kurman, 2016).

The advances in deep learning began with a machine vision competition for ANN to correctly identify the entities in the images. In 2010 the winning ANN had a 28% error rate on 100,000 test images (where the human error rate is about 5%). The error rate decreased every year with improved technology, until the winning ANN in 2015 had an error rate of 3.6%, which was the first time an ANN surpassed the human ability to correctly recognize entities in images (Lipson and Kurman, 2016).

Now new deep learning architectures and algorithms are being developed, including dynamic deep learning systems which can recognize entities from their movement, not just their appearance. Two problems are that deep learning and ANN have "black box" architectures so the analyst cannot usually determine why the system provides the output (e.g., makes the decision) that it does, and ANN may "decide" on their own what to detect during their training, thereby biasing their decisions. For example, if an ANN were shown thousands of pictures of dogs to learn how to recognize a dog, and if many of the dogs in the images had collars, the ANN might decide that the existence of a collar is the sole determinant of whether the entity is a dog.

Advances in deep learning technology drive advances in robot sensory perception, but other fields of AI, such as natural language, expert systems, and genetic algorithms will help form the evolving robot brain and its emergent mind. Advances in sensor technology will contribute significantly to enhanced robot perception. Advances in mechanical engineering and control (cybernetics) technology will greatly improve robot dynamics, including that of legged and humanoid robots.

While autonomous ground robots are beginning to demonstrate their potential value in many sectors of the economy, including transportation, agriculture, energy, manufacturing, construction, commercial services, military, and law enforcement, we are focusing here on healthcare.

Healthcare

Robots will revolutionize patient care in hospitals and elder care in nursing homes, assisted living facilities, and in-home care. Healthcare jobs will be performed by autonomous robots better and cheaper than by humans, and their value is especially apparent in times of a pandemic because they are immune to infection and can relieve the burden on overworked healthcare personnel. There are a number of healthcare functions in hospitals and other medical facilities that autonomous robots are now able to perform quite satisfactorily, including: greeting and escorting visitors; cleaning floors and bathrooms; providing facility security; delivering laundry, meals, pharmaceuticals, and mail; comforting and entertaining patients; and providing information to nurses and doctors (Finkelstein, 2019). We are excluding from this discussion the non-mobile surgical robots, which are largely teleoperated, and stationary pharmacy robots used to store and retrieve a hospital's pharmaceuticals.

Current robots are able to accomplish other tasks at least partially and will likely become fully proficient in these tasks within the next few years, including: lifting, repositioning, and transporting patients; performing toilet choirs such as providing and emptying bedpans; delivering and handling all supplies. Within the next five to ten years robots will be able to accomplish more complex tasks, safely, efficiently, and effectively, including: changing bedding; performing all laundry tasks; dressing, grooming, and bathing patients; administering medication to patients; and obtaining patient vital signs directly from patients and monitoring machines (Finkelstein, 2019).

In the 1990s an early generation of autonomous robots named HelpMate served as mobile couriers to deliver meals, pharmaceuticals, laboratory specimens, equipment and supplies, medical records, and radiology films back and forth between support departments and nursing floors in hospitals. They were even capable of autonomously summoning elevators and determining if there was sufficient available space for them to enter the elevator. They were about 5 ft. tall, weighed about 600 lbs., and carried a payload of about 200 lbs. About 100 such robots worked 24 hours per day, 7 days per week. Hospital courier robots are now larger, more efficient, and more effective, but they are not yet highly intelligent (Finkelstein, 2019).

Presently, there are four companies active in the delivery robot marketplace for hospitals: Aethon, Panasonic, Swisslog and Vecna. A fifth, CCS Robotics, sold its delivery robot line to Swisslog. For example, the TUG autonomous robot, developed by Aetheon Inc. carries materials and clinical supplies throughout hospitals 24/7 while reducing cost-per-delivery by as much as 80 percent. It relieves overworked nurses from having to constantly call for deliveries of medications, meals, supplies and tests, and which the robots deliver on time. The robot also provides a safer, more efficient way to manage waste removal. The system integrates with automated dumping equipment so the cart can empty itself of waste without employee assistance, which is especially useful after hours. The robot also provides scheduled and on-demand delivery of linens to nursing units. Housekeeping staff do not have to leave the unit for linen supplies because the robot delivers wherever the supplies are needed. On average, the robots move hospital materials and supplies over 370 miles every week (Finkelstein, 2019).

As another example, Diligent Robotics of Texas is testing the autonomous Moxi nursing assistant robot in several Texas hospitals to assist overworked nurses and clinical staff with logistical tasks, such as gathering and delivering supplies for new patient admissions, restocking supply rooms, delivering lab samples, and carrying out heavy linen bags as part of cleaning patient rooms. It can undertake about 30% of the nurse's (non-care related) workload. Moxi displays the beginnings of social intelligence through modes of expression with its head and eyes, to "clearly and purposefully communicate what it is doing so people feel comfortable," although it does not yet verbally communicate. The robot navigates using cameras and lidar. It has two arms. When a nurse addresses it, the robot moves its head in their direction and displays heart and rainbow symbols to convey pleasure (Finkelstein, 2019).

The Xenex autonomous LightStrike Germ-Zapping robot is designed to kill microorganisms that cause healthcare associated infections (HAI). Xenex claims the robot is user-friendly and can deliver up to 4,300 times more germicidal UV pathogen-killing intensity than a UV-C mercury vapor system. It can disinfect an entire patient room in as little as 20 minutes (Finkelstein, 2019).

The AIST PARO therapeutic robot, the eighth version in use since 2003, is modeled after a cuddly baby harp seal, covered with soft artificial fur to make people feel comfortable, as if they are touching a real animal. While a live baby harp seal spends most of the day sleeping, the robot is active during the daytime, but gets "sleepy" at night. It has five kinds of sensors: tactile, light, audition, temperature, and posture, with which it can perceive people and its environment. It can recognize light and dark, and feel being touched or held. It can also hear and recognize the direction of voice and words, such as its name, greetings, and praise. The robot provides the verified benefits of animal therapy for patients in hospitals and extended care facilities where live animals may not be suitable. The robot: reduces patient and caregiver stress; stimulates interaction between patients and caregivers; improves patient motivation; and improves the socialization of patients with each other and with caregivers. The PARO robot can learn to behave in a way that the patients prefer, such as responding to a new name, or remembering how a patient stroked it and then encouraging that behavior. In its interaction with patients, the robot responds as if it were alive, imitating the voice of a real baby harp seal, moving its head and legs, making sounds, and displaying behavior preferred by the patient (Finkelstein, 2019).

The Japanese Riken RIBA-II care-giving robot is designed to help lift patients onto a wheelchair. A Japanese survey of nursing homes found that lifting patients was the most tiring task for workers, who do it an average of 40 times each day. The robot's technology includes high-precision tactile sensors and flexible motor control. The robot can lift a patient weighing up to 80kg (176 lb.) from floor-level bedding into a wheelchair (Finkelstein, 2019).

Various autonomous security robots are available to provide inside and outside security for healthcare facilities. For example, the Knightscope K3 robot patrols the interior of facilities. It is 4.3 ft. tall, weighs 340 lbs. and can move at speeds up to 3 mph. The K3 captures video from 360 degrees. It has speakers and microphones for verbal interaction when needed between control center personnel and people it encounters; it does not engage in its own natural language conversations. It has thermal imaging for night vision, monitoring temperature-sensitive equipment, and to detect fire (Finkelstein, 2019).

Taski Intellibot, one of several companies with commercial floor cleaning robots, is selling three types of autonomous robots for cleaning the floors of medical facilities and other commercial spaces: the Swingbot 1650, a scrubber for hard surfaces, the Duobot 1850, a sweeper/scrubber for hard surfaces, and the Aerobot 1850, a vacuum cleaner. The robots all communicate their working status to a remote monitoring station, and can have their system diagnosed remotely. They all use less water and chemicals than manually-operated machines. The robots are controlled by a software package that can handle many different environments (Finkelstein, 2019).

Autonomous wheeled and tracked robots are available now to perform many of the arduous tasks needed by medical facilities. But within 15 years autonomous bipedal humanoid robots, with requisite physical strength and agility, higher cognition and empathy, will be able to perform many of the tasks and functions in hospitals, nursing homes, assisted living facilities, and in-home care now performed by medical personnel. They will be safe, reliable, and economical. They will be able to provide the services needed by the increasing demographic of the elderly. By mid-century autonomous robots, humanoid and other forms, will be able to replace physicians, nurses, and technicians in medical facilities and, especially, for in-home care. The ability of robots to care for the growing numbers of elderly in their own homes will save considerable family and societal resources and enhance the quality of life for the vast majority of an ageing population (Finkelstein, 2019).

Robots and The Economy

The economy is usually decomposed into three sectors: (1) the primary sector, which is manufacturing; (2) the secondary sector, which is agriculture, fishing, and extraction, such as oil drilling and mining; (3) the tertiary sector, which is service (Zeder, 2020). Over time, the various sectors in the U.S. economy have altered in relative importance and human labor required, such as the historical shift in the U.S. from labor-intensive agriculture (from 75% to 3% of population) to labor-intensive services.

There is a compelling need for robots in most subsectors of a national economy, where many are already active, including the military, industry, agriculture, forestry, construction, commercial facilities, entertainment, law enforcement, and transportation of goods and people. A new generation of robots is essential for national security and the national economy: an economic necessity for the future wealth and health of a nation. But increased robots will increase human unemployment, as they have over the years in manufacturing. In the near future about 3.5 million U.S. professional drivers (trucks, taxis, buses, delivery vans, Uber, etc.) will lose their jobs with the advent of autonomous vehicles. As the autonomy technology in vehicles is replicated over the coming decades in ubiquitous autonomous robots throughout the economy, how are people to make a living and support their families?

There are three commonly proposed solutions to the expected robot-caused unemployment apocalypse: (1) Do nothing and assume that new technology will engender new jobs as it always has, i.e., jobs are lost in certain sectors but new jobs elsewhere will more than compensate. But the disruptive and transformative nature of this technology (autonomous intelligent machines) may mean that the old employment processes will no longer hold. (2) The government establishes a national mutual fund with the new wealth created by the increased productivity of autonomous robots, to invest in the robotic companies, and everyone gets an annual income from the fund sufficient for a basic sustainable life. (3) The government receives increased tax revenues from the increased national wealth and everyone gets a Universal Basic Income (UBI) (Finkelstein, 2016).

There are problems with the mutual fund or UBI prospective solutions. To many in the U.S. these solutions are redolent of socialism and will stimulate a political backlash and postpone a feasible solution until the economy collapses into a depression. And these solutions, to be fully implemented for everyone, require that the highly productive autonomous robots be ubiquitous, in all sectors of the economy, which may take several decades and not be completed until near the end of the century.

There is another solution, which, for example, can be implemented immediately for jobs lost to autonomous vehicles by drivers and then others displaced by autonomous robots in the coming decades. Allow erstwhile drivers (and then others made unemployed by robots) to own the means of production – to become entrepreneurs or business owners employing robots. This is a blatant capitalist solution, but one requiring major government assistance. The government should: (1) provide loan guarantees to allow the former workers to purchase and employ autonomous robots, such as autonomous cars and trucks; (2) provide training so that former workers can become owners and managers. Not all former workers may qualify for loan guarantees or management training, but if many do it will help solve the unemployment problem (Finkelstein, 2016).

Robot Ethics

Robot ethics, to guide and constrain robot behavior, seems like something that might be useful sometime in the distant future. But they will be needed sooner than most people think. Military robots will proliferate beyond 2030 or so, but autonomous robotic vehicles will be available in your neighborhood showroom by 2023 or so. It seems obvious why robotic cars would need to obey the law. But why would autonomous cars also need ethics and morality? Well - why do human drivers need ethics and morality? There are manifold decisions that drivers make where the law is not apparent or irrelevant. A car moving along a road encounters a squirrel in the middle of the lane. Should it run over the squirrel, which would be legal, or swerve to miss it, which would be humane, even if swerving increases, however slightly, the risk of an accident? What if it encounters a dog instead of a squirrel? What if it encounters a person standing in the middle of the road and there is a non-trivial 0.5 probability of an accident causing injury or death to the car's occupants, if it swerves to miss the person in the road? At what level of risk, and for what sort of entity in the road, should it avoid swerving? And who decides: the manufacturer or the owner?

There are two approaches to robot ethics being contemplated: the top down approach, in which ethical theory or rules are encoded in the robot's software, and the bottom-up approach, in which the robot acquires ethics over time through experience interacting with the world, as do human children (Lin, et. al., 2012). Unlike humans, robots can be "born" with a huge set of ethical rules, where humans have very few such rules genetically encoded by evolution. But any set of rules can fail in certain situations, as Isaac Asimov demonstrated in his "I, Robot" stories, where the robots would inexplicably violate one of the rules and the central mystery was to find out why (Asimov, 1950). It is not clear which approach is better, but most likely a hybrid combination of both top-down and bottom-up approaches would be best.

The top-down approach would specify an ethical theory, such as utilitarianism or Asimov's laws for robots, and analyze the computational requirements to guide the design of algorithms and subsystems capable of implementing that theory. The bottom-up approach creates an environment where the robot explores courses of action and is rewarded for behavior that is ethically praiseworthy, or it is allowed to interact with the world and its observed behavior is praised or chastised to reinforce the desired ethical behavior. Thus the robot learns through its experiences and ethical principles are discovered or constructed by the robot. All useful lessons can be shared rapidly by all current robots around the world, and encoded in future robots as part of the top-down rules.

Visions of The Robot Mind

The mind is what the brain does. Robots, no different than organisms, could have emergent minds. If we define self-awareness as the ability to mentally simulate oneself (e.g., in a world model), then it is apparent that a robot can be considered to be self-aware if it is able to predict what it will perceive in the future based on its current behavior, and then use those predictions to plan its future behavior.

There is no agreement among cognitive scientists as to the nature or meaning of consciousness (Edelman and Tononi, 2000; Graziano, 2013). One definition of consciousness is being aware of oneself and one's relationship to the world. Most of what transpires within the human body occurs unconsciously, e.g., the functioning of organs, cells, digestive tract, etc. Even decisions are often made subconsciously, with a retroactive delusion that they were consciously made decisions. Our preferred definition of consciousness is: the perception of perception (i.e., meta-perception). Self-awareness can emerge in any system, organic or machine, that includes a self-object in its representation of the world. Any system that can build and maintain a world model that represents objects, events, entities, situations, and relationships, should be able to include itself as one of the entities (Albus and Meystel, 2001). A self-aware system can keep track of itself, i.e., its external and internal states and dynamics. It is possible to build a machine that appears to an observer to be conscious. If a machine behaves as if it were intelligent, conscious, and self-aware, then who is to say it is not?

Isaac Beshevis Singer sardonically said that we have to believe in free will because we have no choice (Brainy Quote (3), 2020). There is no experiment or scientific evidence, as yet, to show that humans have free will. While there many different notions of free will, one definition is: the ability to choose among alternatives, or to act in certain situations, independently of natural, supernatural, or social constraints (Gazzaniga, 2011; Harris, 2012). It might be that robots, as intelligent machines, are more likely to have free will than Homo sapiens.

REFERENCES

Albus, J. S., & Meystel, A. M. (2001). *Engineering of Mind*. John Wiley & Sons, Inc.

Asimov, I. (1950). *I Robot*. Doubleday.

Bower, J. L., & Christensen, C. M. (1995, January–February). Disruptive Technologies: Catching the Wave. *Harvard Business Review*, 43–53.

Buehler, M., Iagnemma, K., & Singh, S. (Eds.). (2009). *The DARPA Urban Challenge: Autonomous Vehicles in City Traffic*. Springer-Verlag. doi:10.1007/978-3-642-03991-1

Burns, L. D., & Shulgan, C. (2018). *Autonomy: The Quest to Build the Driverless Car—And How It Will Reshape Our World*. Ecco Publ.

Christensen, C. M. (1997). *The Innovator's Dilemma*. Harvard Business Review Press.

Congressional Research Service. (2019), Defense Primer: U.S. Policy on Lethal Autonomous Weapon Systems Updated, Washington, DC: Author.

Davenport, C. (2015). Military Pushes for Emergency Robots as Skeptics Worry About Lethal Uses. *Washington Post*.

Edelman, G. M., & Tononi, G. (2000). *A Universe of Consciousness*. Basic Books.

Finkelstein, R. (2006), Military Robotics: Malignant Machines or the Path to Peace? Presented at the *Military Robotics Conference: Advancing Autonomy to Reduce Human Casualties, Institute for Defense and Government Advancement*, Georgetown University Conference Center, Washington, DC.

Finkelstein, R. (2010), Military Robotics: Malignant Machines or the Path to Peace? Presented at the *First AUVSI-IEEE Joint Conference on Unmanned Vehicles, Ben-Gurion University*, Beer-Sheva, Israel.

Finkelstein, R. (2016). *Autonomous Cars: Opportunities and Threats.* Presented at the University Seminar On Reflexive Systems, The George Washington University, Washington, DC.

Finkelstein, R. (2016). *Autonomous Robots: The Next Big Thing That Will Change The World.* Presented at the Czech Technical University, Czech Institute of Informatics, Robotics, and Cybernetics, Prague, Czech Republic.

Finkelstein, R. (2019). *Autonomous Intelligent Robots for Hospitals, Nursing Home, and Assisted Living Facilities.* White Paper, Potomac, MD.

Finkelstein, R. (2020). Evolutionary Psychology and Robotics. In T. K. Shackelford (Ed.), *SAGE Handbook of Evolutionary Psychology*. Sage Publications.

Finkelstein, R., & Shaker, S. (1994). *Unmanned Vehicle Systems: Military and Civil Robots for the 21st Century & Beyond.* Pasha Publications, Inc.

Gazzaniga, M. S. (2011). *Who's in Charge: Free Will and the Science of the Brain.* Harper Collins.

Graziano, M. S. (2013). *Consciousness and the Social Brain.* Oxford University Press.

Harris, S. (2012). *Free Will.* Free Press.

Holley, P. (2019). George Mason Students Have a New Dining Option: Food Delivered by Robots. *Washington Post.*

Huang, H.-M. (Ed.). (2008). *Autonomy Levels for Unmanned Systems (ALFUS) Framework, Volume I: Terminology, Version 2.0.* NIST Special Publication 1011-I-2.0, National Institute of Standards and Technology.

Lin, P., Abney, K., & Bekey, G. A. (2012). *Robot Ethics: The Ethical and Social Implications of Robotics.* MIT Press.

Lipson, H., & Kurman, M. (2016). *Driverless: Intelligent Cars on the Road Ahead.* MIT Press.

Mattingly, M., & Finkelstein, R. (1992), Combat Robotics: From Kaiser to the New World Order. In Defence Yearbook. Royal United Services Institute and Brassey's Publisher.

Meystel, A. M., & Albus, J. S. (2002). *Intelligent Systems: Architecture, Design, and Control.* John Wiley & Sons, Inc. doi:10.6028/NIST.IR.6576

Mixon, T. (2020). *USAF to Use Four-Legged Robots for Security at Florida Base.* The News Herald.

QuoteB. (2020a). https://www.brainyquote.com/quotes/alfred_north_whitehead_108058

QuoteB. (2020b). https://www.brainyquote.com/quotes/thomas_bailey_aldrich_141405

QuoteB. (2020c). https://www.brainyquote.com/quotes/isaac_bashevis_singer_141018

Schranz, M., Umlauft, M., Sende, M., & Elmenreich, W. (2020). Swarm Robotic Behaviors and Current Applications. *Frontiers in Robotics and AI*, *7*, 36. Advance online publication. doi:10.3389/frobt.2020.00036 PMID:33501204

Shaker, S. M., & Wise, A. R. (1988). *War without Men: Robots on the Future Battlefield*. Pergamon-Brassey's International Defense Publishers, Inc.

Shuttleworth, J. (2019). Automated-Driving Graphic Update, *SAE. Standardization News, J3016*. https://www.sae.org/news/2019/01/sae-updates-j3016-automated-driving-graphic

U.S. General Accounting Office. (2003). *Issues Facing the Army's Future Combat Systems Program*. Author.

Zeder, R. (2020). *The Sectors of the Economy*. https://quickonomics.com/category/microeconomics/

Chapter 10
Toward Benevolent AGI by Integrating Knowledge Graphs for Classical Economics, Education, and Health:
AI Governed by Ethics and Trust-Based Social Capital

Don MacRae
HIAlba, Australia, Australia

ABSTRACT

This chapter considers the contributions that may be made to the evolution of an ethical and compassionate form of artificial general intelligence (AGI) by integrating knowledge graphs for classical economics, education, health, and space science. Classical economics offers pathways to the evolution of a form of AGI characterised as moral, ethical, and providing a capacity for building trust-based social capital of societies. Pathways that can be elucidated by applications of machine and deep learning to knowledge graphs of the fields of classical economics, ethics, and social capital. A network platform based on the application of distributed ledger technology is proposed to provide the basis for eliciting insights from interdependences between an ever expanding digital-quantum cloud hosting similarly empowered knowledge graphs for an ever increasing myriad of advancing fields. Knowledge graph sketches for education, health, space science, and other fields germane to building social capital serve to illustrate this proposed process and attendant business opportunities.

DOI: 10.4018/978-1-7998-6772-2.ch010

INTRODUCTION

Goertzel (2020), one of the world's leading developers of software and robots seeking to achieve Artificial General Intelligence (AGI), visualises at least three plausible scenarios for the evolution of AGI:

1. *"Corporate Techno-Fascism"* leading to "a dark future of corporate-totalitarian hegemony";
2. *Decentralised-Digital-Democracy (DDD)* leading to "a future of diverse flourishing creativity";
3. "*Human Free*" leading to "a future in which advanced AGI tech leaves biological humanity entirely by the wayside".

In chapter 3 the author refers to decentralisation and digitalisation of remote, rural, and regional communities worldwide. By harnessing breakthroughs in AI-based technology and business opportunities these communities could become increasingly capable of establishing and operating their own processes in the target sectors of food production, education and healthcare achieving high levels of self-sufficiency and resilience post-covid-19. MacDonald & MacRae (2019) also consider AI-based building of social capital focussed on reducing inequality in target sectors for housing, employment and belonging – DDD leading to "diverse flourishing creativity".

To maximise benefits, AI could be trained to not only determine the effective utilisation of technology to achieve high levels of self-sufficiency and resilience in the target sectors but also in so doing develop an understanding of and empathy with human desires to achieve such outcomes. Chapter 3 proposes further AI advances within the sectors while this chapter offers a view of how AI in progressing towards AGI can be directed towards fulfillment of the latter. This view also considers how to minimise the threats of Goertzel's dark options of unimpeded "corporate techno-fascism" and "human free" sidelining.

First up is an annotated list of organisations guiding the evolution of benevolent AI. Consideration then turns to a platform to integrate Knowledge Graphs (KGs) using distributed ledger technology (DLT) such as blockchain. KGs can be constructed from a knowledge base of entities and the relationships among them. Perhaps the World Bank (2018) has published the most concise DLT definition:

Distributed ledgers use independent computers (referred to as nodes) to record, share and synchronize transactions in their respective electronic ledgers (instead of keeping data centralized as in a traditional ledger).

The integration of KGs involves exchanging insights and facts stimulated by requests between KGs that are DLT facilitated, accountably and securely. Analytical techniques such as machine learning (ML) and its subset deep learning (DL) encompassing artificial neural networks (ANN) can generate insights and extract or abstract new facts from the content of KGs. Beyond the current focus in the AI world on advancing and using these powerful analytical techniques, applications are proposed to filter exchanges for trustworthiness (encompassing authenticity, logic, and empathy), truthfulness, adherence to codes of ethics, common sense, causality, the need to negotiate terms of exchanges, and rating the efficacy and impact of the exchange of insights and new facts. Applications of APIs are proposed to transform the content of a request for exchange and/or the response to the exchange in forms comprehensible to the respective KGs.

Consideration is then given to the formulation of guiding KGs describing pathways to the attainment of global well-being. These would be available to operate on request as a guide to the processes of formulating and developing KGs. Early candidates for guiding KGs, sketched only in the following, consider roles for AI in generating insights and extracting or abstracting new facts from building trust-based social capital, reinvesting in classical economics and the contributions of leaders in AI.

Finally, consideration is given to the formulation of KG sketches designed to explore AI inspired pathways to achieving post-covid self-sufficiency and resilience in education and health of remote, rural and regional communities worldwide. Emerging roles for AI in outer space and their interplay with the foregoing is also outlined in a final KG sketch.

Organisations Guiding The Evolution of Benevolent AI

Many public institutions and private sector organisations have been established to regulate, guide, and assess the rapidly increasing growth in the application of AI.

Governmental and Public Institutions

- **GDPR:** EU General Data Protection Regulation "is the most important change in data privacy in 20 years, fundamentally reshaping the way in which data is handled across every sector, from healthcare to banking and beyond". View at https://eugdpr.org/.
- **UK Information Commissioner's Office**: Big Data, artificial intelligence, machine learning and data protection. View at https://ico.org.uk/media/fororganisations/documents/2013559/big-data-ai-ml-and-data-protection.pdf.
- **House of Lords Select Committee on AI**: report AI in the UK: ready willing and able? makes the case that "the UK is in a strong position to be a world leader in the development of AI", particularly by putting "ethics at the centre of AI's development and use". View at https://publications.parliament.uk/pa/ld201719/ldselect/ldai/100/100.pdf.
- **Université de Montréal**: Montréal Declaration for a Responsible Development of AI (2017). Arose from a deep concern over dark side and harmful purposes of AI as well as beneficial developments. This led to a University of Montréal forum to draft the declaration based on consulting with individuals on the internet and in bookstores, gathering opinions across disciplines – including philosophers, sociologists, jurists, and AI researchers.
- **Future of Life Institute**: Currently focusing on keeping AI beneficial and exploring ways of reducing risks from nuclear weapons and biotechnology. Based in the Boston area, and welcomes the participation of scientists, students, philanthropists, and others worldwide. Activities can be viewed at https://futureoflife.org/.
- **AI Ethics Lab**: Based in Boston & Istanbul - "Our goal is to enhance technology development by integrating ethics from the earliest stages of design and development for the mutual benefit of industry and communities. Our work aims to provide ethics guidance to researchers, developers, and legislators." Activities can be viewed at https://aiethicslab.com/.

Private Sector Organisations

- **Association for the Advancement of Artificial Intelligence** (AAAI) - founded in 1979 as a non-profit scientific society devoted to advancing the scientific understanding of the mechanisms underlying thought and intelligent behaviour and their embodiment in machines by promoting: (a)

AI research and its responsible application; (b) increased public understanding of AI; (c) improving the teaching and training of AI practitioners; and (d) providing guidance for research planners and funders concerning the importance and potential of current AI developments and future directions. Activities can be viewed at https://www.aaai.org/.

- **IEEE Global Initiative on Ethics of Autonomous and Intelligent Systems**: Institute for Electrical and Electronics Engineers: An incubation space for new standards and solutions, certifications and codes of conduct, and consensus building for ethical implementation of intelligent technologies. Activities can be viewed at https://ethicsinaction.ieee.org/.
- **OpenAI**: Discovering and enacting the path to safe artificial general intelligence by ensuring that AI discoveries are publicly available and preventing private companies from doing evil. Established in 2015 by Elon Musk based on the prospects of an "existential threat from machine superintelligence" with support expressed by Bill Gates and Stephen Hawking. View at https://openai.com/.
- **International Joint Conferences on Artificial Intelligence:** a non-profit corporation founded in California in 1969 for scientific and educational purposes, including dissemination of information on AI at conferences in which cutting-edge scientific results are presented and through dissemination of materials presented at these meetings in the form of proceedings, books, video recordings, and other educational materials. View at https://www.ijcai.org/.

Towards Benevolent Artificial General Intelligence by Integrating Knowledge Graphs

Consideration is given to the evolution of an ethical form of Artificial General Intelligence (AGI) by establishing a platform to integrate knowledge graphs (KGs) using distributed ledger technology and applications to analyse, filter, and transform the content of KGs.

Knowledge Graphs

A KG can be built from a knowledge base of entities and the relationships among them. Entities are readily identifiable objects related to people, places, processes, organisations, applications, data, and things. They can also include concepts/insights, colours, and feelings. The relationships between entities are expressed by nodes and the edges between these nodes in the KG.

A history of KGs and various forms of knowledge bases and representations from the 1950s – provided at the website *knowledgegraph.today* – reveals that they meet the original aim of seeking to create intelligent computer-based systems capable of integrating knowledge and data. This history organises hyperlinked-cited research under data, knowledge, and data + knowledge headings for the periods: "advent of the digital age (1950s and 1960s), data and knowledge foundations (1970s), managing data and knowledge (1980s), data, knowledge and the Web (1990s), data and knowledge at large scale (2000s), and where are we now?". In this history, KGs originate from advances in fields such as knowledge representation, and more recently from the development of technologies for the "semantic web", machine learning, and natural language processing. This integration of widely differing technologies has led to the rapid uptake of KGs by academia and industry. To derive the full benefit of this integration of knowledge and data it is reasoned that understanding the evolution of these technologies from the earliest advances in computer science will facilitate in not only maximising their utility but also avoiding the much-quoted aphorism of George Santana: *"Those who cannot remember the past are condemned to repeat it."*

Semantic web advances are germane to the utility of KGs. To quote the principal developer of the web and the Semantic Web, Berners-Lee:

The Semantic Web is not a separate Web but an extension of the current one, in which information is given well-defined meaning, better enabling computers and people to work in cooperation.

And from Berners-Lee et al (2001):

Most of the Web's content today is designed for humans to read, not for computer programs to manipulate meaningfully." "The Semantic Web will bring structure to the meaningful content of Web pages, creating an environment where software agents roaming from page to page can readily carry out sophisticated tasks for users.

Building on a foundation of standards for interoperability, such as:

Unicode (IT standard for consistent encoding, representation, and handling of text expressed in most of the world's writing systems),

URI (a string of characters used to identify a resource on a computer network, of which the best-known type is the web address or URL), and

RDF (Resource Description Framework – a set of World Wide Web Consortium specifications originally designed as a metadata data model that organizes elements of data and standardizes how they relate to one another and to the properties of real-world entities) central to the Semantic Web which has always applied semantics to instil meaning in web content and now by extension in KGs. This meaning is achieved by extracting or abstracting web-based content, including the sphere of rapidly increasing web-based KGs, other publicly or privately held KGs in cloud platforms and the expanding zone of open platform ventures such as *OpenAI*, *AI Ethics Lab* and *LFI,* referred to above in the list of AI organisations seeking to guide the evolution of benevolent AI. Other open platforms include:

1. *GitHub.com* – a Microsoft subsidiary providing a curated list of AI courses, books, video lectures and papers and software development hosting supporting 50 million users discover, fork, and contribute to over 100 million projects.
2. *AI2.org* – AI for the common good with the mission to contribute to humanity through high-impact AI research & engineering.
3. *SingularityNet*.io – blockchain-based decentralised AI marketplace letting anyone create, share, and monetise AI services at scale with the goal of forming the basis for the emergence of the world's first true AGI – led by Ben Goertzel.
4. *OpenCog.org* - open-source software project with the vision of creating an open-source framework for AGI – led by Ben Goertzel.

Considering the many profound contributions to AI from 1950 to 2000 in advancing the state of AI art, science and engineering would be a daunting task for researchers deeply expert in and committed to analytical AI techniques, including ML, DL, ANN and XAI. Yet this is what is being called for by those who question adherence only to forms of AI of the past decade based on using ever-expanding massive databases crunched by equally massive and growing computing power and for that matter rapidly growing electrical power needs. One way of accelerating the hybridisation of yesteryear and today prominently articulated by Marcus and Davis (2019) in their call to "reboot AI" would be to utilise the semantic-based extractive and abstractive technology developed at AI2, founded by Microsoft co-founder Paul Allen – see *allenai.org*. At this stage, the focus is on scientific literature with the development of Semantic Scholar. This semantic-based facility for conducting highly advanced literature searches combines natural language processing, data mining and computer vision *"to create a rich academic search experience that helps scientists discover and understand research papers more efficiently than ever"*. It is suggested that Sematic Scholar could be used to search for purposeful linkages with past AI research findings and algorithmic advances and current technologies and engineering capabilities by using the categorisation of research profiled in the AI-KG-based scientific history provided by *knowledgegraph.today.* A further hybrid approach is proposed by Futia & Vetro (2020) when they outline three challenges for future research in integrating KGs into deep learning models.

There has been increasing awareness of KGs since the advent in 2012 of Google providing link-laden KGs to profile prominent entities alongside their search results for the same. Google has not publicised the code for the extractive method it has developed for this purpose. Burdick et al (2019) report at GitHub on their research to generate KGs by automatically sorting through large volumes of web-based data and meaningfully connecting real world events and relations with entities that their system extracts from images and videos. Wilcke et al (April 10, 2017) report that KGs are a suitable candidate for operating as a deep data model with the capability of learning from sources of heterogeneous knowledge, namely information of different types and from different domains, where existing models are generally domain specific.

Integrating Platform Based on Distributed Ledger Technology

As referred to in the foregoing a platform to integrate (KGs) using DLT is envisaged. The integration of KGs involves exchanging insights and facts stimulated by requests between KGs that are DLT facilitated, accountably and securely. Goertzel (2020) considers that blockchain technology would evolve with different roles for the three scenarios for the evolution of AGI quoted at the outset of this chapter. "Further, the ways in which blockchain technology is developed right now will be part of the nexus of factors influencing what sort of outcome eventuates for humanity generally".

The next section outlines part of one way in which DLT may evolve by facilitating the operation of a platform of integrating applications analysing, filtering, and transforming the content of KGs. Two other parts are considered in the following two sections.

Integrating Platform Applications

The integration of KGs involves exchanging insights and facts stimulated by requests between KGs that are DLT facilitated, accountably and securely. It is proposed in the following that these exchanges can be supported by analysis, filtering, and transformation of the content of the exchanges, requests, and responses.

Applications Analysing Content of Knowledge Graphs

Analytical techniques such as machine learning (ML) and its subset deep learning (DL) encompassing artificial neural networks (ANN), and expert systems can generate insights and extract or abstract new facts from the content of KGs. These can be viewed within branches of AI software and algorithms considered as requirements for a machine to be intelligent, that include:

1. **machine learning** and the subset **deep learning** encompassing **artificial neural networks**.
2. **expert systems** are knowledge-based decision support systems replacing human experts.
3. **natural language processing** encompassing classification & clustering and information extraction.
4. **automated planning, scheduling & optimisation** also comprising intelligent agents, autonomous robots, and unmanned vehicles.
5. **speech** comprising speech to text and text to speech.
6. **vision** comprising image recognition and machine vision.
7. **robotics** integrating much of the foregoing, generally learning to perform tasks through ML.

ML can provide to KGs the ability to learn automatically and improve from experience without being explicitly programmed. DL enhances this learning capacity where KGs comprise unstructured or unlabelled data. It can also support KGs to detect objects, recognise speech, translate languages, and make decisions using artificial neural networks mimicking the workings of the human brain. Expert systems dating back to the 1950s were coded manually and while still in use today they scale poorly, leading to automated systems to extract semantic relationships in unstructured texts of KGs. The adoption of structured data KGs in natural language processing (NLP) is leading to NLP approaches to produce coherent human-readable text.

Applications Filtering Content of Knowledge Graphs

Applications to filter the content of KGs are proposed from the following perspectives:

1. Trust – starting with "who can you trust", trust encompassing authenticity, logic, and empathy, and "rebooting AI: building AI we can trust".
2. Truth – detecting fake news and bias.
3. Ethics – developing a framework for codes of ethics to deal with various forms of concern.
4. Common Sense – loosely structured open-text descriptions of knowledge.
5. Causality – applied in identifying entities in the responses to a KG's requests for information that display causality.
6. Participatory Negotiation – "Getting to yes" in negotiating terms of exchanges between KGs.
7. Rating – the efficacy and impact of the exchange insights and new facts.

Trust

Botsman (2017) in asking "who can you trust" and "how technology brought us together and why it might drive us apart" relates to what individuals all know and experience:

From government to business, banks to media, trust in institutions is at an all-time low. Widespread corruption, elitism and economic disparity have led to a worldwide upsurge of anti-establishment movements.

Conceivably this has led to the emergence and increasing acceptance of the sharing economy such as AirBnB, Uber and peer-to-peer exchange cryptocurrency largely transacted among strangers. Many are moving from dictatorship oriented institutional enterprises characterised as

Opaque — Closed — Centralised — Licensed — Top-down

to distributed network enterprises respectively characterised as –

Transparent — Inclusive — Decentralised — Accountable — Bottom-up.

The orientation to dictatorship is characterised by the prevalence of enterprise DLTs that are

Closed — Centralised — Permissioned — Censorable

in contrast to democracy enhancing DLT networks, such as the Blockchain-based Bitcoin, that are

Open — Decentralised — Permissionless — Censorship-Resistant.

In the view of Frie (2018), Harvard Business School professor of technology and operations management, “trust is the foundation of everything we do”. In a widely viewed TED Talk (nearly 5 million views) she depicts trust as three sides (edges) of a triangle (graph) linking authenticity, logic, and empathy nodes. A wobble (perturbation) in any node threatens trust. According to Frie, achieving trust among all stakeholders in negotiations can lead to “unprecedented human progress”. Achieving authenticity can be the most vexing because “we as human beings can determine in an instant whether or not someone is being authentic with us.” So, the challenge in the negotiation processes is to establish conditions which fosters the contributions of stakeholders as their “authentic selves”. Achieving empathy can be the wobbliest because many will doubt that others with all their distractions are in it for them. Identifying where, when, and to whom distractions are likely to be offered traces to where, when, and with whom empathy is likely to be withheld. Logic wobbles can occur if either the quality of the logic is at risk or the ability to communicate it is lacking. Promising initiatives leading to the creation of computer-based applications for Authenticity & AI, Empathy & AI and Logic & AI follow.

Authenticity and AI

Tan (2018), in asking if trust AI machines can be trusted, describes a process to verify authenticity by AI and DLT-Blockchain.

Empathy and AI

In recent times studies of human relations have investigated the utility of computational modelling of empathy. Yalçın & DiPaola (2020) in reporting that modelling empathy involves linking affective and cognitive processes, provide a comprehensive background on its theoretical foundations, including working definitions, components, and models of empathy proposed by various fields. They provide a detailed review of research to model empathy.

Logic and AI

Based on the theory of tractable Boolean circuits (directed acyclic graphs whose internal nodes are Boolean functions and are tractable if obtained from neurons as in ANN), Darwiche (2020) proposes three roles for AI logic: as a basis for computation; to learn from combinations of data & knowledge; and to reason about the behaviour of ML models.

Marcus & Davis (2019) in their call to reboot AI relate this to building AI that humanity can trust. They view this as a need in advancing AI to the next level, obviating the "Human Free" scenario of machine overlords. Endowing machines with common sense and deep understanding, rather than ever more sophisticated statistical analysis applied to burgeoning data collections, will position humanity to create an AI that can be trusted.

Truth

And what of fake news? Fortuitously there are many algorithms to detect fake news ranging from the simple quick-and-ready DIY to the more advanced. Dyson & Golab (2017) have applied NLP methods to machine identification of misleading news sources, that is to the detection of 'fake news'. More recently research has led to DL models that can detect ever more sophisticated approaches to manipulating multi-media content, such as speech, images, and video, resulting in ever-more natural and realistic "deepfake" content (Terri, 2020).

Beyond detection, Brody & Meier (2018) have developed a mathematical approach to assessing the impact of fake news. They use the mathematical theory of communication to simulate the way fake news influences elections and referendums, indicating ways to mitigate its effects.

Truth is lacking in the many forms of bias found in AI applications. There has been a slew of articles on the need to deal with bias in AI in recent years. Manyika et al (2019) ask what must be done about the many and varied forms of bias that creep into AI algorithms. The data used to train these algorithms can contain biased human decisions or mirror inequities, historical or social, even when common causes such as gender, race, or sexual orientation sensitivities are removed.

AI and Ethics

The scientific literature is replete with views on AI and ethics. The *Future of Life Institute* website – futureoflife.org – provides an impressive range of articles covering topics such as:

1. AI, ethics, and society – mechanisms to ensure moral behaviours in AI systems.
2. Ethical questions behind AI – dealing with concerns about poorly designed AI.
3. Towards a code of ethics applied to AI – for researchers to consider in all aspects of their work.
4. Ethical priorities for neurotechnology and AI – augmenting and manipulating brain functions.
5. Law and ethics of AI – major concerns include autonomous weapons and self-driving cars.
6. Machine ethics & AI governance – social & political mechanisms for mitigating risks of AI.
7. Applying AI safety & ethics today – AI supporting humanity achieve desired futures.

Chander & Ming (2019) explore how ethics must keep pace with exponential change. While moral guidelines generally define existing anomalies, technology is not waiting around for ethics to catch up. In a *McKinsey Podcast,* Chui & Wigley (2019) discuss the emerging field of AI ethics and the key steps companies can take to ethically deploy AI.

Rolls-Royce (2020) has announced two breakthroughs in AI ethics, possibly leading to society's trust of AI, accelerating the next generation of industrialisation. The first breakthrough is an AI ethics framework for use by entities seeking to ensure the decisions it takes to use AI in its applications are ethical. The second, a first of its kind, is a step-by-step process within the framework to ensure outcomes of AI algorithms can be trusted.

As noted earlier, AI Ethics Lab seeks to enhance AI tech development by integrating ethics from the earliest stages of the design and development process for the mutual benefit of industry and communities. And in so doing they aim to provide ethics guidance to researchers, developers, and legislators. Perhaps a case could be made for the core of this proposed filter app to be developed as a framework of codes of ethics for various issues of concern, including the code for AI researchers. In this way a transference of this framework may be sufficient for the emergence of benevolent AGI.

Common Sense

Without exhibiting common sense, could AI machines ever be trusted? DL with artificial neural networks (ANN) combined with natural language processing algorithms to simulate machine-endowed common sense are built on automatically generating KGs from knowledge bases comprising a vast range of patterns of *language use* but not *language understanding*. As such they fare significantly worse than humans in tests set up to test understanding. The larger the test the worse they perform. However, advancing computing-powered DL integrated with a semantic network providing word-based meaning to language patterns is closing the gap, as reported by Bosselut et al (2019) in "COMET: Commonsense Transformers for Automatic Knowledge Graph Construction". Notwithstanding, many AI researchers are beginning to realise that endowing machines with common sense is likely to be the most intractable as well as important problem to solve on the road to AGI. It has been so since the earliest AI research endeavours in the 1950s. Accordingly, Gary Marcus co-founder of Robust.ai has suggested that hybrid approaches combining the most promising advances from today's DL-ANN grunt with yesteryear's semantic finesse may be more fruitful. And from left field, Hao (2020) reports on the prospect of giving "sight" to existing large-scale language models to facilitate better understanding of the real world. It is considered that the resulting visual language could lead to a breakthrough in building common sense AI. A novel approach to fusing image recognition systems with natural-language processing algorithms provides the visual-language outcome.

Causality

Dhar (2020) and a team of co-researchers consider understanding causality is a challenge for ML and DL. They reason that AI machines and robots able to understand the reason for a set task will improve their transfer of knowledge to other environments. Accordingly, if a KG understands causalities in its content and the corresponding content of a query from another KG, it is likely to generate a superior response. The team's tool "CausalWorld" is being developed to facilitate reasoning, planning, capturing causality, and generalising. Technology breakthroughs that strengthen the application of tools of this type will enhance the reach of KG integration.

Participatory Negotiation

Well worth including as content for this guiding KG would be the following works from the Harvard Negotiation Project:

1. The best-selling book *Getting to Yes: Negotiating Agreement Without Giving In*, by Fisher & Ury (1981). Editions in 1991 and 2011 added Bruce Patton as co-author.
2. The eBook on the Harvard Business Review's ten must reads on negotiation produced by Nobel Prize winning economist Kahneman et al (2019).BOOK

It is likely that bots trained on the insights of the foregoing will eventually match or better the best human negotiators. Perhaps the terms of exchanges between KGs could be conducted by bots representing the KGs that are trained on the foregoing theory and purposes of the KGs. Win-win outcomes could be achieved by "Getting to Yes" on exchanges, subject to the terms of access to the content of KGs.

Rating

Rating the quality of a response of a KG to a query from another KG could provide feedback that would be automatically considered by both KGs. Then responding to assess the appropriateness of its answers to such requests. The querying in providing a rating could use this to determine whether it would be worth (a) querying the responding KG in future on queries of this kind or (b) reformulating the query. A simple 0-10 rating system could be just as effective for these purposes as attempting to develop systems to rate the response for efficacy or the impact it would have on the objectives of the querying KG.

Applications Transforming Content of Knowledge Graphs

Applications of APIs are proposed to transform the content of queries-responses and exchanges between KGs to mutually comprehensible forms. While a case could be made to use APIs or open code sources for fully developed forms of all the foregoing filters of KG content building, attention is restricted to the contributions to transforming KG content using the following transformers: (a) the gargantuan GPT-3 natural language processor; (b) MISIM to extract the meaning of a piece of code leading to machines programming themselves, and (c) Conscious AI.

GPT-3

Developed by OpenAI, GPT-3 (Generative Pre-trained Transformer 3) is the latest and by far the largest NLP system of pre-trained language representations. As for GPT-2, OpenAI has continued its policy of not releasing the source code for GPT-3 as in the hands of actors with malicious intent it could be used as a tool for evil, including the creation and viral dissemination of utterly convincing and highly damaging fake news, comprising text, images, speech, and video forms. As the most powerful language model ever built it can produce streams of convincing text for a range of varied styles when prompted with an opening sentence. It comprises a massive 175 billion DL-ANN parameters, more than two orders of magnitude greater than GPT-2 and an order of magnitude greater than the Microsoft's transformer-based language model. (Parameters are the values that an ANN seeks to optimise during training.) The result is that GPT-3 can be applied to any language task, such as language translation, writing poetry, music, and social media commentary, as well as completing sentences and writing convincing articles in a wide variety of writing styles. It can also read and generate code. While experiments include generating code for Google's home page, JavaScript XML, and other impressive demos it is a long way from competing with software developers coding real-world applications.

Its contribution to building KGs probably lies in accessing the API requesting draft text and image-based vignettes for possible inclusion in natural language KG sketches such as those guiding the emergence of benevolent AGI considered below from the perspective of seeking to achieve the global well-being of humanity.

MSIM

Invented by Intel (2020) and MIT's labs, MISIM (Machine Inferred code Similarity) is a machine programming system that is intended to be a precursor to machines programming themselves. The system is an automated engine with a context-aware semantic structure allowing it to determine what a piece of software does and assessing the software structure to enable analysis of differences in the syntax of similar software code – operating like natural language processing (NLP) systems reading text written in English. However, machine programming exponents are eventually aiming for use by non-programmers with a good software idea. Used in conjunction with NLP this fully evolved form of MISIM could respond to verbal cues allowing anyone to create computer programs based on text or a verbal description – perhaps expanding the 26.4 million software developers in the world today (as estimated by Evans Data Corporation for 2019) to billions. In this way the MISIM API could be used in communication between KGs with different supporting software systems and in time communicating to express text-based insights generated from responses to queries in software code. It is worth noting that the MISIM developers are planning at this early stage to deal with the malevolent intent of a small fraction of the billions.

Conscious AI

In providing a theoretical computer science study of consciousness, eminent computer scientists working at the leading-edge of consciousness research, Blum & Blum (2020) formalise an established theory, based on a simple cognitive architecture, developed to account qualitatively for a large set of matched pairs of conscious and unconscious processes. Their "major contribution lies in the precise formal definition of a …. Conscious AI" (CAI), constituting a simple model of the complex concept of consciousness. After defining CAI formally, they proceed to provide a definition of consciousness in CAI. This leads to their view of why CAI has "the feeling of consciousness", with their definitions and explanations judged for reasonableness depending on the degree of their concurrence with (a) intuitive concepts of human consciousness that are commonly accepted, (b) various related concepts that the CAI model can explain easily and naturally, and (c) "the extent of its agreement with scientific evidence." If this research effort towards endowing machines with consciousness passes the rigours of peer review, it augurs well for progress towards such endowment. Add to this the prospects considered earlier for the emergence of visual-language building common sense in machines, and machine programming eventually delivering code-writes-code outcomes and the role of AGI looms larger. It is worth noting that if machine programming eventually enables billions to develop software the basis to build artificial common sense will be exponentially enriched.

Knowledge Graph Sketches Towards Human Well Being

KG sketches for AI and social capital, classical economics and thought leaders are proposed as a guide for the emergence of benevolent AGI. A guide that articulates pathways to achieve human well-being on a global scale.

AI and Social Capital

This KG sketch begins a proposed process of digging deep into how AI can contribute to building social capital. The concept of social capital was coined by Hanifan (1916) and described "as those tangible assets [that] count for most in the daily lives of people: namely goodwill, fellowship, sympathy and social intercourse among individuals and families who make up a social unit". As the analysis of Putnam (2000) of social capital in the USA demonstrates, in his best-selling book *Bowling Alone*, understanding changes in social capital has important implications for a nation's overall well-being, including its productive capacity and consequently its economic growth. Putnam argues that although Americans have become wealthier in aggregate in the last century their sense of community has declined and depreciated. 'Cities and traditional suburbs have given way to 'edge cities' and 'exurbs' – vast anonymous places where people sleep and work and do little else'. Putnam uses the example of bowling to argue that the decline of the community networks that once led Americans to bowl together represents a loss of social capital. Coleman (1988) in his article "Social Capital in the Creation of Human Capital" has argued that social capital is an important component of human capital and therefore a key determinant in the productivity/economic growth nexus.

Fukuyama (1995) in *The Social Virtues and the Creation of Prosperity* pushes this further by noting that the ability of people to associate and create social capital is a function of Trust which can be demonstrated to have measurable economic value. The decline of trust and sociability in many advanced countries has led to the breakdown of social capital in terms of the family and a wide range of intermediate structures. Fukuyama presents an interdisciplinary perspective on many of the topics that economists have come to claim as theirs only. He reasons that there are many issues in societal behaviour that cannot be satisfactorily explained by the assumption of rational economic agents. He believes that social capital is a major contributing factor in building economic prosperity around the world. In the *End of History?,* Fukuyama (1989) argues that the convergence of modern-day economies into market-based economies and their integration into the global economy suggests that the Marxist-Hegelian sense of history as a 'broad evolution of human societies advancing toward a final goal' has arrived.

AI and Classical Economics

The emergence of the discipline of economics in the 18th century was the result of people trying to explain something that had never happened before. At the time a handful of countries were becoming fabulously rich, while others remained dirt-poor. In 1500 the world's richest country was twice as well-off as the poorest one; by 1750 the ratio was five to one. It is no coincidence that the most famous book in economics, published in 1776, inquired into "The Nature and Causes of the Wealth of Nations".

To explain such a divergence between rich and poor countries, *The Economist* (2020) reports that the early economists were obsessed with culture, a catch-all term encompassing a society's beliefs, preferences, and values. Adam Smith, the author of "The Wealth of Nations", explored the ways in which culture helped or hindered capitalism. He argued that certain norms were required for market economies to thrive—most importantly, that people would be self-interested, but that they would satisfy their self-interest by adapting to the needs of others. Karl Marx, a few decades later, worried that a culture of "oriental despotism" prevented the emergence of capitalism in Asia.

The *Financial Times* (August 2018) reporting on "How Adam Smith predicted the power of Big Tech" and noting Smiths' highly elaborated concerns in the *Wealth of Nations* (WN) about the rise of corporate power, oligopoly, and monopoly, finds an echo in the *IMFBlog* (2018) lamenting the marked rise in the "economic wealth and power" of corporations in Big Tech, airlines, pharmaceuticals, and auditing. Of primary concern is that the significant increases in mark-ups averaging 43% in advanced economies since the 1980s have been trending higher in the past decade, mostly driven by Big Tech – principally the FAANGs (Facebook, Apple, Amazon, Netflix, Google). If the WN anticipates the pre-eminence of Big Tech then Smith's *Theory of Moral Sentiments* (TMS) provides a potentially profound contribution to achieving the benevolence of AI as the driver of this pre-eminence – even maximising the well-being of humans and the global environment.

Writing in *The Economist*, Penn (2018) of the University of Cambridge paraphrases Nobel economist Herbert Simon in his title "AI thinks like a corporation – and that's worrying: AI was born of organisational decision-making and state power; it needs human ethics." Simon as a coinventor of the virtual machine the *Logic Theorist*, heralded as the first working AI prototype, identified Frederick Taylor, the father of the scientific management movement, and Charles Babbage, originator of the concept of a digital programming computer, as intellectual predecessors, and that "Physicists and electrical engineers had little to do with the invention of the digital computer". Penn then reports that Simon infers:

The real inventor was the economist Adam Smith. He explained the connections that Gaspard de Prony, a French civil engineer, set out to "manufacture" logarithms using techniques drawn from WN. Babbage, inspired by Prony, converted this insight into mechanical hardware. In the mid-1950s, Simon transmuted it into software code.

If, as Simon deduces, Smith played a role in the process of inventing the digital computer, then as noted above, TMS, especially the concluding Part VII: Systems of Moral Philosophy, provides a process that could contribute to reaching agreement on human ethics for the governance of corporations and agencies of states exploiting AI advances in the exponential advancement of their businesses and administrations.

More recently there has been a slew of articles on the role of Smith's "invisible hand" in the laissez faire operation of markets in which the bottom-up, substantially unregulated, emergence of advances in AI technologies in the west is compared with the top-down, state-governed aspirations of China's national plan to achieve global leadership in AI: "Is Xi Jinping's iron grip better than Adam Smith's invisible hand for technological innovation?" asks Jing (2019) in the *South China Morning Post* (January 2019). Earlier Feng (May 2018) reports in *The Washington Post* that "AI will spell the end of capitalism":

The most momentous challenge facing socio-economic systems today is the arrival of artificial intelligence. If AI remains under the control of market forces, it will inexorably result in a super-rich oligopoly of data billionaires who reap the wealth created by robots that displace human labor, leaving massive unemployment in their wake. But China's socialist market economy could provide a solution to this. If AI rationally allocates resources through big data analysis, and if robust feedback loops can supplant the imperfections of "the invisible hand" while fairly sharing the vast wealth it creates, a planned economy that actually works could at last be achievable. The more AI advances into a general-purpose technology that permeates every corner of life, the less sense it makes to allow it to remain in private hands that serve the interests of the few instead of the many. More than anything else, the inevitability of mass unemployment and the demand for universal welfare will drive the idea of socializing or nationalizing AI.

In reporting "Artificial Intelligence must avoid a backlash", the *Financial Times* (2018) expresses the view that "AI is the ultimate invisible hand", one that is already at work in almost every aspect of our lives."

What the foregoing and much of economics in general misinterpret is the role of Smith's "invisible hand", referred to only once in WN and once in TMS. In WN, Smith, notwithstanding his pre-capitalist support of free market systems and small government, provides an enlightened and nuanced case against what is now referred to as neoliberal globalisation. In effect he argues that merchants and manufacturers while aware of the benefits to themselves of offshoring production of goods to cheaper labour markets would know that this would have deleterious impacts on labour at home and as such as if by an "invisible hand" they probably would not do so. In the case of TMS, the usage provides a case for equality in an essentially agrarian society. Smith provides the example that if a landowner accumulated very large land holdings on which people were dependent then just as by his natural sympathy for the welfare of others he would distribute the wealth so that it would end up with all concerned being relatively equal as if by an "invisible hand". These interpretations, made a few years ago in a video by Noam Chomsky (2015) (linguistic scholar, cognitive scientist, philosopher, and social critic), reinforce the foregoing proposal to return to the philosophical basis and underpinning moral sentiments of classical economics in the role it can play in benevolating AI. It accentuates the case for the application of AI to assessing the contribution of TMS to benevolating AI. Of relevance to such an endeavour is the recent pathbreaking partnership of AI companies, Franz (2018) and Semantic Web Co., to create a KG of the gargantuan body of the writings of Chomsky, thereby turning the linked information into a searchable and retrievable resource for the public. Also promising is the application of animation to widen community involvement in public discourse in fundamental debates such as provided over 30 years ago (but never out of print) *Manufacturing Consent: The Political Economy of the Mass Media* by Herman & Chomsky (1988) in which they contend that the media creates the illusion of democracy. With over 1.8 million views the recent video "The Five Filters of the Mass Media" is revitalising interest in their analysis of the ways corporate mass media "manufactures consent" for a status quo most people do not actually want. Perhaps similar animation of TMS would lead to its wider comprehension and appreciation, including an appraisal of its contribution to benevolating AI.

AI Thought Leaders

The debate on AI and the technological singularity – when AGI surpasses human intelligence and exponentially increases the gap between the two – sees diametrically opposed reasoning from MIT intellectual titans Chomsky and Minsky? Chomsky (2013) takes the view that "the singularity is science fiction" and quips that for the past sixty years at MIT he has been exposed to leading AI researchers continuously claiming that it is in the foreseeable future. In an interview with Ray Kurzweil, Minsky (2016) responds to the question of whether the singularity is near, "Yes, depending on what you mean by 'near'…It may well be within our lifetimes." Around the same time Minsky is recorded as saying that "Chomsky's theories of language were irrelevant".

As a cognitive scientist credited with remodelling the scientific study of language, Chomsky has sought to extend his early reasoning on (a) the innate nature of the linguistic capacity of humans (not acquired gradually through natural selection in the evolution of syntax, but rather due to a rewiring of the brain caused by a chance mutation) and (b) the development of the interplay between rules of language and the mind in efforts to (c) understand the mechanisms that the brain uses to create these rules. He comes from a point of profound comprehension to reason that humans are nowhere near understanding the interplay between the preconscious mind (comprising, inter alia, memories, values, beliefs, emotions, sentiments – moral and otherwise) and the conscious mind and hence there is little prospect of AI bridging this gap in the foreseeable future. Notwithstanding, in responding to the question of whether he believes the mind to be computable Chomsky (2016) is in accord with the view that the mind is "a computing machine of some sort".

As the cofounder in 1959 of the MIT Computer Science and AI Lab, Minsky earlier built the first neural network machine, pioneered the analysis of neural networks, and created a framework for representing knowledge leading to a new paradigm in computer programming. His major contribution was the construction of a model of human intelligence systematically built up from the interactions between a vast "society" of simple "mindless" processes, referred to as agents, constituting what he called a "society of mind". In espousing a philosophy that "minds are what brains do" he responds to his own query "What magical trick makes us intelligent?" with "The trick is there is no trick. The power of intelligence stems from our vast diversity, not from any single, perfect principle." In the same interview noted above, Minsky considers that the "brain is 400 different computers" and in modelling and building a brain "You can't do it from the bottom up."

Perhaps the application of Knowledge Graphs and attendant AI and machine learning could be used to "mine" new insights from the collective works of Chomsky and Minsky related to language, brain, mind, and AI. Alongside this, insights for benevolating AI may be derived from applying Knowledge Graphs to probe collectively TMS and Chomsky's prodigious output as a political and social critic – including his writings, burgeoning video-based commentary, and a legendary online body of email responses to queries from individuals.

Recent advances in the speed and scope of digitisation, connectivity, big data, and AI are now probing deeply into places and powers that humanity has never experienced and governments have never regulated. As such deep AI covers an increasing range of deep innovative drives including deep learning, deep artificial minds, deep insights, deep automation, deep surveillance, deep facial recognition, deep voice recognition, deep artificial neural networks, deep brain photoreception, and deep models. While some of these advances and associated technologies offer great promise and some dire peril, they are here now and must be managed. Too often presaged with darkness, deep drives can be bereft of transparency and enlightenment. This is a reality which Shoshana Zuboff (2019) has searchlighted in *The Age of Surveillance Capital: The Fight for a Human Future at the New Frontier of Power:*

Surveillance capitalism unilaterally claims human experience as free raw material for translation into behavioral data. Although some of these data are applied to service improvement, the rest are declared as a proprietary behavioral surplus, fed into advanced manufacturing processes known as 'machine intelligence,' and fabricated into prediction products that anticipate what you will do now, soon and later. Finally, these prediction products are traded in a new kind of marketplace that I call behavioral futures markets. Surveillance capitalists have grown immensely wealthy from these trading operations, for many companies are willing to lay bets on our future behavior.

The depth of assessments of the issues to be resolved and the proposals for their resolution proffered by Zuboff makes her body of work on this subject a worthy candidate for inclusion in the KG project of the TMS and work of social critic Chomsky proposed above, subsequently designing and conducting incorporation of the seminal works of 16th century enlightenment thinkers along with the seminal social capital works considered in the above. A useful starting point for inclusion of the works of the classical economists in this KG would be the overviews of the contributions of Smith, Malthus, Ricardo, Marx, Stuart Mill, George, Schumpeter and Keynes by Robert Heilbroner in the seventh edition of *The Worldly Philosophers* published by Simon & Schuster.

AI in Education, Health and Outer Space

KG sketches are provided for the achievement of post-covid self-sufficiency and resilience in education and health of remote, rural and regional communities worldwide through AI-based technology. The chapter concludes with a KG sketch for the interplay between the foregoing sketches and AI-based technologies for the exploration of outer space currently used and projected over the next 20 years, 20-30 years and 50-100 years.

AI in Education

Academic research on AI in education is conducted by the International Artificial Intelligence in Education Society and its official journal *The International Journal of Artificial Intelligence in Education* (IJAIED), focussed on the application of AI techniques and concepts to the design of systems to support learning. IJAIED: (a) spans several fields of research and the interplay between them offering a broad spectrum of topics; (b) considers the application of complex methodologies intrinsically related to education: (c) provides an international perspective of principles used for the design of computer-based learning systems.

The *Institute of Electrical and Electronics Engineers* (IEEE) is the world's largest technical professional organization dedicated to advancing technology for the benefit of humanity. The mission of *IEEE Brain* is to facilitate cross-disciplinary collaboration and coordination advancing research, standardisation and development of technologies in neuroscience to help improve the human condition.

Speaking at a recent Singularity University Summit on exponential technologies, Vivienne Ming, theoretical neuroscientist, technologist, and serial entrepreneur considers "the capability to enhance and accelerate our abilities, helping us to reach our full potential as human beings by combining neuroscience and artificial intelligence complementing a growing understanding of the human brain to increase dramatically our ability to learn and to achieve". As precursors Ming created *Muse,* an AI app for parents to interact with and understand their children as reported by Kamenetz (June 8, 2016) in *Apps That Aim To Give Parents 'Superpowers'*. Ming also established *Socos Labs* (see https://socos.org/) to explore the future potential of humans, through explorations of the potential of machine learning, AI, and even the eventual power of augmented intelligence and neuroprosthetics. *Socos Labs* is an independent think tank experimenting with whole new visions of work, education, innovation and inclusive economies to inform more human-centered policy. In considering "The Workforce of the Future", Ming (2015) sets the initial goal as robot-proofing kids by developing their "meta-learning skills – the deeper abilities that enable learning – to produce a generation of problem-solvers".

In relation to AI education Andrew Ng takes centre stage. Founder of the Google Brain Deep Learning Project and former Baidu Chief Scientist, Ng is a business and industry leader in the AI field. Founder of the AI education companies *Coursera Deep Learning Specialization* (an online learning platform) and *Deeplearning.ai*, he is often quoted as AI being the new electricity: *"Just as electricity transformed almost everything 100 years ago, today I actually have a hard time thinking of an industry that I don't think AI will transform in the next several years."* His website at *deeplearning.ai* leads with his quote: *"Deep Learning is a superpower. With it you can make a computer see, synthesize novel art, translate languages, render a medical diagnosis, or build pieces of a car that can drive itself. If that isn't a superpower, I don't know what is."* On the other hand, concerning the fears that AI will replace humans Ng holds the view that: *"Despite all the hype and excitement about AI, it's still extremely limited today relative to what human intelligence is."* Research firm IDC (2020) estimates that worldwide spending on AI Is expected to double in the next four years reaching $110 billion in 2024.

AI in Health

Doraiswamy (2020) writing in Scientific American describes the increasing reach of digital medicine in diagnosing the ailments of patients and supporting the delivery of prescribed treatments. Existing and emerging apps can autonomously detect and monitor mental and physical ailments and deliver therapies. Such digital medicines enhance traditional medical care provide support to patients where there is limited access to treatments. A wide range of therapeutic and ever more effective and efficacious and diagnostic apps and bots are emerging rapidly.

The McKinsey Global Institute (MGI) report by Remes et al (July 8, 2020) makes a convincing case for how prioritising health improvement at the population level would boost long-term global economic growth. They stress that the silver lining of the Covid-19 black cloud provides the stimulus for this to be achieved by collaborations among governments, companies, and communities worldwide. The imperative is to deal with gaping weaknesses in healthcare, supply chains, and social organisation, and immense inequities. MGI's model estimates that the resulting economic benefits would add $12 trillion or 8 percent to global GDP in 2040, some 0.4 percent increased growth year on year. The benefits arise through healthier populations leading to fewer early deaths increasing employment, fewer health ailments, and productivity gains enabled by physically and cognitively healthier workers.

This expansion of the supply of labour could add an estimated 0.3 percent to global employment growth. High-income economies would generate a fifth of new labour-market entrants, fully countering the projected slowdown in labour-force growth. Low- and middle-income countries would generate the balance of 80 percent of new labour-market entrants.

MGI then uses a standard economic approach to measuring welfare resulting in an estimate of the social benefits from reduced illness and averted deaths to be about $100 trillion without adjustments for income levels, some eight times estimated GDP benefits. *"Promising innovations include genomics to deliver more targeted prevention and treatment; data science and AI to detect and monitor disease and enhance research; tech-enabled delivery to expand and reimagine access; and advances in the understanding of the biology of aging."*

A team from Stevens Institute of Technology (2020) has developed an AI tool that can diagnose Alzheimer's with more than 95 percent accuracy, eliminating the need for expensive scans or in-person testing. The team uses XAI to build trust in the use of AI-based tools and processes.

AI In Outer Space

AI systems are playing wide ranging roles in the exploration of outer space from the design and planning of interplanetary missions to clearing junk orbiting Earth. Bandivadekar & Berquand (2021) outline various contributions currently being made by AI to space ventures:

1. Astronaut assistants: Early advances are being made in the development of AI assistants to perform tasks designed to reduce stress on astronauts. Companion robots designed to perform risky tasks on the ISS are being developed by NASA.
2. Mission design & planning: ML applications powering AI assistants to draft (a) mission designs and plans and their revisions much more rapidly and (b) designs for satellite systems and constellations observing and communicating with Earth, the Moon, Mars and beyond.
3. Satellite data processing: AI is well suited to in-depth analysis of the voluminous data generated by satellites to (a) generate Earth observations, (b) assess heat island effects of urban areas, (c) estimate wind speeds by combining met data with satellite imagery, (d) estimate solar radiation, and (e) predict the health performance of satellites themselves.
4. Space debris: ML applications have been developed to support satellites to avoid collision with the presence of over 30,000 objects greater than 10 cm and on completing missions to disintegrate on re-entry.
5. Navigation systems: Simulate the imagery of having navigation satellites around the Moon, Mars and so on by using ML to construct navigation systems from the vast troves of images generated by observation-satellite missions to the planets in recent decades.

The technological breakthroughs necessary to fulfill the vision of astrophysicist Neil deGrasse Tyson of creating the conditions for building world peace resulting from the superabundance of riches flowing from mining the rare minerals of the planets and moons of the Solar System is somewhat more futuristic, say 50 to 100 years, than extensions of the above AI-based technologies over the next 20 to 30 years to keep the Earth safe from the existential threat of asteroid strikes and to mine the many asteroids passing within one lunar distance every year.

Challenges include:

1. Establishing AI-powered monitoring systems that vastly improve on current capacities of predicting/tracking when asteroids will pass Earth within one lunar distance (of the 106 asteroids of 1-200 metres in 2020 – 60% were detected after passing, 13% less than one day before, 25% up to 7 days before, and 2% greater than one week before) Wikipedia (2021). Surely there is a high priority case for using AI and satellite technology to ensure 100% detection greater than one year before.
2. Serious investment in start-up asteroid mining companies that began in 2012 with the aim of commencing operations by 2025 evaporated by 2019 with the realisation that a further two decades of technological development is likely required (Abrahamian, 2019).
3. Determining more than one month before passing the rare mineral payloads of asteroids, including water to produce hydrogen fuel.
4. Rockets that go beyond the current capacity of launching hundreds of satellites into orbit to AI controlled space craft capable of (a) landing on asteroids, (b) dispatching robotised workforces and equipment (3D printers to produce mining equipment from asteroid minerals and eventually

replication of robots) to mine pre-determined sources of valuable minerals and (c) returning the payload to Earth, possibly via Moon-based processing.

Before these long and medium term prospects are the prospects for AI-based advances in the next 20 years to accelerate self-sufficiency and resilience of remote, rural and regional communities worldwide not only in education and health but also in food production, housing, employment and belonging as reported by MacDonald & MacRae (2019). See also reporting by the UNOOSA - *United Nations' Office for Outer Space Affairs* (n.d.) on the "benefits of space for humankind" (in Agriculture, Global Health, Environment, Sustainable Development, Disasters, Education, Human Settlements, Research & Development, Transportation, Communication, Humanitarian Assistance, International Peace & Security). As climate change is anticipated to create harsher conditions for crop growth on Earth, decades of science is proving that the even harsher space environments may be able to produce bacteria and microbes, or grow seedlings, more quickly than on Earth. These efforts are being commercialised by space services company Nanoracks (2020): Space Growth Crops for Earth Farms. This and other similar initiatives, including the transportability of robotic systems for greenhouse maintenance in space, are reported by Pultarova (2021) writing in Space.com.

Without doubt the harmonised interplay between advancements in AI-based technologies accelerating the harnessing of these benefits and AI-based technologies enabling space exploration over the next 20 years could accelerate outcomes on both fronts and with it the eradication of poverty on Earth. Indeed it could be argued AI-based advancements that are guided not to prioritise such outcomes would not augur well for the evolution of benevolent artificial super-intelligence. Fortuitously institutional arrangements as reported by *UNOOSA* (n.d.) on *Space Law* – particularly the *United Nations Treaties & Principles on Outer Space* (1967) – the activities of 37 national space agencies and three international agencies provide a basis to reach for benevolent AI.

REFERENCES

Bandivadekar, D., & Berquand, A. (2021). Five ways artificial intelligence can help space exploration. *The Conversation.* https://theconversation.com/five-ways-artificial-intelligence-can-help-space-exploration-153664

Berners-Lee, T., Hendler, J., & Lassila, O. (2001). The Semantic Web. *Scientific American, 284*(5), 34–43. doi:10.1038cientificamerican0501-34 PMID:11323639

Blum, M., & Blum, L. (2020). *A Theoretical Computer Science Perspective on Consciousness.* https://arxiv.org/ftp/arxiv/papers/2011/2011.09850.pdf

Bosselut, A., Rashkin, H., Sap, M., Malaviya, C., Celikyilmaz, A., & Choi, Y. (2019). *COMET: Commonsense Transformers for Automatic Knowledge Graph Construction.* https://arxiv.org/pdf/1906.05317.pdf

Botsman, R. (2018). *Who Can You Trust: How Technology Brought Us Together and Why It Might Drive Us Apart.* Penguin. https://www.penguin.com.au/books/who-can-you-trust-9780241296189

Brody, D. C., & Meier, D. M. (2018). How to model fake news, *Cornell University*, https://arxiv.org/abs/1809.00964

Burdick, L., Wang, M., Ignat, O., Wilson, S., Zhang, Y., Wei, Y., Mihalcea, R., & Deng, J. (2019). *Building a Flexible Knowledge Graph to Capture Real-World Events*. https://lauraburdick. github.io/papers/AIDA_UMich_Princeton_2019.pdf

Chander, D., & Ming, V. (2019). Exponential Ethics. *SingularityU Portugal Summit Cascais*. https://www.youtube.com/watch?v=KchPtF09PZ0

Chomsky, N. (2010). Some simple evo devo theses: how true might they be for language? In *The Evolution of Human Language* (pp. 45–62). Cambridge University Press. doi:10.1017/CBO9780511817755.003

Chomsky, N. (2013). *Noam Chomsky on AI: The Singularity is Science Fiction!* https://www.youtube.com/watch?v=0kICLG4Zg8s

Chomsky, N. (2015). *Noam Chomsky on Adam Smith's Invisible Hands*. https://www.youtube.com/watch?v=_qfX0ASvg3Y

Chomsky, N. (2016). *Noam Chomsky – Mind, Consciousness & AI*. https://www. youtube.com/watch?v=0ORHGa-vQp0&t=3s

Chomsky, N. (2017). *The Five Filters of the Mass Media*. https://www.openculture. com/2017/03/an-animated-introduction-to-noam-chomskys-manufacturing-consent.html

Chui, M., & Chris, W. (2019). AI Ethics in today's world. *McKinsey Podcast*. https://www.mckinsey.com/featured-insights/artificial-intelligence/ai-ethics-podcast

Coleman, J. S. (1988). *Supplement: Organizations and Institutions: Sociological and Economic Approaches to the Analysis of Social Structure*. American Journal of Sociology, 94. https://www.jstor.org/stable/2780243

Darwiche, A. (2020). Three Modern Roles for Logic in AI, *Cornell University*. https://arxiv.org/abs/2004.08599

Dhar, P. (2020). Understanding Causality Is the Next Challenge for Machine Learning. *IEEE Spectrum*. https://spectrum.ieee.org/tech-talk/artificial-intelligence/machine-learning/understanding-causality-is-the-next-challenge-for-machine-learning

Doraiswamy, P. M. (2020). Therapeutic and diagnostic apps and bots are almost here. *Scientific American*. https://www.scientificamerican.com/article/digital-medicine-can-diagnose-and-treat-what-ails-you/

Dyson, L., & Alden, G. A. (2017). *Evaluating Machine Learning Algorithms for Fake News Detection, Final Project for CAPP 30255 Advanced Machine Learning for Public Policy*. U of Chicago.

Feng, X. (2018). AI will spell the end of capitalism. *The Washington Post*. https://www. washingtonpost.com/news/theworldpost/wp/2018/05/03/end-of-capitalism/

Financial Times. (2018a). *How Adam Smith predicted the power of Big Tech*. https://www.ft.com/content/3a1b0632-a61e-11e8-a1b6-f368d365bf0e

Financial Times. (2018b). *Artificial Intelligence must avoid a backlash*. https://www.ft.com/content/d94d07e4-b278-11e8-8d14-6f049d06439c

Fisher, R., & Ury, W. (1981). *Getting to Yes: Negotiating Agreement Without Giving In*. Penguin Group.

Franz. (2018). *Franz and Semantic Web Co. Partner to Create a Noam Chomsky Knowledge Graph*. https://allegrograph.com/franz-and-semantic-web-company-partner-to-create-a-noam-chomsky-knowledge-graph/

Frie, F. (2018). How to build trust. *TED*. https://www.ted.com/talks/frances_frei_how_to_build_and_rebuild_trust

Fukuyama, F. (1989). The End of History? *The National Interest*, (16), 3-18.

Fukuyama, F. (1995). *Trust: The Social Virtues and the Creation of Prosperity*. Free Press. https://books.google.com.au/books/about/Trust.html?id=BaS2AAAAIAAJ

Futia, G., & Vetro, A. (2020). Information, On the Integration of Knowledge Graphs into Deep Learning Models for a More Comprehensible AI—Three Challenges for Future Research. *Information (Basel)*, *11*(2), 122. doi:10.3390/info11020122

Goertzel, B. (2020). *Say Hello to the Singularity*. https://www.coindesk.com/say-hello-to-the-singularity

Hanifan, L. J. (1916). The Rural School Community Center. *The American Academy of Political and Social Science*. https://journals.sagepub.com/doi/10.1177/000271621606700118

Hao, K. (2020). This could lead to the next big breakthrough in common sense AI. *MIT Technology Review*. https://www.technologyreview.com/2020/11/06/ 1011726/ai-natural-language-processing-computer-vision/

Herman, E. S., & Chomsky, N. (1988). *Manufacturing Consent: The Political Economy of the Mass Media*. Pantheon Books. https://books.google.com.au/books/about/ Manufacturing_consent.html?id=Up5sAAAAIAAJ&redir_esc=y

IDC. (2020). Worldwide Spending on AI. *IDC*. https://www.idc.com/getdoc.jsp?containerId=prUS46794720

IMFBlog. (2018). *Chart of the Week: The Rise of Corporate Giants*. https://blogs.imf.org/ 2018/06/06/chart-of-the-week-the-rise-of-corporate-giants/

Intel. (2020). *Intel, MIT and Georgia Tech Deliver Improved Machine-Programming Code Similarity System*. https://newsroom.intel.com/news/intel-mit-georgia-tech-machine-programming-code-similarity-system/#gs.r77d21

Jing, M. (2018). Is Xi Jinping's iron grip better than Adam Smith's invisible hand for technological innovation? *South China Morning Post*. https://www.scmp.com/tech/article /2173128/xi-jinpings-iron-grip-better-adam-smiths-invisible-hand-technology-innovation

Kahneman, D., Malhotra, D., Meyer, E., & Bazerman, M. H. (2019). HBR's 10 Must Reads on Negotiation. *Harvard Business Review*. https://www.booktopia.com.au/hbr-s-10-must-reads-on-negotiation-harvard-business-review/book/9781633697751.html

MacDonald, R., & MacRae, D. (2019). Maximising Scotland's Well-Being by Bravely Innovating. *HIAlba-IDEA*. http://box5817.temp.domains/~hialbaor/wpcontent/uploads/2019/04/HIAlba-IDEA-Overview-16a.pdf

Manyika, J., Jake, S., & Presten, B. (2019). What Do We Do About the Biases in AI? *Harvard Business Review*. https://hbr.org/2019/10/what-do-we-do-about-the-biases-in-ai

Marcus, G., & Davis, E. (2019). *Rebooting AI: Building Artificial Intelligence We Can Trust*. Penguin Random House. https://www.penguinrandomhouse.com/books/603982/rebooting-ai-by-gary-marcus-and-ernest-davis/

Ming, V. (2015). The Workforce of the Future. *TedMegBlog*. https://blog.tedmed. com /tag/vivienne-ming/

Ming, V. (2016). The Hidden Tax on Being Different. *HR Magazine*. https://www.hrmagazine.co.uk/article-details/the-hidden-tax-on-being-different

Minsky, M. (2016). *Kurzweil Interviews Minsky: Is Singularity Near?* https://www.kurzweilai.net/kurzweil-interviews-minsky-is-singularity-near

Minsky, M. (2016). *Chomsky's theories of language were irrelevant*. https://www.youtube.com/watch?v=wH98yW1SMAo

Nanoracks. (2020). *Space Growth Crops for Earth Farms*. https://nanoracks.com/space-grown-crops-earth-farms/

Penn, J. (2018). AI thinks like a corporation - and that's worrying. *The Economist*. https://www.economist.com/open-future/2018/11/26/ai-thinks-like-a-corporation-and-thats-worrying

Pultarova, T. (2021). *Could space greenhouses solve Earth's food crisis?* https://www.space.com/space-greenhouses-nanoracks-food-crisis

Putnam, R. (2000). *Bowling Alone*. New York: Simon & Schuster. http://bowlingalone.com/

Remes, J., Linzer, K., Singhal, S., Dewhurst, M., Dash, P., Woetzel, J., Smit, S., Evers, M., Wilson, M., Rutter, K., & Ramdorai, A. (2020). *Report - Prioritizing health: A prescription for prosperity*. McKinsey Global Institute. https://www.mckinsey.com/industries/healthcare-systems-and-services/our-insights/prioritizing-health-a-prescription-for-prosperity

Rolls-Royce. (2020). Rolls-Royce Has Achieved A Breakthrough in Ethics & Trustworthiness. *Rolls-Royce*. https://www.carandbike.com/news/rolls-royce-has-achieved-a-breakthrough-in-ai-ethics-trustworthiness-2291937

Stevens Institute of Technology. (2020). *A.I. Tool Promises Faster, More Accurate Alzheimer's Diagnosis*. https://www.stevens.edu/news/ai-tool-promises-faster-more-accurate-alzheimers-diagnosis

Tan, J. (2018). In Machines We Trust? – Verifying Authenticity by AI and Blockchain. *10 Plus Brand*. https://10plusbrand.com/2018/10/21/in-machines-we-trust-verifying-authenticity-by-ai-and-blockchain-part-1-of-2/

Terri, P. (2020). *Advancing artificial intelligence research*. MIT Schwarzman College of Computing. https://news.mit.edu/2020/advancing-artificial-intelligence-research-1118

The Economist. (2020). *Economists are turning to culture to explain wealth and poverty*. https://www.economist.com/schools-brief/2020/09/03/economists-are-turning-to-culture-to-explain-wealth-and-poverty

United Nations Office for Outer Space Affairs. (n.d.). *Benefits of Space for Humankind.* https://www. unoosa.org/oosa/ en/benefits-of-space/sustainable-development.html

United Nations Treaties & Principles on Outer Space. (1967). https://www. unoosa.org/pdf/publications/ STSPACE11E.pdf

UNOOSA. (n.d.). *Space Law.* https://www.unoosa.org/oosa/en/ourwork/spacelaw/index.html

Wikipedia. (2021). *List of close asteroid approaches to Earth in 2020.* https://en. wikipedia. org/wiki/ List_of_asteroid_close_approaches_to_Earth_in_2020

Wilcke, X., Bloem, P., & de Boer, V. (2017). The knowledge graph as the default data model for learning on heterogeneous knowledge. *Data Science, 1*(1-2), 39-57. https://content.iospress.com/articles/ data-science/ds007

World Bank. (2018). *Blockchain & Distributed Ledger Technology.* The World Bank. https://www. worldbank.org/en/topic/financialsector/brief/blockchain-dlt

Yalçın, Ö. N., & DiPaola, S. (2020). Modeling empathy: Building a link between affective and cognitive processes. *Artificial Intelligence Review, 53*(4), 2983–3006. doi:10.100710462-019-09753-0

Zuboff, S. (2019). *The Age of Surveillance Capital: The Fight for a Human Future at the New Frontier of Power.* Profile Book. https://www.bookdepository.com/Age-Surveillance-Capitalism-Shoshana-Zuboff/9781781256848

Section 3
Output-Oriented Developments

Chapter 11
The Transition to Online Teaching and Learning: 'Like Ducks to Water'

Greg Whateley
Universal Business School Sydney, Australia

Alan Manly
Universal Business School Sydney, Australia

ABSTRACT

The onset of the COVID-19 pandemic in early 2020 provided both a need and an opportunity for educational institutions in Sydney to explore new ways of providing teaching and learning for their students. UBSS was able to respond quickly because it had already prepared to introduce online learning. In mid-2019, the institution had decided to offer an online option and a senior staff member and support staff had been working since then to convert existing product into suitable online format. By the time the demand for online teaching and learning arrived, the institution was already well placed for the conversion.

INTRODUCTION

Group Colleges Australia (GCA) is the umbrella organisation for a set educational and training institutions that host close to 5,000 students, both Australian and international. It employs about 100 staff and contractors. GCA's higher education division - Universal Business School Sydney (UBSS) - provides undergraduate degrees in Accounting and Business as well as a multi-streamed postgraduate degree in Business Administration (MBA). Its undergraduate business school is ranked #1 for teaching quality among the more than 60 business schools in Sydney.

GCA's paramount goal is to ensure that its students receive the highest-quality academic experience. It recognises that new technologies can make an important contribution to improving educational outcomes, monitors closely the development of new technologies, and is pro-active in identifying and exploiting the latest and best technologies. This focus has provided GCA with a significant advantage

DOI: 10.4018/978-1-7998-6772-2.ch011

in the COVID-19 environment where face to face teaching has had to be replaced by online teaching and at the same time maintain student engagement.

The onset of the COVID-19 pandemic in early 2020 provided both a need and an opportunity for educational institutions in Sydney to explore new ways of providing teaching and learning for their students (Bolton, 2020). UBSS was able to respond quickly because it had already prepared to introduce online learning. In mid-2019, the institution had decided to introduce an online option and a senior staff member and support staff had been working since then to convert existing product into suitable online format. By the time the demand for online teaching and learning arrived, the institution was already well placed for the conversion (Nerlich, 2020). Higher education in Australia is regulated by an independent national body, the Tertiary Education and Quality Standards Agency (TEQSA). Approval must be obtained from this Agency for any "material change" in a provider's offering to students. GCA believed that its planned move to online delivery was a material change, and in October 2019 had applied to TEQSA for approval to convert to online delivery. The submission was detailed and comprehensive and demonstrated both a willingness and a propensity for a part conversion to online learning and teaching.

In March 2020, when TEQSA gave notice that all courses were to be delivered online in response to COVID-19, UBSS simply accelerated implementation of the plan it had been developing prior to submitting the request for a Material Change. Some in the industry said that UBSS took to online delivery '*like ducks to water*'. They were correct; the institution had done so because the "ducks" had studied the mechanics of swimming carefully, had ensured the access to the water was disciplined and appropriate, and had made thorough preparations before getting into the water.

The Introduction of Online Education

The transition from on-campus to online teaching occurred in Trimester 1, 2020. The speed and agility (given the foresight) with which UBSS made the transition is shown in the following sequence of events:

- In Week 8, UBSS delivered its final week of on-campus teaching to students in its undergraduate and postgraduate programs.
- In Week 9, it provided intensive training to staff using Microsoft's TEAMS platform – the training was thorough and extensive throughout the week.
- In Week 10, UBSS commenced online teaching with staff onsite. The decision to have staff deliver their subjects in the UBSS classrooms rather than in their homes was made with the quality of the student experience in mind. Staff were required to dress professionally, and tracking cameras were installed so they could move around the classroom, providing students with the feeling that they were sitting in their normal classrooms.
- In Weeks 13 and 14, end-of-trimester examinations were held online.

This focus required an enormous focus and commitment by several very skilled personnel. It is important to remember that none of the lecturing staff at UBSS had signed up for online delivery – it was imposed upon them and done rather quickly and with some urgency. In many ways their response to the challenge was magnificent.

Initial Impacts of The Shift to Online Education

Despite the sudden shift in delivery and assessment modes, the initial outcomes were pleasing. Examination submissions increased from 81% in Trimester 3, 2019 (the last full trimester of on-campus delivery and assessment) to 84% in Trimester 1, 2020 (the first trimester that included online delivery and assessment). It might be expected that the more highly motivated and capable students would participate in the assessment process whatever the mode, and that the increase in participation with online assessment would capture less motivated and/or less capable students. However, grade distributions were the same in Trimester 1, 2020 (online) as in Trimester 3, 2019 (on-campus). The aggregated progression rate was also the same in both trimesters (79%).

Table 1. Grade Distribution in Trimester 1, 2020 (% shares)

Grade	BACC	BBUS	MBA	UBSS
High Distinction	16	2	9	6
Distinction	16	12	13	14
Credit	25	24	24	24
Pass	37	40	31	36
Fail	16	22	23	20
Total	110	100	100	100

Further Improvements

While the initial move to online education went smoothly, it soon became apparent to UBSS Management that some immediate changes to practice were required, including the formal adoption of a new assessment model, a move to more robust platform, and investment in (and enhancement of) classrooms settings. It was also recognised that some of these changes needed to be made more quickly than was provided for in the normal operating schedule.

In Week 15, Trimester 1, the Chair of the UBSS Academic Senate executively approved the change to an assessment model better suited to online delivery. The three on-campus invigilated examinations regime (which had operated and worked effectively with regard to face to face teaching and learning) was replaced by a 40% assignment and 60% final examination model. This decision was ratified by the full UBSS Academic Senate at its next meeting.

Initially, UBSS used Microsoft TEAMS as its delivery platform since time was important and the platform was already in place. TEAMS is designed primarily for online meetings and it has many strengths in this area. It can also be used to deliver online learning; however, it does not support some of the software packages that are purpose-designed to improve the educational experience. During Weeks 1-3, Trimester 2, 2020 UBSS substituted Blackboard Collaborate, a platform designed specifically for online education. Blackboard Collaborate provides two-way audio, multi-point video, an interactive whiteboard, application and desktop sharing, breakout sharing and session recording as well as an attractive environment for hosting club meetings and creating a collaborative workspace for students. In general, staff took to Blackboard Collaborate quite quickly and efficiently given it was teaching/learning specific software.

The investment in technology and the focus on sustaining a high standard of teaching and learning was evidenced in the conversion of classrooms into lecture studios. With the likelihood that online teaching would continue throughout 2020 and beyond, GCA started to invest heavily in converting classrooms into lecture studios (not unlike television studios) from the beginning of Trimester 2, 2020.

First Survey of Student Attitudes toward On-Campus and Online Learning Modes

In Week 5, Trimester 2 a survey was conducted to gain a better understanding of student response to the sudden shift from traditional on-campus face to face delivery to an online operation and to ascertain their attitude toward returning to campus for their learning. The students were made aware that any proposed on-campus delivery would be part of a hybrid model (both on-campus and online classes) with regulated class numbers, and that on-campus delivery would be conducted in a manner that observed the workplace regulations and conditions set by Safe Work Australia to reduce the spread of COVID-19. Incidentally, GCA and UBSS had worked quickly and efficiently towards achieving NSW COVID Safe Certification. The survey was sent to all of UBSS' active 1,380 students and 495 students (36%) responded, with 72% indicating a preference for remaining wholly with online learning and 28% indicating a desire to return, at least partly, to on-campus learning.

With the latter group in mind, UBSS sent a memo to staff and students in Week 6, Trimester 2 advising them that it intended to introduce a hybrid model, with online classes continuing to be provided and a small number of students being invited to on-campus classes in a setting that strictly observed COVID19 protocols. It also planned to hold invigilated examinations at a designated examination centre for units accredited by peak accounting bodies. UBSS sought and received NSW Government COVID19 Safe certification to ensure that the processes in place were appropriate and of a standard required for maximum health and safety.

In Week 9, the hybrid model was introduced. However, attendance was exceedingly low, with evidence that some of the students who had indicated a desire to return to campus had either misinterpreted or mis-stated their feelings. Circumstances in NSW also changed, with the threat of a second wave of COVID-19 becoming more likely, and UBSS cancelled the on-campus classes after only one week of delivery.

Student Feedback on Units

At the end of every trimester, UBSS conducts a survey of the students' evaluation of each subject in every discipline. The survey comprises 11 questions focusing on teaching quality and the learning needs of students. Students give a score to each question, ranging from 1 (extremely bad) to 5 (extremely good). Individual subject scores are made available to staff on their Moodle sites as well as to the three Program Directors who then meet with individual staff to analyse the scores and explore ways of improving performance.

The overall aggregated score during the 11 preceding teaching periods (T2, 2016 – T3, 2019) was 4.2 out of a possible 5. Because of the sudden introduction of online teaching and the many other problems faced by students due to COVID-19, UBSS had set a target of 4.0 for Trimester 2. However, as is shown in Table 2, the overall score was 4.2, above the target score and in line with historical scores. It supported the conclusion from the earlier survey that the transition from on-campus to online learning had been successful.

Table 2. Student Feedback on Units, Term 2, 2020

	Survey Questions	BBus	BAcc	MBA
Q1	The subject provided useful knowledge and skills	4.30	4.13	4.23
Q2	The learning outcomes were achievable	4.18	4.03	4.13
Q3	The subject workload was manageable	4.13	4.07	4.03
Q4	The subject helped to develop relevant professional skills such as problem solving and critical thinking	4.17	4.17	4.14
Q5	The lecturer was well prepared for each class	4.31	4.24	4.30
Q6	The lecturer provided useful feedback	4.25	4.21	4.18
Q7	The lecturer had a good knowledge of the subject matter	4.32	4.32	4.33
Q8	The lecturer used e-learning resources eg smartboard Moodle in a way that aided learning in the subject	4.20	4.20	4.26
Q9	The lecturer was available to discuss learning problems outside of class time	4.12	3.99	4.02
Q10	The assessment requirements were clearly explained	4.26	4.09	4.20
Q11	Overall, the teaching in the subject was of a high quality	4.26	4.13	4.16
	Average	**4.23**	**4.14**	**4.18**

Internal Audit

UBSS has a two-pronged approach to online learning: it focuses, first, on the factors that directly influence the students' learning experience; and second, on ensuring that this experience is in line with or exceeds accepted industry practice.

UBSS has invested heavily in IT resources to ensure that students are not disadvantaged by the transition to online learning. It believes that its position as a market leader in the industry is due to its combination of state-of-the-art resources and passionate lecturers. The Executive Dean publishes daily updates for all staff on the particular nuances and experiences of online delivery. In addition, UBSS remains one of the few providers to insist that lecturers deliver their subjects from campus rather than their homes. This ensures that lectures are delivered in a scholastic manner and in an environment with which students are familiar. It also ensures transmission is of a high standard and technology glitches are minimised. Limited bandwidth, local peak usage issues, and inferior equipment are no longer considerations nor hurdles to overcome.

The Executive Management Team ensures that the student online experience is in line with accepted industry practice in the following manner. Firstly, the IT Director provides ongoing IT updates to the Board of Directors on the state of resources. Secondly, UBSS conducts internal audits with reporting lines back to Senior Management and the Academic Senate on the progress of online delivery. These updates and reports are comprehensive and detailed.

The institution's audit tool is derived from TEQSA's Online delivery key considerations for Providers (April 2020) https://www.teqsa.gov.au/latest-news/publications/online-delivery-key-considerations-providers. It also conducts a watching brief on the TEQSA website to alert Management when the resources are updated or changed. The outcomes are best captured in the following traffic light report that is reported regularly to the UBSS Academic Senate:

Table 3. Traffic Light Report

Criteria	TEQSA Reference Guide	Action	Status	Responsible Officer
A set of questions is provided before class		Lecturers provide homework and questions for students before class		Lecturers
Glossary of terms is used for online learning (in class)		Glossary is provided in the unit outline		Lecturers
Students have access to technology (tablet, phone, laptop)		Students have access to computers. Learning material is also accessible on phones		Student Services
Online readings available		Readings are available in Moodle		Lecturers
Attendance is recorded promptly		Attendance is recorded in Moodle via the login page		Lecturers
FAQ, YouTube, Ted Talk links, tables are used		These are included in the Moodle page for each unit		IT/Lecturers
Warm Up activities are used for students to get to know each other		Students are put into small groups and given meet-and-greet activities		Lecturers
Students are provided information on support services		Support services are available on Moodle and the UBSS website. Students also provided information during orientation		Lecturers
Lecturers are available to students after the structured learning		Lecturers use the Chat forum and queries are addressed at the end of the class		Lecturers

In Week 12, Trimester 2 UBSS conducted an internal review of its online resources, practice, and experience against the TEQSA good practice guidelines (Table 3). The findings indicated that UBSS met fully all the TEQSA criteria.

External Benchmarking

In May 2020, approximately 2 months after the COVID 19 restrictions on education delivery were put in place, UBSS participated in a benchmarking project with members of the Higher Education Private Providers Quality Network (HEPP-QN) to assess the responses of independent higher education providers to COVID19. HEPP-QN is a consortium of 35 higher educational institutions that aims to strengthen the status of private providers in the higher education industry by encouraging and supporting cooperation with each other in developing and implementing quality assurance frameworks. Benchmarking, then, is an important aspect of the work.

The HEPP-QN project involved two focus groups. One group comprised *business schools* and included the following 5 providers - Australian Institute of Business, Polytechnic Institute Australia, Russo Business School, Universal Business School Sydney, and Western Sydney International College. The main topics covered by this group were:

- Governance and decision making to facilitate the required changes
- Support provided to students
- Support provided to staff
- Measures to ensure that quality is maintained

UBSS compared its responses to those of the four other members of the group in the areas of decision-making, staff and student support and measures implemented for quality assurance. It noted that one of the measures that had been introduced by UBSS but not by the other providers was the provision of practice online exams for students. It believed that this measure may have had a positive impact on the academic performance of students in Trimesters 1 and 2. It may also account for the increase in examination completions.

Ongoing Monitoring

In Week 15 of Trimester 2, UBSS commenced a study of options for invigiliatating online assessments. In the same week, the results of student performance in Trimester 2 became available. The distribution of grades was similar to that of Trimester 1, and there was a slight increase in the overall pass rate, to 81% (Table 4). The ability of UBSS to not only sustain the improvement in progression achieved in Trimester 1 but to increase it further in Trimester 2 suggests that the institutions' early planning for online delivery, generous provision of online resources, extensive staff training and flexible response to challenges and opportunities achieved the institution's overarching goal of providing a quality teaching and learning environment for its students.

Table 4. Grade Distribution in Term 2, 2020 (% shares)

Grade	BACC	BBUS	MBA	Total
High Distinction	5	2	4	4
Distinction	15	12	14	14
Credit	24	24	26	26
Pass	38	40	37	37
Fail	18	22	19	19
Total	100	100	100	100

Second Survey of Student Attitudes Towards On-Campus and Online Learning Modes

In Week 3, Trimester 3 UBSS conducted a second survey of the attitudes of students toward online delivery and of their enthusiasm for a return to on-campus learning albeit it in a hybrid mode. The survey separated postgraduate students (who comprise 60% of UBSS' student population) from undergraduate students to elicit a more refined, granular set of responses. The survey was sent to 1,380 students, of whom 455 (33%) replied.

The key result was that 93% of postgraduate students and 86% of undergraduate students reported that they 'prefer to continue studying remotely from home'. The overall preference for online learning, of 90%, was 18 percentage points higher than for the survey conducted in Trimester 2. This came as no great surprise given the qualitative feedback received throughout the online teaching and learning period and the T2 Student Feedback on Units (SFU) data suggesting high levels of satisfaction.

CONCLUSION

The conclusions that can be drawn from the UBSS experience with online learning include:

1. Careful planning for the introduction of online education, provision of adequate resources to support online delivery and assessment, and appropriate training of staff in online teaching methods and use of online resources can contribute to the successful transition from on-campus to online learning.
2. Though experience with the new mode of education is limited, students prefer quality online learning to similar-quality on-campus learning (Kidd, 2020).
3. The large jump in preference for online learning in only its second trimester of delivery suggests that the preference will increase more, perhaps to nearly 100%, as delivery and assessment methods improve and become more student-friendly and as students become more accustomed to the online mode of delivery (Nott, 2020).
4. The preference for online learning is stronger for postgraduate than undergraduate students. This may be due to postgraduate students having clearer goals for their careers and greater need for efficiency in their learning activities. It may also reflect a greater focus among undergraduate students, especially first-year students, on building a social network.
5. The student experience is central to success. If onlin opportunities can faciliatate a postive and enriching experience for student at both undergraduate and postgraduate levels online learning could be here to stay – or at least a blended model (Rogers, 2020).

REFERENCES

Bolton, R. (2020). Uni students prefer online learning: study. *Financial Review.* https://www.afr.com/work-and-careers/education/uni-students-prefer-online-learning-study-20201011-p563yr

Kidd, E. (2020, May 12). My staff don't want to return to work – coming back after COVID-19. *Human Resources Director.*

Nerlich, S. (2020). 2020 by the numbers. *International Education Association of Australia.* https://www.ieaa.org.au/blog/2020-by-the-numbers

Nott, W. (2020). UK: Face-to-face teaching vital for international students. *Professionals in International Education*. https://thepienews.com/news/uk-face-to-face-teaching-vital-for-intl-students

Rogers, C., & Cook, C. (2020). Where to now? Blended futures. *Deloitte*. https://www2.deloitte.com/au/en/pages/public-sector/articles/where-now-blended-futures.html

Chapter 12
Crisis in US Accounting Education:
Twenty Years Later

Vincent J. Shea
St. John's University, USA

Nina T. Dorata
St. John's University, USA

Kevin E. Dow
University of Auckland, New Zealand

ABSTRACT

In 2001, Professors W. Steve Albrecht and Robert J. Sack wrote in their 2001 CPA Journal article that the accounting practice community perceived accounting education to be completely outdated. Fewer students majored in accounting, many accounting firms perceived accounting students higher educational training as outdated and hired fewer accounting majors. In 2021, these issues are still prevalent. This chapter discusses what has changed over the last 20 years and what issues are faced with current accounting graduates. A more agile accounting profession, a new CPA exam approach in 2024, and the use of more advanced analytical and technological tools in the profession only make the challenge to higher education even harder. The discussion then concludes with mitigating strategies to not only improve requisite skills that the profession demands, but also discusses strategies to curtail declining enrollment trends.

INTRODUCTION

The Accounting profession is a constantly changing organism that molds itself to the meet the needs of the clients it serves. Over the last century the Accounting profession has changed dramatically from providing basic bookkeeping, tax filing services to sophisticated, complex client consultation services.

DOI: 10.4018/978-1-7998-6772-2.ch012

The accounting profession went from focusing on transactions style business to knowledge service business. These changes are a result of meeting the needs of the clients and to remain as efficiently and effective as possible to satisfy the clients' needs.

Accounting education in universities and colleges have helped to provide accounting education and foundational knowledge to prepare the students for entering the profession. In the early days this education was basic bookkeeping but as the profession has changed and modified itself to meet the needs of its clients, so too has higher education for accounting.

Over the last two decades academics and the profession have questioned whether academia is keeping up with the needs of the profession. In the early 2000s, both academic and practitioner articles had begun mentioning that accounting education is outdated. The point of this chapter is that accounting education needed to change its focus and approach to *teaching* accounting to ensure future accountants' skills are relevant for the twenty-first century. Today, in 2021 these concerns are still relevant (some would say even more pronounced), and academics and practitioners still question the relevancy of accounting education. As the profession continues to change so must the approach to accounting education. Accounting educators lay the foundation to the knowledge base of future accountants, thus, to have a continuing strong accounting profession, accounting education must adapt as well.

The purpose of this chapter is to address the gap between accounting educational curriculum at universities and colleges in the United States and the expectations of the profession. This chapter discusses how has higher education have failed to address the changes in meeting the needs of the profession and what gaps exist to meet the future professional needs. In addition, this chapter discusses the technical and technological skillsets are needed now in accounting education based on the tools currently being used in the profession.

BACKGROUND

20 Years Ago People Questioned The Relevancy of Accounting Education

At the beginning of the new millennium academic and practitioner articles stated there is a knowledge gap between what was taught in the classroom and what is required by the profession. Professors W. Steve Albrecht and Robert J. Sack wrote in their 2001 CPA Journal article (Albrecht and Sack 2001) that the accounting practice community perceived accounting education to be completely outdated. They highlighted that practicing accountants in their study would not major in accounting if they had to redo their own education. In addition, student enrollment decreased dramatically in the 1990s as students did not see the true value of an accounting education. As depicted in Figure 1, accounting firms began the process of hiring more non-accounting majors as the profession continued to change from firms that provide accounting services to professional services firms (AICPA 2019).

This bleak picture of the future of accounting education raised major concerns about the relevancy of education and the output of quality education. Some argued that that the decrease in quality students was a result of lower starting salaries and opportunities for graduates (Gabbin 2002). Students at that time were choosing majors in Finance and Information Systems because of the perceived better job opportunities and job growth. The perception of accounting was that accounting is already outdated and stagnant.

The American Accounting Association (AAA) introduced benchmarking programs to guide accounting programs to implement relevancy and needed changes. For example, in its *Accounting Education*

Figure 1.

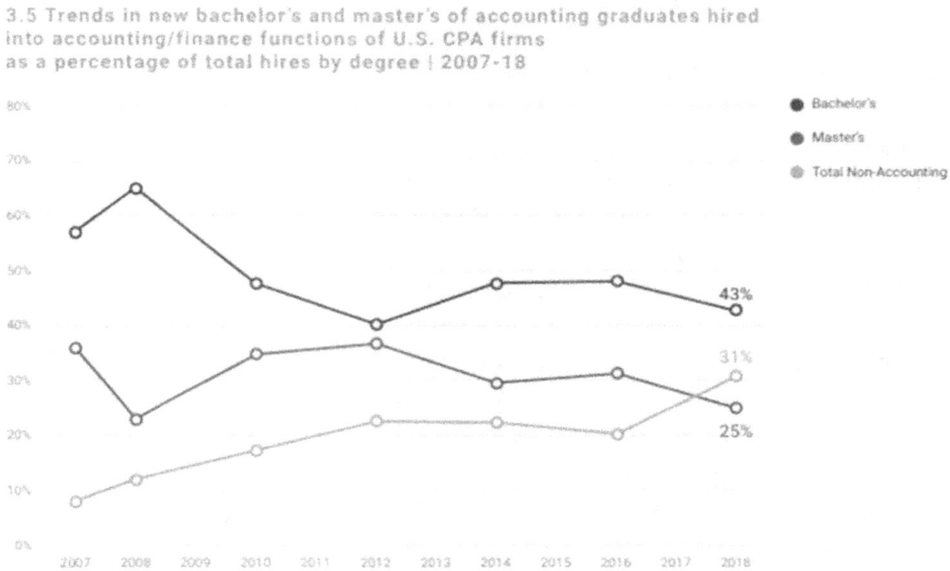

Change Commission Great Experience thirteen schools ran proactive, curriculum experiments that surveyed employers and alumni (Flaherty 1998). These thirteen schools would modify their school's curriculum to meet the changing needs of the profession -- including focusing on written and oral communication skills, and a decreased emphasis on the use of textbook to lecture approach in the classroom to more real-world simulations. In essence, the focus was on bringing the real world into the classroom. These real-world simulations would provide students opportunities to enhance their analytical and critical thinking skills. The goal of these programs was to introduce accounting from different perspectives whether as users of accounting information, technical information systems perspective and financial statement preparation perspective. The curriculum experiment these thirteen schools attempted was considered successful even though many of the schools failed to modify their curriculum to the changing stakeholders' needs (Gabbin 2002).

Accounting Education Has Not Changed Over The Last Twenty Years

The *great experiment* completed by the thirteen schools are considered a success, but most of the colleges and universities made minor modifications across their curriculum by emphasizing general education and cultural diversity over the changing nature of the accounting profession - and only some minor modifications were made in accounting education.

Accounting academics have consistently argued that their accounting curriculum lacks agility because of many reasons. First, bureaucracy within the accounting department, business college and the university level does not allow timely changes to meet the profession's needs. In addition, state regulatory and CPA exam eligibility regulations requires reviews of most changes and that takes time for approval in the academy (Albrecht and Sack 2001). Second, faculties may be challenged to update knowledge requirements due to inadequate funding for professional development or an unwillingness to change.

Curriculum changes are difficult and challenging as they are awash with many established barriers, but any potential loss of relevance is unbearable.

Why do Accounting Curriculums Need To Change Quicker?

In the United States, traditionally many accountants attempt and obtain the Certified Public Accountant (CPA) license. Although National Association of State Board of Accountants (NASBA) and the American Institute of Certified Public Accountants (AICPA) Education prescribe knowledge content requirements for the CPA exam, specific education, and experience requirements for sitting for the exam and attaining the license CPA license is determined and regulated by licensing jurisdictions. Starting 2024 the CPA *Evolution* CPA exam takes effect. Evolution will not only significantly change the CPA exam model, but also create seismic changes to the education model. The goal of Evolution is to prepare newly licensed CPAs with the requisite knowledge to succeed in the profession. Although the exam will keep its four-part model, the exam will focus on core knowledge and a discipline. The core, which represents three parts, focuses on accounting, auditing, tax, and technology. Candidates will choose a fourth part from currently three disciplines, that include: Information systems and controls, business analysis and reporting, and tax compliance and planning. Evolution will present challenges for colleges and universities to adapt to its model. Without the ability to be agile to modify the accounting curriculum students may lack knowledge to pass the exam. Additionally, the three-discipline approach will strain accounting programs who do not have adequate resources to offer courses in all three disciplines. Smaller accounting programs will be forced to limit programs, seek partnerships with other programs with available resources, or worse yet, eliminate offerings in accounting.

Many Colleges and Universities are now faced with recovering from COVID-related budgetary issues combined with recent declining enrollment trend in accounting programs, makes implementing Evolution program changes insurmountable. These challenges will only hamper the academy to meet the needs of the profession who may continue its increasing trend to hire from non-accounting programs (AICPA 2019) to secure talent. The hiring trend is apparent as displayed in figure one. CPA firms are hiring fewer accounting majors and more non accounting majors. The CPA *Evolution*'s goal is to reverse that trend and make accountants relevant to the profession.

Accounting firms have been big adopters of technology over the last twenty years. The profession is using more tools such as data analytics, artificial intelligence, and bots to improve their firms' performance. Failure to implement emerging technology tools in the accounting curriculum will result in graduating students not possessing skills demanded by the profession. For example, the Big 4 audit firms have expressed that as technology is more prevalent in the profession, firms will reduce accounting graduates hires by fifty percent (Welker 2017). The profession will continue to hire from fields outside of accounting, especially gradates with skills in data analytics (Huerta and Jensen 2017). Education programs need to urgently adapt to the Evolution accounting curriculum; otherwise, accounting students may become even less relevant to the profession.

Technologies have changed dramatically over the last twenty years and will continue to change in the future. In 2000, a major focus in the accounting profession included XBRL, Enterprise Resource Planning (ERP) and statistical analysis. Today in 2021, the accounting profession continues to use these technologies but take them a step further. For example, ERP traditionally were stored on servers but today many firms are shifting to using ERPs on the cloud. Accountants need to understand the differences in risk associated with servers versus the cloud. Another growing area is data analytics. In 2000, accountants mainly used descriptive analytics, describing the data over a historical period of time. For example, what are the sales for division B in the third quarter. Today, descriptive analytics is not enough. The accounting profession is using predictive and prescriptive analytics. Understanding statistics is not enough to succeed in the accounting profession. Accountants are expected to provide their skillset to provide more insist and analysis to their firm or clients.

Mitigating The Crisis

The accounting profession is a constantly changing and morphing to meet the global marketplace, so must too accounting education. The biggest hurdles are bureaucracies within the colleges and universities as well as the State Accounting Boards. Programs need to be nimbler and more innovative to meet the challenging business world. Some steps have been taken by academia.

One starting point for addressing these changes is the updated AACSB standards. A goal of the AACSB accounting accreditation is to advance accounting as a learned profession that produces excellent graduates, impactful scholarship, and mutually dependency between the academy and the profession (AACSB 2018). CPA Evolution is a prime example of that mutually dependency. Current AACSB accounting standards signals that an attribute of a quality accounting program is the level of engagement between the academy and the professional community. Accredited programs requirement to adhere to standard A5 evidences the professional demand for emerging technological skills.

Professional engagement prepares students for meaningful professional experiences, as well as societal and personal lives in the view of the AASCB. That engagement encourages and documents continuous improvement and innovation in programs. The recently enacted AACSB business school standards emphasize "…quality business education cannot be achieved when either academic or professional engagement is absent, or when they do not intersect in meaningful ways" (AACSB 2020). The AACSB charge of professional engagement is not novel. Yet, unless the academy rises to the challenge of not only intersecting but acknowledging the dependency it has on the profession, the accounting academy may find itself without its vital consumer of its product.

Accounting programs need to be more flexible. Programs should require less specified courses in the curriculum and more electives that supports the knowledge content necessary for the three disciplines of the new CPA exam. Currently, many accounting curricula at colleges and universities are rigid, allowing students few electives, if any within their major. By focusing on broad core components of the CPA exam, students have the opportunity to take timelier, technology and analytics intense courses. These courses would help the students to be more marketable and ready to enter the new accounting profession. The courses would provide technology and cases to allow the student to gain hand's on experience, simulating the profession (Dow, Jacknis and Watson 2021).

The goal of accounting education is not to turn students into accounting experts, but to provide the foundation to enter the accounting profession. The foundation of any new accounting graduate should include the knowledge and understanding of how transactions are processed and flow through the business and how to analyze the transactions and processes. These graduates should also have experience in the necessary tools to adapt to a changing profession. Accounting firms are hiring more non accounting graduates as results of the non-accounting graduate's data analytics and technological skills (Huerta and Jensen 2017). However, accounting graduates have specific knowledge and skillsets to help management to create and analyze specific questions of the business and to interpret and communicate the results of the issue at hand (IMA 2018). Thus, by ensuring new accounting graduates have the right knowledge and understanding of technological tools, this ensures accounting graduates have a place in the changing accounting profession.

Integration of technology requires application to the content knowledge. As firms continue to supply ample support for technology use, filling the application void can be challenging to complete without costly professional development. As the Evolution exam progresses in establishing topics and objectives, there will be a demand that emerging technology application be spread across the core exams for financial, tax, and audit. This will be a tall order for many in the academy that typically rely on publisher and standard setters' resources for content knowledge. It is conceivable that professional resources will be paramount as the curriculum adjusts to Evolution resulting in the profession taking a dominant role in delivering accounting education.

The profession's influence and support are nothing new for the academy. Ample firm established resources for faculty already exist in the format of software, case material, targeted conferences, funding, and the like. However, a more direct approach to ensure that accounting programs and the respective graduates meet the needs of the profession is to partner with programs. That is the case with KPMG in establishing partnerships with nine graduate accounting programs. KPMG lends their branding, funding in all but one case, and more importantly, sharing and infusing their knowledge expectations in programs that produce newly licensed CPAs. In return, graduates streamline into KPMG workforce possessing the requisite skills to succeed in the firm and in the profession. Further, if firm program partnerships are viewed by college entrants as a direct pathway to permanent employment, declining enrollment trends may curtail.

Firm partnered programs sound like a win-win; university administrative welcome the funding, students value the experience, and firms ensure a skilled pipeline. But curriculum changes are fraught with barriers that the academy needs to reconcile or potentially suffer the loss of relevancy to the profession. These alarms for the academy are not unique but, now the ringing is deafening as seismic changes may be also needed for the academy to survive.

CONCLUSION

Twenty years ago, the profession questioned the relevancy of accounting higher education. Today, this question is still relevant. Higher education in accounting is faced with more challenges than it did in 2000. Today, the education system is not only being challenged by professional and technology requirements, but also global challenges, including COVID-19 and the Global economy. Higher education needs to adapt quicker to the changing markets to ensure graduating students are prepared to enter a new accounting profession and thus, remain relevant. The academy is at the cusp of shaping the crystal ball. Failure could be an option for those who view these challenges as formidable, including the challenge of changing imbedded and antiquated academic processes. The outcome could be fewer programs, perhaps elite in nature, yet adeptly producing what the profession requires. On the other hand, success could be an option. Accounting higher education needs to be more of a stakeholder in the profession than just an entry point. Accounting higher education needs to be nimble to meet the profession's changing needs. As the accounting profession changes, so to must accounting higher education. The question of relevancy of accounting higher education has been asked multiple times since the creation of accounting education (Chu & Man 2012). Skeptics may argue that the relevancy question does not matter at this point because accounting higher education still exists. Perhaps, though, relevancy of accounting higher education is more prevalent than any time in history. Accounting enrollments continue to decrease, and non-accounting graduates hiring at accounting firms continue to increase. The canary in the coal mine is here today and if accounting higher education does not adjust to the changing profession, accounting education will go down in the history books as a major colleges and universities once offered.

REFERENCES

AACSB. (2018). *2018 Eligibility Procedures and Accreditation Standards for Accounting Accreditation*. Available on-line on March 1st, 2021 at: https://www.aacsb.edu/-/media/aacsb/docs/accreditation/accounting/standards-and-tables/2018-accounting-standards.ashx?la=en&hash=8DCDA6CE3B0CEF6AB82D39CBF53995DA96111196

AACSB. (2020). *2020 Guiding Principles and Standards for Business School Accreditation*. Available on-line on March 1st, 2021 at: https://www.aacsb.edu/-/media/aacsb/docs/accreditation/business/standards-and-tables/2020%20business%20accreditation%20standards.ashx?la=en&hash=E4B7D8348A6860B3AA9804567F02C68960281DA2

Albrecht, W. S., & Sack, R. J. (2001). The Perilous Future of Accounting Education. *The CPA Journal*, *71*(3), 16–23.

American Institute of Certified Public Accountants (AICPA). (2019). *Accounting Graduates Supply and Demand Report*. Available on-line on December 15, 2020 at: https://www.aicpa.org/content/dam/aicpa/interestareas/accountingeducation/newsandpublications/downloadabledocuments/2019-trends-report.pdf

American Institute of Certified Public Accountants (AICPA). (2020). *CPA Evolution*. Available on-line on December 12, 2020. https://www.evolutionofcpa.org/

Chu, S., & Man, H. (2012). History of Accounting Education. *Journal of Higher Education Theory and Practice*, *12*(1), 119–128.

Dow, K., Jacknis, N., & Watson, M. (2021). A Framework and Resources to Create a Data Analytics-Infused Accounting Curriculum. *Issues in Accounting Education*, ISSUES-19-071. Advance online publication. doi:10.2308/ISSUES-19-071

Flaherty, R. E. (Ed.). (1998). Accounting Education Series: Vol. 14. *Accounting Education Change Commission Grant Experience: A Summary*. AAA.

Gabbin, A. L. (2002, April). The Crisis in Accounting Education. *Journal of Accountancy*, *193*(4), 81–86.

Huerta, E., & Jensen, S. (2017). An Accounting Information Systems Perspective on Data Analytics and Big Data. *Journal of Information Systems*, *31*(3), 101–114. doi:10.2308/isys-51799

Institute of Management Accountants (IMA). (2018). *Management Accounting Competencies: Fit for Purpose in the Digital Age?* Available on-line on January 21, 2021 at: https://www.imanet.org/insights-and-trends/the-future-of-management-accounting/management-accounting-competencies---fit-for-purpose-in-a-digital-age?ssopc=1

KPMG Master of Accounting with Data and Analytics Program. (n.d.). Available on-line on January 21, 2021 at:https://audit.kpmg.us/kpmg-master-accounting-data-analytics-program.html

Welker, B. (2017). Will the Big Four change hiring practices in the coming years? *CPA Practice Advisor*. Available on-line on January 26, 2021 at: https://www.cpapracticeadvisor.com/article/12296978/will-the-big-four-change-hiring-practices-in-the-coming-years

Chapter 13
Removing the Constraints of Disability:
How New Technology Is Transforming the Experience of Disabilities

Wadad Kathy Tannous
https://orcid.org/0000-0002-3153-5652
Western Sydney University, Australia

Laney McGrew
Western Sydney University, Australia

ABSTRACT

One billion people globally live with disabilities that are physical, sensory, psychiatric, neurological, cognitive, or intellectual. Their disabilities are dynamic and can be temporary or permanent, singular or plural, from birth or developed, and can change over time. People with disabilities face barriers to economic, social, political, and cultural participation. Assistive technology, artificial intelligence, and broader technology can amplify their inclusion, participation, and independence. This chapter will highlight emerging and evolving technologies, rooted in machine learning and neural networks, which assist across different disabilities and seek to improve the user's sense of ability and independence. These include Seeing AI app, OXSIGHT, OrCam, Envision smart glasses, and Dot Watch for vision impairment; Ava app and cognitive hearing aid for hearing impairment; Liftware self-stabilising utensils for limited hand mobility; Eyegaze and Tobii – assistive technologies that allow users to control computer and smartphone screens with their eyes; and 3D printed prosthetics.

DOI: 10.4018/978-1-7998-6772-2.ch013

INTRODUCTION

Conceptualizing Disability

Disability is not easy to define. The boundaries are elusive, changing across environments, circumstances and individuals. A contested concept, the nebulous nature of disability is one of the reasons that basic human rights for people with disabilities (PWD) can be so difficult to secure. For the purposes of this chapter, disability is understood as any condition impacting a person's mobility, neurological functioning, communication, cognition, intellect, physical ability, mental health, or social engagement (Government of Western Australia, n.d.). Roughly one billion people, or 15% of the global population, live with a disability (The World Bank, 2020). In Europe, Australia and the United States, this is as high 20% and is expected to continually increase over the next few decades (Belton, 2016; Australian Network on Disability, 2020; Disabled World, 2020). Disabilities are dynamic and multifaceted: they can be temporary or permanent, visible or invisible, singular or plural, present since birth or developed later in life. Some of the most prevalent disabilities include hearing loss, vision loss, dyslexia, Autism Spectrum Disorders (ASDs), musculoskeletal disorders, and nervous system disorders (World Health Organization, 2011).

Disability is generally understood and analysed through one of two lenses: the medical model or the social model. The medical model posits disability as an impairment of the individual and seeks to correct disabilities through medical intervention under the assumption that this will improve the life of the disabled person. Under this framework, the disabled body is constructed as the pathologised antithesis to the normal, superior able body (Johnstone, 2012). The disabled body, and by extent, the disabled individual, are deviant and problematic and must be 'fixed' to conform with the norms of abled-ness (Berghs et al., 2016). In contrast, the social model of disability regards disability as a socially constructed phenomenon that stems from barriers and norms that centre the lived experience of able-bodied persons. Emerging from British disability activism of the 1960s–70s which identified the limitations of the medical model and its tendency to remove agency and autonomy from PWD, the social model challenges the assumption that PWD are inherently broken or wrong. As outlined in the manifesto penned by the Union of the Physically Impaired against Segregation (UPIAS) (1976), "it is society which disables physically impaired people. Disability is something imposed on top of our impairments by the way we are unnecessarily isolated and excluded from full participation in society" (p. 14). Further thinking about disability has seen the development of new models that recognise both physical and sociocultural impacts on the experiences of PWD and posit disability as "an inability or limitation in performing socially defined roles and tasks expected of an individual within a socio-cultural and physical environment" (Nagi, 1991, p. 315). In addition, these hybrid models take into account identity politics, psychosocial factors and human rights issues as components that inform experiences of disability (Brewer et al., 2012; Masala & Petretto, 2008).

Efforts to expand and reimagine models of disability have led to substantive analysis of the relationship between capitalism and disability. Russell and Malhotra (2002) explain that, as capitalism and industrialisation began to take hold in the 17th Century, "people's bodies were increasingly valued for their ability to function like machines" (p. 213). This shift towards mechanical labour established rigid norms and expectations that were difficult for many PWD to achieve. In turn, PWD were posited as unfit and excluded from the labour market. The construction of the 'working body' informed the pathology of the disabled body as an instrument that could not fulfil its duties. Disability is thus the "central contradiction of capitalism" (Russell and Malhotra, 2002, p. 212) that both constructs and penalises disabled bodies. It is worth noting that the oppression of PWD is not exclusive to capitalism; the rights of PWD are frequently compromised under any political or economic system, as is evidenced by the treatment of PWD in Nazi-occupied Germany and the USSR (Friedlander, 1995; Phillips, 2009). Thus, it is not capitalism specifically that constructs and, consequently, impedes on the rights of PWD but the social value system that posits the white, heterosexual, cisgender, able-bodied male as the norm around which deviance is constructed (Schalk, 2017).

PWD face myriad impediments to economic, social, political and cultural participation which are often a result of environmental factors such inadequate policies, poor public attitudes, deficient services, and lack of funding, data and consultation. In turn, PWD frequently experience poorer health, lower educational achievements, higher rates of poverty, and restricted social and community engagement (World Health Organization, 2011). In understanding the history behind "capitalist modes of production and their relationship to structural inequalities and social exclusion" (Berghs et al., 2016, p. 36), we can better grasp the interconnectedness of these themes and how they take shape under neoliberal capitalism. For example, poverty and disability have a cyclical relationship: PWD experience poverty and unemployment at twice the rate of able-bodied people, and people living in poverty are more likely to develop a disability, making it increasingly difficult to escape poverty (Disabled World, 2020; Belton, 2016; World Health Organization, 2011). It comes as no surprise, then, that two thirds of the global disabled population lives in relative poverty (Degener, 2016). Furthermore, as work is a fundamental aspect in the development of identity under neoliberal capitalism, the impediments to the labour market faced by PWD have material impacts on not only economic but also social well-being (Budd, 2011; Galer, 2012). Understanding disability justice as a multifaceted issue that stretches across economic, social and political realms is vital to developing technology that will provide adequate, meaningful assistance to PWD.

Reimagining (Assistive) Technology

The emergence of the digital era has seen technology put in the back pocket of virtually every individual. Daily tasks and interactions are routinely informed by artificial intelligence (AI) and an invisible web of networks and algorithms, including machine learning, deep learning and neural networks (Saravanan & Sujatha, 2018; Hosny et al., 2018; Islam et al., 2017). While often invisibilised, these models and processes govern nearly every aspect of everyday life in the digital era. A subset of technology that often overlaps with digital technology is assistive technology (AT), which refers to any device or piece of equipment "whose primary purpose is to maintain or improve an individual's functioning and independence to facilitate participation and to enhance overall well-being" (World Health Organization, 2020a). While an estimated one billion people worldwide require AT, only 10% of those in need actually have access to it. Historically, AT has been widely inaccessible due to poor policies, high prices, and a lack of awareness and availability, and even when obtained, abandonment rates of AT are as high as 30% due to routine failures in meeting the needs of users (World Health Organization, 2018). While this has been changing in recent years due to "the confluence of technological development, human rights progress, and progressive public policy" and a push for AT to achieve "the twin goals of universal health coverage and sustainable development" (Buchanan & Layton, 2019, p. 1), much of it remains clunky, dated, and mono-functional (Horsley, 2019). Furthermore, the gaps in technological accessibility and innovation for PWD exemplify how the digital field acts as a microcosm of larger society, emulating the same norms, hierarchies and standards that comprise the wider sociocultural fabric (Goggin & Newell, 2003). In the same way that the 'abled/disabled' binary works to categorise people, so too does it categorise technology. AT is necessarily second to mainstream technology, which is informed both by the dynamic and productive nature of technology as well as the way AT is regarded as a 'fringe' market (Hall, 2019). Foley & Ferri (2012) explain: "by the time an accessible version is developed, mainstream technology has often been updated or otherwise changed" (p. 197). As AT is positioned as the antithesis to mainstream technology, it is not afforded the same care and significance around design and aesthetics. While aesthetics and function are of equal importance in mainstream technology, design takes a back seat to function when it comes to AT. This lack of care around innovative design and failure to consider a nuanced user experience mean many PWD experience stigma and social exclusion as a result of using AT, which contributes to high abandonment rates and the reluctance to get AT in the first place (dos Santos et al., 2020). Further, it is not just AT that is inaccessible to PWD but technology as a whole. Even something as commonplace as the Internet is significantly under-utilised by PWD, despite the fact that online communities offer a way for PWD to connect to one another and experience greater social inclusion (Moon et al., 2019).

For too long, there has been a disconnect between assistive technologies and their intended users. PWD are rarely consulted when it comes to designing AT; rather, AT is developed by people who do not imagine themselves as the user. As Sheldon (2004) explains, "technology is not neutral. It is created by the same oppressive society that turns those with impairments into disabled people… It is no surprise, then, that disabled people have a complicated relationship with technology" (p. 155). This disconnect means that AT often relies on "assumptions about disabilities and disabled users that may or may not be accurate" (Foley & Ferri, 2012, p. 198). The Amazon Kindle, for example, was promoted as a breakthrough e-Reader that was also equipped with text-to-speech software, ostensibly allowing it to be accessed by people with visual impairments and learning disabilities. When put to the test, however, the Kindle failed to meet the needs of these groups as the interface itself was unable to be operated by voice: users could not purchase books, select texts to read, or even activate the text-to-speech function by voice, instead having to rely the Kindle's touchscreen interface (Foley & Ferri, 2012). While the Kindle contained technology that could be helpful for PWD, those users were never actually consulted nor were their realities imagined in the design of the product. This lack of consideration came back to bite Amazon, as ultimately what made this well-intentioned product fail in this regard was its awareness of sociocultural and political factors without fully considering and consulting with the very people it intended to help.

Despite the recent ubiquity of technology, barriers to accessibility mean that PWD are frequently excluded from this new normal. However, this accessibility gap can also be understood as an opportunity gap. For example, the wearable technology ('wearables') market is expected to be worth $54 billion by 2023, yet it is still regarded as existing on the outskirts of rather than within the mainstream technology market (Barkho, 2019; Hall, 2019). A huge range of possibilities exist for AI and broader technology to help amplify "inclusion, participation and independence for people with disabilities" (Essl Foundation, 2019), and this untapped potential affects not only PWD but also companies and entrepreneurs that could expand into a new market desperate for innovative, cutting-edge technology.

This chapter will highlight some of the emerging assistive technologies that are in line with the rapidly evolving technological landscape of our current age. These technologies harness the power of modern developments such as machine learning, neural networks and 3D printing to assist across a range of disabilities and areas of human activity. In recognising the changing role of AT and the increasing prevalence of disintermediation, or the ability to access AT on one's own instead of through a professional, these technologies break down the stigmas and assumptions that PWD are inherently indeed of help and instead seek to improve the sense of ability and independence, rather than interdependence, for people living with disabilities (Buchanan & Layton, 2019). Most notably, the companies developing these technologies move away from the medical/social dichotomy and instead utilise a hybrid model of disability that acknowledges the sociocultural, physical and environmental factors that create and inform disability. Through consultation and employment, the voices and lived experiences of people with disabilities are centred rather than regarded as an afterthought. This multifaceted approach to technological research and development can be summarised by the slogan of the disability rights movement of the 1990s: "Nothing About Us Without Us" (Charlton, 1998). While there is an endless amount of emerging AT worth discussing, this chapter will cover a select few standouts in the areas of visual impairment, hearing loss and physical disabilities.

It is important to note that inclusion in both the development of AT and the wider labour market is, of course, not the be all and end all of disability rights. A multitude of factors inform the sociocultural, economic and political ostracisation of PWD, and to provide each a meaningful and nuanced consideration would be outside the scope of this chapter. Merely centring the experiences of PWD in the research and development of AT will not break the cycles of poverty disproportionately endured by PWD, nor will it guarantee PWD equal and unproblematic access to AT or opportunities in the labour market. The technologies described in this chapter focus on increasing users' autonomy, independence and self-determination, concepts which are posited as unquestionably desirable under neoliberal capitalism (Kafer, 2013). Drawing on Mia Mingus' (2011) idea of "access intimacy", this chapter highlights the importance of involvement with and access to AT as merely one facet of disability justice, not in a quest to reach an end goal of equitable labour participation but rather of self-actualisation.

Assistive Technologies For Visual Impairment

An estimated 2.2 billion people suffer from visual impairment to some degree (World Health Organization, 2019). Ranging from blurred vision to complete vision loss, people with visual impairment harbour a diverse set of needs which are reflected in the variety of available assistive technologies. The most common wearable for people with visual impairment has become normalised to the extent that it is nearly invisible as a piece of AT: eyeglasses. Eyeglasses are so routine that moderate vision loss is no longer regarded as a disability, whereas moderate hearing loss is. This is an example of the way in which disability is dynamic, socially constructed and shifts across time and place. Until the 1930s, eyeglasses were regarded as medical equipment in the UK. It was not until eyeglass developers put user experience and desires at the forefront of design that they began to emerge as a fashion accessory (Pullin, 2009). This shift highlights the way in which disability can be reimagined not as something to be hidden but something to be celebrated, helping to destigmatise disability and assistive technologies.

A step up from traditional eyeglasses, smart glasses are a popular wearable for people with more significant vision impairment. OXSIGHT, a company developed by Oxford University researcher Stephen Hicks, designs and creates glasses specifically for people who experience peripheral vision loss, or tunnel vision, which can be caused by the degenerative eye disease retinitis pigmentosa (RP) as well as strokes. These glasses utilise small cameras that project HD imagery of the wearer's peripheral view into the lenses, expanding their horizontal visual field to up to 68 degrees and allowing for easier and more thorough engagement with one's surroundings. OXSIGHT glasses are sleek, lightweight and come in two styles: OXSIGHT Prism™ or OXSIGHT Crystal™ (Oxsight, 2020). OXSIGHT is also developing glasses for central vision loss which will be released in late 2020 (OxSight, 2019). Another example of innovative wearable technology is OrCam MyEye, a voice-activated device that attaches to any pair of glasses to offer visually impaired people greater independence and mobility. OrCam MyEye uses digital sensors, speakers and machine learning algorithms to capture the user's surroundings and convert it into audio. The device can recognise faces, read text, or identify objects, and narrate them in real time. It is hands-free and does not rely on the Internet to function (OrCam, 2020; Zonshine, 2020). Innovations such as OXSIGHT and OrCam glasses expand the potential of traditional eyeglasses and seek to provide people with substantial vision impairment greater independence, social inclusion and access to modern technologies.

While glasses may be the most obvious choice of wearables for visually impaired people, they are not the only option. South Korean company Dot Incorporation, in partnership with Serviceplan Korea, launched the Dot Watch in 2017. Touted as "the world's first braille smartwatch" (Murillo, 2020), the watch pairs to an app on the user's smartphone via Bluetooth, allowing them to receive phone notifications in braille on the screen of the watch. Users can answer or reject calls, read and respond to texts and notifications (the device can display notifications from any app on the user's smartphone), and access all traditional smartphone clock features such as the date and time, stopwatch, timer and alarm clock (Dot, 2020). The device has nine language settings and is also equipped with a teaching program that allows users to learn braille if needed. Furthermore, the Dot Watch app, which is required in order to use the smartwatch itself, is completely accessible to blind or visually impaired people and does not require sighted assistance (Murillo, 2020). In contrast to the aforementioned Kindle debacle, it is clear that Dot has thoroughly considered the needs of blind and visually impaired people in order to make every aspect of the Dot smartwatch accessible. Dot identified an accessibility gap for visually impaired people who felt excluded from modern technological norms such as texting and social media. Innovative technology for visually impaired people was lacking, and existing braille devices were clunky, expensive and inaccessible (Serviceplan Group, 2018). The Dot smartwatch is sleek and modern, coming in a range of fashionable styles that rival other leading, non-AT smartwatches on the market today. It is lightweight and durable, a stark contrast to much of the bulky AT that comes overfitted with protection devices, and its stylish design is a testament to Dot's commitment to its users' desires, needs and preferences.

Case Study: Microsoft

In recent years, Microsoft has demonstrated a strong commitment to accessibility in both product development and hiring. This has been largely informed by CEO Satya Nadella's experience of having two children with disabilities, one with significant learning disabilities and another who is quadriplegic, visually impaired and has limited communication abilities (Francisco, 2017). The company launched AI for Accessibility in 2018, a $25 million grant program that provides funding for AI developers whose products seek to help PWD and "amplify human capability" (Microsoft, 2020a). They also have an autism hiring program, provide disability training programs for employees, offer a range of accommodations for interviewees with disabilities, and host various 'Ability Hiring Events' for candidates with disabilities as well as an annual Ability Summit (Microsoft, 2020b). Microsoft has also developed several assistive technologies in recent years, most notably the Seeing AI app, which uses deep machine learning and neural networks to narrate descriptions of the world via the user's smartphone camera. This software was developed with input from visually impaired programmers, allowing the software to be extraordinarily adept at meeting the needs of the blind community (Microsoft, 2020c; Microsoft, 2016). Seeing AI can read printed and handwritten text, identify currencies, locate barcodes and recognise products, identify people and describe their emotions, and describe one's surroundings, including colours and light (Microsoft, 2020c). Microsoft has also developed breakthrough eye-tracking technologies, which will be covered in depth further in this chapter. Overall, Microsoft is an excellent model for any company wishing to implement changes at a structural level that will promote inclusivity and accessibility for PWD.

Assistive Technologies For Hearing Loss

Approximately 466 million people worldwide live with a hearing disability (World Health Organization, 2020b). Many deaf and hard of hearing (HOH) people who use traditional hearing aids cite difficulty hearing properly in group scenarios or in crowded places such as restaurants as a primary concern. While typical hearing aids can filter background noise, they do not have the ability to separate and amplify one voice out of many. In order to combat this issue, a team of researchers at Columbia University are currently developing a new kind of cognitive hearing aid. Spearheaded by associate professor of electrical engineering, Nima Mesgarani, the device identifies which voice in a room a user is attempting to listen to and amplifies that voice while quieting the other noises. This shift in development relies on two primary components: a new style of microphone and the brain's own audio filtration abilities. The new microphone, developed by Mesgarani, uses deep neural networks to separate the inputs in real time. It then compares these inputs to the user's neural signals to identify which voice the user is focusing on and amplifies that voice, all of which takes under ten seconds. The product is currently still in development but is anticipated to be on the market by 2023 (Columbia University's Zuckerman Institute, 2018).

A more accessible and immediately available option is Ava, is a captioning program for live conversations. What started as a smartphone app developed by three students at the University of California, Berkeley, one who was born the only hearing person in his family, has transformed into an app and computer software that provides live captioning for individual or group conversations (University of California, Berkeley, 2020; TEDx Talks, 2017). Ava uses an individual's smartphone or computer microphone to listen to what is being said and translate that into written text. Driven by neural networks and machine learning, the program has an average accuracy rate of 95% and allows deaf or HOH people to follow conversations in real time. The interface is simple and functional and tailors the user's experience to whether they are hearing, HOH, or deaf (Matney, 2016). In order for the app to function, Ava must be downloaded by every person in a conversation; each smartphone is assigned a colour which allows the user to more easily differentiate the various streams of text. When being used by a business, the software creates an 'Ava Room' in which one user hosts and other members join (Ava, 2020). Additionally, the company has recently developed Ava Scribe, their premiere captioning service that uses machine learning and human interference to ensure captions of 99% accuracy. Ava has a variety of options for individuals and businesses, ranging from free to USD $9.95 per month plans for individuals and USD $99 to $145 per month for businesses which also have the option for a custom plan to suit the needs of a specific company (Ava, 2020).

Communication is a two-way street, meaning deaf and HOH people face difficulties not only with hearing others but also with being understood. Approximately 70 million people worldwide use sign language as their primary language but only 1.4 million have access to AT for communication, and the majority of 'hearing' people, or those without aural impairments, do not speak sign languages (Knowles, 2018). To help tackle this problem, designer and programmer Hadeel Ayoubt created BrightSign, a smart glove that uses sensors to translate the wearer's sign language into speech and written text via a companion smartphone app. While many iterations of smart gloves existed before BrightSign, none of them achieved the level of functionality and accuracy needed to become marketable. During the development process,

Ayoub recognised that many of her predecessors' shortcomings were due to poor style and functionality. While the technology was innovative, the design was not. The gloves were bulky and unappealing; they *looked* like AT and failed to break away from the normative assumptions and stigmas surrounding PWD and assistive devices. Furthermore, previous smart gloves were designed with a 'one-size-fits-all' approach, using pre-installed gesture libraries that limited the potential users and did not account for individual variations in signing. Ayoub consulted with members of the HOH and deaf community, who all gave the same feedback: "the glove had to be as weightless and invisible as possible" (BrightSign, 2020). The idea of creating a companion app was Ayoub's way to move away from bulky screens and allow for a more slim, sleek and fashionable design that could be worn on its own or underneath another pair of gloves. Furthermore, while BrightSign can understand over 100 different sign languages and over thirty spoken languages, it also uses a machine learning algorithm and neural networks that allows its users to train the glove to understand their particular idiosyncrasies and even create their own library of signs (BrightSign 2020; Jewell, 2019; Rony & Alam, 2019). This shift in approach has given the glove a 97% accuracy rate (Knowles, 2018). Smart gloves such as BrightSign provide an unprecedented level of independence, social connectedness and communication to people in the deaf and HOH community. While most comparable assistive devices retail at approximately £2,000, BrightSign is available for purchase for only £945, making it a practical, affordable, and appealing wearable for deaf and HOH people.

Another initiative that assists with communication is Google's Project Euphonia, an automatic speech recognition (ASR) software trained to recognise non-typical speech. ASR technology allows users to speak to a device instead of using the keyboard and includes functions such as speech-to-text, voice search, voice dialing, and speaker recognition (Singh, 2019). While ASR is not new for PWD, its recent technological ubiquity due to virtual assistants such as Apple's Siri and Amazon Alexa that are not specifically developed for PWD has created an accessibility deficit. Most ASR systems are only familiar with 'typical' speech patterns, causing people with speech impairments or accents to be left out of using this technology. Speech impairments can be caused by a range of disabilities including Parkinson's, multiple sclerosis, spinal cord injury, and amyotrophic lateral sclerosis (ALS) (Cattiau, 2019; Shor & Emanuel, 2019). Recognising this gap in accessibility, Project Euphonia was developed to expand the range of voices and speech patterns understood by ASR systems. The software was created by using two neural networks, RNN-Transducer (RNN-T) and Listen, Attend and Spell (LAS), to fine-tune a high quality ASR system and train data on atypical speech patterns. In order to gather enough data to be effective, Project Euphonia offers open participation that allows people to submit their own voice profiles if they consider their speech difficult to understand (Shor & Emanuel, 2019; Tweed, 2019). As Project Euphonia was developed in collaboration with PWD, including deaf people and those with ALS, it centres the experiences and opinions of those for whom it was created and helps to bridge the gap between creator and consumer (Cattiau, 2019).

Assistive Technologies For Physical and Mobility Disabilities

As physical and mobility disabilities encapsulate such a broad range of conditions, AT must be specialised to suit the diverse needs of people living with these disabilities. Many people who lose partial or complete functioning of their limbs due to conditions such as Cerebral Palsy or ALS retain the ability to move their eyes. With the development of eye-tracking technology, people who cannot use their limbs can instead control devices with their eyes, allowing for greater inclusion in modern technology (Tobii AB, 2020). Microsoft's Eye Control and Eye Drive technologies are prime examples of innovative eye-tracking technology. Eye Control allows Windows users to access and control devices using just their eyes, and Eye Drive acts as a "virtual joystick" that gives users the ability to control other devices such as wheelchairs or remote-control cars. Microsoft also released their Eye Control application programming interfaces (APIs), "which allow any developer to harness the power of Eye Control for their own applications to add accessibility for eye-tracking" (Microsoft, 2018). Similarly, Tobii Dynavox, a subset of the Tobii Group, is a company that specialises in developing AT for people with physical and cognitive disabilities (Tobii Dynavox, 2020b). When developing their latest series of eye-tracking tablets, the I-Series, Tobii engineers sought out customer feedback and learned that their users wanted the new devices to be more durable, have more features, and better enable social inclusion. Thus, the I-Series has better sound quality, less weight, new technology that allows users to turn the device on or off with their eyes, a longer battery life, a rear-facing partner window for easier two-way conversations, and no extra parts, eliminating the need for assembly (Stern Law, PLLC, 2020; Tobii Dynavox, 2020a). Eye-tracking technology is a groundbreaking innovation that assists across a range of areas and allows PWD to have increased independence as well as higher rates of technological and social inclusion.

For able-bodied people, feeding oneself may not seem like a big deal but losing that ability can be one of the most dehumanising, embarrassing and infantilising events a person can experience. In an effort to help those in need "eat with confidence" (Liftware, 2020), Liftware creates self-stabilising and self-levelling utensils that allow people with tremors or limited arm or hand mobility to feed themselves. Liftware utensils come in two models: Liftware Steady™ and Liftware Level™. The former uses Active Tremor Cancelling technology to help people with tremors be able to feed themselves more easily and effectively. It features an electronic handle with four possible attachments: a soup spoon, an everyday spoon, a fork and a spork. A study of the Lift Steady™ device found significant improvements in users' ability to eat and transfer food as well as a 71% to 76% reduction in shaking compared to the user's hand (Pathak et al., 2013). Liftware Level™, on the other hand, is a computerised device created for individuals with restricted arm and hand mobility. It uses electronic sensors and self-levelling technology to accommodate for the user's range of motion, adjusting to any twist or bend, and is available with fork and spoon attachments (Liftware, 2020). The Liftware devices are modern and practical: they are compact and rechargeable, the attachments are dishwasher-safe, and the devices resemble normal eating utensils rather than flaunting the fact that they are AT. With each device retailing at USD $195.00 for the starter pack, Liftware technologies are excellent options for anyone wishing to regain the independence, confidence and freedom experienced in being able to feed oneself.

Prostheses, which are artificial body parts that replace a missing body part, have traditionally been limited and expensive (Zuniga et al., 2016). In 2004, then-engineering student Mat Bowtell had the opportunity to try on a bionic arm while studying at Japan's Chiba University. Bowtell was amazed by the technology but horrified at its USD $1 million price point. Thirteen years later, when he was made redundant from his job at one of Toyota's Victorian manufacturing facilities, he decided to pursue his dream of creating affordable prostheses for people in need. Thus, Free 3D Hands was born, a charity that 3D prints prostheses at a fraction of the cost of standard production and distributes them across the world. Bowtell has created a kinetic finger for AUD $0.90, a prosthetic hand for $8, and is currently developing a bionic arm that will cost between $50-$100 to produce, rivalling other prosthetic arms on the market which cost anywhere from $40,000 to $95,000 (Monash, 2019; Ben-Achour, 2020; Jeffrey, 2019). Furthermore, as accessibility is at the heart of Free 3D Hands, all of their designs are free to download, allowing anyone with a 3D printer to produce the items. The organisation uses crowdfunding to source capital and sends the finished products out to people across the globe. The prostheses are often fun colours and resemble something out of a superhero movie, making them perfect for children (Monash, 2019). Similarly, e-NABLE, a decentralised, online global community of volunteers, design and create free 3D printed bionic fingers, hands and arms for people in need and are currently developing prostheses that cater more towards adults rather than children (e-NABLE, 2020). Other organisations and companies also exist that have more of a focus on personalising prostheses to the children's specific needs and personalities. These include Limbitless in the United States, a non-profit organisation that 3D prints prosthetic arms for children, and Open Bionics, a UK-based for-profit company that creates affordable, bionic 'Hero Arms' for children over the age of 8 (Limbitless Solutions, 2019; Open Bionics, 2018). In creating prostheses that are fun and accessible rather than dull and expensive, people who need prostheses can more easily access AT and rewrite the narratives around disability as something to be celebrated rather than hidden.

CONCLUSION

The development and creation of AT has long been informed by outdated models and assumptions around disability. Inclusive, innovative and thoughtful design is imperative for AT and mainstream technology moving forward. If AT research and development does not include input from the very people that the technology is intended to assist, both the technology and its intended clients will suffer unnecessarily. In the spirit of true inclusivity in both technology and the wider world, understandings of disability and impairment must be reinvigorated and all assumptions in this area must be critically examined, deferring instead to the lived experiences of PWD. The possible applications of emerging technology to support PWD are endless, and by once again referring to the disability rights slogan of the 1990s, "Nothing About Us Without Us", we can begin to build environments and futures that provide everyone with access to the tools and resources necessary to live the life they want to live.

REFERENCES

Australian Network on Disability. (2020). *Disability statistics.* Australian Network on Disability. https://www.and.org.au/pages/disability-statistics.html

Ava. (2019, April 4). *How to use Ava? The accessibility mobile app for deaf and hard-of-hearing people* [Video]. YouTube. https://www.youtube.com/watch?v=PNoSWNUz1Z0

Ava. (2020). *Pricing.* Ava. https://www.ava.me/pricing

Barkho, G. (2019, August 12). *The wearable tech industry is expected to hit $54 billion by 2023.* Observer. https://observer.com/2019/08/wearable-tech-industry-hit-54-billion-by-2 023/

Belton, P. (2016, January 28). *The tech giving people power to deal with disability.* BBC. https://www.bbc.com/news/business-35427933

Ben-Achour, S. (2020, January 9). *Bringing down the cost of bionic arms.* Marketplace. https://www.marketplace.org/2020/01/09/bringing-down-the-cost-of-bionic-arms/

Berghs, M., Atkin, K., Graham, H., Hatton, C., & Thomas, C. (2016). Implications for public health research of models and theories of disability: A scoping study and evidence synthesis. *Public Health Research*, *4*(8), 1–166. doi:10.3310/phr04080 PMID:27512753

Brewer, E., Brueggemann, B. J., Hetrick, N., & Yergeau, M. (2012). Introduction, background, and history. In B. Brueggemann & G. L. Albrecht (Eds.), *Disability key issues and future directions: Arts and humanities* (pp. 1–62). SAGE Publications.

BrightSign. (2020). *Why BrightSign?* BrightSign. https://www.brightsignglove.com/

Buchanan, R., & Layton, N. (2019). Innovation in assistive technology: Voice of the user. *Societies (Basel, Switzerland)*, *9*(48), 1–10. doi:10.3390oc9020048

Budd, J. W. (2011). *The thought of work*. ILR Press. doi:10.7591/9780801462658

Cattiau, J. (2019, May 7). *How AI can improve products for people with impaired speech.* Google: The Keyword. https://www.blog.google/outreach-initiatives/accessibility/impaired-speech-recognition/

Charlton, J. (1998). *Nothing about us without us*. University of California Press. doi:10.1525/9780520925441

Columbia University's Zuckerman Institute. (2018). *A new party trick: A hearing aid that reads minds* [Video]. YouTube. https://www.youtube.com/watch?v=8P3QalFi73U

Degener, T. (2016). Disability in a human rights context. *Laws*, *5*(3), 1–24. doi:10.3390/laws5030035

Dillet, R. (2017, July 12). *The OrCam MyEye helps visually impaired people read and identify things.* TechCrunch. https://techcrunch.com/2017/07/12/the-orcam-myeye-helps-visually-impaired-people-read-and-identify-things/

Disabled World. (2020, March 30). *Disability statistics: Information, charts, graphs and tables.* Disabled World. https://www.disabled-world.com/disability/statistics/

Dos Santos, A. D. P., Ferrari, A. L. M., Medola, F. O., & Sandnes, F. E. (2020). Aesthetics and the perceived stigma of assistive technology for visual impairment. *Disability and Rehabilitation. Assistive Technology*, 1–7. doi:10.1080/17483107.2020.1768308 PMID:32501732

Dot Incorporation. (2020). *Dot Watch.* Dot. https://www.dotincorp.com/

e-NABLE. (2020). *FAQs.* Enabling the Future. https://enablingthefuture.org/faqs/

Essl Foundation. (2019, June 10). *Artificial Intelligence and the potential to increase inclusion, participation and independence for people with disabilities*. Panel presented at the United Nations Conference of States Parties to the Convention on the Rights of Persons with Disabilities, New York, NY. https://www.un.org/development/desa/dspd/wp-content/uploads/sites/22/2019/05/10-June-4.45_Essl-Foundation_CN_AI-and-the-Potential.pdf

Foley, A., & Ferri, B. A. (2012). Technology for people, not disabilities: Ensuring access and inclusion. *Journal of Research in Special Educational Needs*, *12*(4), 192–200.

Francisco, J. (2017, September 27). *Satya and Anu Nadella open up about their family life.* Good Housekeeping. https://www.goodhousekeeping.com/life/inspirational-stories/a46221/satya-anu-nadella-microsoft/

Friedlander, H. (1995). *The origins of Nazi genocide: From euthanasia to the final solution*. University of North Carolina Press.

Galer, D. (2012). Disabled capitalists: Exploring the intersections of disability and identity formation in the world of work. *Disability Studies Quarterly*, *32*(3). https://dsq-sds.org/article/view/3277/3122

Goggin, G., & Newell, C. (2003). *Digital disability: The social construction of disability in new media*. Rowman & Littlefield.

Government of Western Australia. (n.d.). *What is disability?* Government of Western Australia Department of Communities Disability Services. http://www.disability.wa.gov.au/understanding-disability1/understanding-disability/what-is-disability/

Hall, J. (2019, November 18). *Wearable tech is failing people with disabilities*. OneZero. https://onezero.medium.com/wearable-tech-is-failing-people-with-disabilities-673f67d65724

Hicks, S. (Speaker)., & Lenzi, M. (Host). (2019, August 8). *HealthRedesigned: Oxsight: Hacking reality for the visually impaired* [Audio podcast]. Hanno. https://hanno.co/healthredesigned/oxsight-smart-glasses/

Horsley, P. (2019, September 13). *The 3 big trends influencing assistive technology*. Remarkable. https://remarkable.org.au/trends-influencing-assistive-technology/

Hosny, A., Parmar, C., Quackenbush, J., Schwartz, L. H., & Aerts, H. J. W. L. (2018). Artificial intelligence in radiology. *Nature Reviews. Cancer*, *18*, 500–510.

Islam, K. T., Mujtaba, G., Raj, R. G., & Nweke, H. F. (2017). Handwritten digits recognition with artificial neural network. In IEEE international conference on engineering technology and technopreneurship (pp. 1–4). Kuala Lumpur, Malaysia: IEEE.

Jeffrey, C. (2019, April 4). *Open Bionics' 3D-printed 'Hero Arm' is now available in the US*. TechSpot. https://www.techspot.com/news/79515-open-bionics-3d-printed-hero-arm-now-available.html

Jewell, C. (2019, October). *BrightSign: a smart glove that gives a voice to those who cannot speak*. WIPO Magazine. https://www.wipo.int/wipo_magazine/en/2019/05/article_0005.html#:~:text=BrightSign%2C%20an%20AI%2Dbased%20smart%20glove%2C%20allows%20sign%20language,photo%3A%20Courtesy%20of%20BrightSign)

Johnstone, D. (2012). *An introduction to disability studies* (2nd ed.). Taylor and Francis.

Kafer, A. (2013). *Feminist, queer, crip*. Indiana University Press.

Knowles, K. (2018). *BrightSign: Incredible sign-to-speech smart glove startup is raising $1.4 million*. Forbes. https://www.forbes.com/sites/kittyknowles/2018/04/30/brightsign-incredible-sign-to-spee ch-smart-glove-startup-is-raising-1-million/#341ca8b773be

Liftware. (2020). *Liftware by Verify*. Verily. https://www.liftware.com/

Limbitless Solutions. (2019). *What we do*. Limbitless Solutions. https://limbitless-solutions.org/ourWork

Masala, C., & Petretto, D. R. (2008). From disablement to enablement: Conceptual models of disability in the 20th century. *Disability and Rehabilitation*, *30*(17), 1233–1244.

Matney, L. (2016, November 22). *Ava gives the deaf and hard-of-hearing a more present voice in group conversations.* TechCrunch. https://techcrunch.com/2016/11/21/ava-gives-the-hearing-impaired-a-more-present-voice-in-group-conversations/

Microsoft. (2016, March 30). *Seeing AI 2016 Prototype - A Microsoft research project* [video]. YouTube. https://www.youtube.com/watch?v=R2mC-NUAmMk

Microsoft. (2018, October 31). *How Eye Control empowers people with disabilities (audio description)* [Video]. YouTube. https://www.youtube.com/watch?v=NXOzDf26FnU

Microsoft. (2020a). *AI for accessibility.* Microsoft. https://www.microsoft.com/en-us/ai/ai-for-accessibility

Microsoft. (2020b). *Inclusive hiring for people with disabilities.* Microsoft. https://www.microsoft.com/en-us/diversity/inside-microsoft/cross-disability/hiring.aspx

Microsoft. (2020c). *Seeing AI.* Microsoft. https://www.microsoft.com/en-us/ai/seeing-ai

Mingus, M. (2011). Access intimacy: The missing link. *Leaving Evidence.* https://leavingevidence.wordpress.com/2011/05/05/access-intimacy-the-missing-link/

Monash. (2019, October 27). *3D-printed prosthetic limbs: how a social entrepreneur is changing lives.* Monash Life Magazine. https://lens.monash.edu/@monash-life/2019/10/27/1377036/3d-printed-prosthetic-limbs-changing-lives

Moon, N. W., Baker, P. M. A., & Goughnour, K. (2019). Designing wearable technologies for users with disabilities: Accessibility, usability, and connectivity factors. *Journal of Rehabilitation and Assistive Technologies Engineering*, *6*, 1–12.

Moritz, C. (2018). A giant step for spinal cord injury research. *Nature Neuroscience*, *21*, 1647–1648.

Murillo, S. (2020). *Review of the Dot, the world's first braille smartwatch.* The Chicago Lighthouse. https://chicagolighthouse.org/sandys-view/review-of-the-dot-the-worlds-first-braille-smartwatch/

Nagi, S. Z. (1991). Disability concepts revised: Implications for prevention. In A. M. Pope & A. R. Tarlov (Eds.), *Disability in America: Toward a national agenda for prevention.* National Academy Press.

Open Bionics. (2018). *FAQ.* Open Bionics. https://openbionics.com/faq/

OrCam. (2020). *OrCam MyEye: For people who are blind or visually impaired.* OrCam. https://www.orcam.com/en/myeye2/

OxSight. (2019, December 20). *OXSIGHT: Best of 2019.* OxSight. https://www.oxsight.co.uk/news/oxsight-best-of-2019/

OxSight. (2020). *OXSIGHT aids for the visually impaired.* OxSight. https://www.oxsight.co.uk/products/

Palermo, A. E., Maher, J. L., Baunsgaard, C. B., & Nash, M. S. (2017). Clinician-focused overview of bionic exoskeleton use after spinal cord injury. *Topics in Spinal Cord Injury Rehabilitation*, *23*(3), 234–244.

Pathak, A., Redmond, J. A., Allen, M., & Chou, K. L. (2013). A noninvasive handheld assistive device to accommodate essential tremor: A pilot study. *Movement Disorders*, *29*(6), 838–842.

Phillips, S.D. (2009). "There are no invalids in the USSR!": A missing Soviet chapter in the new disability history. *Disability Studies Quarterly, 29*(3).

Pullin, G. (2009). *Design meets disability.* MIT Press.

Rony, M. R. A. H., & Alam, M. M. (2019). Hand gesture recognition using image segmentation and deep neural network. *Eleventh International Conference on Machine Vision, 2018. Proceedings of the Society for Photo-Instrumentation Engineers*, *11041*, 1–6.

Russell, M., & Malhotra, R. (2002). Capitalism and disability. *Socialist Register*, *38*, 211–228.

Saravanan, R., & Sujatha, P. (2018). A state of art techniques on machine learning algorithms: A perspective of supervised learning approaches in data classification. In *2018 Second International Conference on Intelligent Computing and Control Systems (ICICCS)* (pp. 949–5). Madurai, India: Academic Press.

Schalk, S. (2017). Critical disability studies as methodology. *Lateral, 6*(1).

Serviceplan Group. (2018, April 6). *DOT: First braille smartwatch* [Video]. YouTube. https://www.youtube.com/watch?v=2ax-NG-YVGw

Sheldon, A. (2004). Changing technology. In J. Swain, S. French, C. Barnes, & C. Thomas (Eds.), *Disabling barriers: Enabling environments* (pp. 155–160). SAGE Publications.

Shor, J., & Emanuel, D. (2019, August 13). *Project Euphonia's personalized speech recognition for non-standard speech.* Google AI Blog. https://ai.googleblog.com/2019/08/project-euphonias-personalized-speech.html

Singh, S. (2019). The role of speech technology in biometrics, forensics and man-machine interface. *Iranian Journal of Electrical and Computer Engineering*, *9*(1), 281–288.

Stern Law, P. L. L. C. (2020). *Eye tracking communication for individuals that are non-verbal.* My Child at CerebalPalsy.org. https://www.cerebralpalsy.org/inspiration/technology/eye-tracking

TEDx Talks. (2017). *Don't silence your anger, find your purpose with it | Thibault Duchemin | TEDxSoMa* [video]. YouTube. https://www.youtube.com/watch?v=YPPUXDMTbuk

Tobii, A. B. (2020). *This is eye tracking.* Tobii. https://www.tobii.com/group/about/this-is-eye-tracking/

Tobii Dynavox. (2020a). *The new I-Series.* Tobii Dynavox. https://www.tobiidynavox.com/products/i-series/

Tobii Dynavox. (2020b). *Tobii Dynavox in brief.* Tobii Dynavox. https://www.tobiidynavox.com/about/about-us/tobii-dynavox-in-brief/

Tweed, A. (2019, December 12). *Ethics, machine learning and disabilities.* AbilityNet. https://abilitynet.org.uk/news-blogs/ethics-machine-learning-and-disabilities

University of California. Berkeley. (2020). *Transcence app makes group conversations between deaf and hearing possible.* Berkeley Innovators. https://innovators.berkeley.edu/spotlights/transcence-app-makes-group-conversations-between-deaf-and-hearing-possible

The Union of the Physically Impaired Against Segregation. (1976). *Fundamental principles of disability.* Retrieved from https://disability-studies.leeds.ac.uk/wp-content/uploads/sites/40/library/UPIAS-fundamental-principles.pdf

The World Bank. (2020, May 15). *Disability inclusion.* The World Bank. https://www.worldbank.org/en/topic/disability

World Health Organization. (2011). *World report on disability.* Author.

World Health Organization. (2018, May 18). *Assistive technology.* World Health Organization. https://www.who.int/news-room/fact-sheets/detail/assistive-technology

World Health Organization. (2019, October 8). *Blindness and vision impairment.* World Health Organization. https://www.who.int/news-room/fact-sheets/detail/blindness-and-visual-impairment

World Health Organization. (2020a). *Assistive devices and technologies.* World Health Organization. https://www.who.int/disabilities/technology/en/

World Health Organization. (2020b). *Deafness and hearing loss.* World Health Organization. https://www.who.int/news-room/fact-sheets/detail/deafness-and-hearing-loss

Zonshine, I. (2020, January 8). OrCam Technologies to debut AI-based, wearable hearing and optical aides. *The Jerusalem Post.* https://www.jpost.com/jpost-tech/business-and-innovation/orcam-technologies-to-debut-ai-based-wearable-hearing-and-optical-aides-613480

Zuniga, J. M., Carson, A. M., Peck, J. M., Kalina, T., Srivastava, R. M., & Peck, K. (2016). The development of a low-cost three-dimensional printed shoulder, arm, and hand prostheses for children. *Prosthetics and Orthotics International*, 1–5.

Chapter 14
Toward Eradicating All Diseases

Qing-Ping Ma
https://orcid.org/0000-0002-4379-0378
The University of Nottingham, Ningbo, China

ABSTRACT

Progress in biomedical sciences and engineering will take mankind closer to a world without disease. The objective of this chapter is to review and forecast research and technologies that contribute to eliminating all diseases. In addition to improved traditional therapeutics such as medicinal chemicals, antibiotics, vaccines, and antibodies, DNA and RNA therapeutics will play key roles in addressing genetic, infectious, and chronic diseases by decreasing disease facilitating proteins or increasing disease suppressing proteins. Nanorobots will find and treat thrombosis and vascular stenosis caused by lipid deposits in the intima of arteries, as well as remove diseased tissues and repair injured tissues. Nanorobots will be much less invasive than keyhole surgical operations, and patients need almost no time to recover from their nanorobotic procedures. Thorough understanding of regeneration and repair mechanisms of injured tissues may make it feasible to regenerate injured or aged tissues and organs and to transplant organs grown in vitro.

INTRODUCTION

A world without diseases and pain has been a dream of people in many nations. In Buddhism, there is a Western Paradise, Sukhavati, where everybody is joyous without any suffering and the land is filled with the music of birds and the tinkling of trees adorned with precious jewels and garlands of golden bells (Pas, 1995). Such a dream is a reflection of the wish to be free from diseases by people who suffer painfully from them. Economic growth and scientific progress since Industrial Revolution have dramatically increased the average living standard and improved people's health. Better nutrition following economic growth improves people's health and strengthens their body's defense against infectious diseases, better public health measures prevent or slow down the spread of diseases, and progress in medicine has made many previously incurable diseases curable. The life expectancy at birth has increased from about 35 years to around 80 years in the late 20th century in developed countries (Ma, 2017; Oeppen & Vaupel,

DOI: 10.4018/978-1-7998-6772-2.ch014

2002; Olshansky, Carnes, & Désesquelles, 2001). The increased longevity demonstrates the great success achieved by medical professionals, biomedical scientists, and biotechnology engineers.

The increase in the life expectancy at birth is attributable firstly to the control of severe infectious diseases and reduction in infant death rate, and secondly to the increased longevity due to improved living conditions and health care (Ma, 2012). In the first half of the 20th century and earlier, the increase in the life expectancy at birth is largely due to the first cause such that longevity has not increased much for people who survive to adulthood. In the second half of the 20th century, the increase in the life expectancy at birth is mainly due to the increased longevity via improved health care. The life expectancy is increasing linearly without a trend to plateau (Oeppen & Vaupel, 2002). Some life scientists think that mankind is at the brink to increase the life expectancy sharply as biomedical sciences have made sufficient progress in understanding the human aging process. It is said that the first person to live 1000 years might have been born (The Economist, 2016). With rapid progress in medicine and biotechnology in the 21st century, people might hope that modern medicine will soon eliminate all diseases and they can live in a disease-free world.

To understand whether mankind will enter a disease-free world, we need to examine, and reflect on, the functions of medicine and the causes of diseases. The two main functions of medicine are to prevent diseases and to treat diseases and injuries. What causes diseases? There are mainly nine causes: 1) biological pathogens, such as bacteria, viruses, fungi, mycoplasma, chlamydia, and parasites including protozoans; 2) defective or mutated genes that underlie hereditary diseases or cause new hereditary diseases; 3) tumors and disorders in immune systems; 4) metabolic disorders; 5) dysfunction in the central nervous system regarding psychological activities; 6) injuries caused by physical and chemical factors; 7) toxic substances; 8) factors associated with long-term overuse or improper use of tissues or organs; 9) aging. To eradicate all diseases, mankind needs to master the knowledge and expertise to prevent and cure diseases caused by all these nine types of cause. Although rapid development in our understanding of human body's physiology and pathology as well as our expanding toolkits for making new medicines bring mankind closer to the eradication of all diseases than ever before, there are still many obstacles to be overcome in the long road to a disease-free world.

The aim of this chapter is to examine the obstacles presented by the nine types of pathological cause, discuss potential solutions to those obstacles, review approaches in dealing with diseases due to different causes, and forecast technologies that might contribute to eradicating all diseases. The rest of the chapter is structured as follows: the next section focuses on diseases caused by biological pathogens, followed by a section on diseases caused by defective or mutated genes, tumors and immune disorders, and metabolic disorders; the fourth section considers mental disorders; the fifth section looks into injuries and poisoning; the sixth section examines issues associated with overuses and aging; the final section concludes the chapter.

DISEASES CAUSED BY BIOLOGICAL PATHOGENS

The battle by mankind against biological pathogens has been the most successful among all human efforts to eradicate diseases. Infectious diseases caused by bacteria and viruses used to be the most deadly and most dangerous threats to mankind. The black death or plague (1347–1351) caused by the bacterium *Yersinia pestis* (Haensch et al., 2010), killed 75-200 million people in Eurasia (DeLeo & Hinnebusch, 2005). Smallpox has also caused many deaths in different countries before Jenner's cowpox vaccine was widely used, and native residents in America has been decimated by the virus brought to the New World by the European colonialists (Patterson & Runge, 2002). The influenza pandemic in 1918 is estimated to have killed somewhere between 20 and 100 million people (Johnson & Mueller, 2002; Selgelid, 2005). Quarantine and public hygiene established during human battles against the plague and other infectious diseases, vaccines, aseptic and sterilization procedures, and antibiotics have got most infectious especially bacterial diseases under control. The sharp increase in the life expectancy at birth since 19th century can be attributed primarily to the success in controlling severe infectious diseases (Ma, 2012).

Vaccines, aseptic and sterilization, and antibiotics as well as dramatically improved living standard make bacterial infection less a serious threat to the life and health, at least in developed countries. However, there are still many obstacles to overcome before mankind can be free from diseases caused by bacteria, viruses, and parasites, etc. The main obstacles to eradication of all diseases caused by biological pathogens today are 1) drug resistance developed by bacteria and protozoans which has been exacerbated by misuse and inappropriate use of antibiotics or anti-protozoan drugs; 2) lack of effective vaccines against many such diseases or failure of vaccines to provide long-term immunity against some diseases; and 3) lack of antiviral drugs for many viral diseases (although most viral diseases are self-limiting in their course).

Superbugs, bacteria resistant to multiple antibiotics, have become a serious public health concern (Morris, Kellner, & Low, 1998). Methicillin-resistant Staphylococcus aureus (MRSA) (Enright et al., 2002), multi-drug resistant Streptococcus pneumonia (MDRSP) (Carter et al., 2005), Vancomycin-resistant Enterococcus (VRE) (Edmond et al., 1995), multidrug-resistant and extensively drug-resistant tuberculosis (MDR- and XDR-TB) (Gandhi et al., 2010), and multi-resistant Acinetobacter baumannii (MRAB) (Koeleman, Parlevliet, Dijkshoorn, Savelkoul, & Vandenbroucke-Grauls, 1997) are some examples of multi-resistant bacteria, which have caused severe clinical problems. Treating them is usually more costly because of the need to find the remaining sensitive drugs and to use multiple courses of antibiotics (McGowan Jr, 2001; Shrestha et al., 2018). Despite effective antimalarial drugs have been found a long time ago, malaria continues to be a serious health threat because of drug resistance. It was estimated that malaria caused 429,000 deaths and 212 million cases occurred globally in 2015 (Varo et al., 2018). Arteminsinin has been the most effective anti-malaria medicine, but there have been reports of resistance in some cases (Dondorp et al., 2009). Tuberculosis causes 1.4 million deaths in 2015 (National Academies of Sciences Engineering Medicine, 2017).

Although drug-resistance is a severe clinical problem, it is surmountable with new antibiotics and other new and emerging therapies as well as better management of their applications. DNA and RNA therapeutics will play key roles in addressing infectious diseases by decreasing disease facilitating proteins or increasing disease suppressing proteins. Surgical operations have been used to remove some inflamed tissues for centuries. The introduction of keyhole surgery (laparoscopy) greatly reduces the invasiveness of surgical procedures. With advances in nanotechnology and robotics, nanomachines or nanorobots might be able to perform operations to clear diseased tissues inside the body, deliver antibiotics to the inflamed tissue directly, and even apply aseptic to the focal infection inside the body.

Except the very successful vaccines against smallpox, poliomyelitis, rabies, measles and mumps, the development of antiviral drugs has lagged far behind antibacterial drugs. There are two reasons for this slow pace in developing antiviral drugs: one is the inherent difficulty in developing antiviral drugs; the other is that most viral diseases are self-limiting in their course such that supportive therapy is enough and antiviral drug is not necessary. Since the emergence of the acquired immunodeficiency syndrome (AIDS), major progress has been made in anti-viral drugs (De Clercq, 2009). With better understanding of the viral multiplication process, more antiviral drugs with high efficacy and high potency will be developed.

Since pathogenesis of diseases caused by biological pathogens has been well understood or relatively easy to be understood, especially with the help of molecular biology technologies to identify new pathogenic viruses, these diseases are most likely to be eradicated as a main threat to human health. Antibiotics, antibodies, antiviral drugs, and RNA and DNA therapies that prevent viral replication and transcription and that remove receptors for virus to enter host cells, etc. will greatly enhance the ability of mankind to treat such diseases. Effective vaccines will provide the ultimate weaponry to eradicate biological pathogens for humans, like the eradication of smallpox. The demand for these therapies provides plenty business opportunities for entrepreneurs. The website Medical Startups (https://www.medicalstartups.org/top/vaccination/) updated on January 8, 2021 its list of top 17 vaccination and vaccine development startups, the No.1 among which is Moderna Therapeutics. Moderna is pioneering the development of a new class of drugs made of messenger RNA (mRNA) and it has developed the second approved COVID-19 vaccine. With better understanding of bacteria and viruses, it is possible to use genetically modified (MO) bacteria and viruses that have the same antigenicity as the wild-type ones without any capacity to cause diseases, to compete with the wild-types and induce immunity in the human body. We may expect that existing pathogen caused diseases will be under control by 2030 and emerging infectious diseases will take less time than COVID-19 to be under control by immunization with new vaccines and treatment with specific antibodies as well as DNA or RNA therapeutics. By 2040, DNA and RNA therapeutics against several types of virus may be developed and used against viruses in the similar manner of antibiotics against bacteria, i.e. one DNA drug may inhibit several different strains of virus.

Although diseases caused by biological pathogens are the most likely to be eradicated as a main threat to human health, it is unlikely that such diseases will completely disappear. There are two reasons: 1) innocuous bacteria become pathogenic due to mutation or immunodeficiency in individual human being; and 2) viruses innocuous to humans mutate to become pathogenic or pathogenic viruses mutate to evade human immunity against them. The appearance of SARS-CoV-2 shows the potential severe threat of new pathogenic viruses (Zhu et al., 2020). Viruses were thought to have originated from genetic material derived from their respective host cells (Forterre, 2003), which implies a possibility of new viruses arising from human and animal genetic material in the future. Therefore, mankind will get all biological pathogens under control and eradicate those most severe ones in the not too distant future with available medicines as well as effective quarantine and public hygiene measures, although new pathogens especially viruses may emerge from time to time and cause new diseases to threat human health. The threat would be much less comparable with historical pandemics, because of not only advances in medicine, but also progress in information technology as well as automation which enables more strict quarantine measures.

DEFECTIVE GENES, TUMORS, IMMUNE DISORDERS AND METABOLIC DISORDERS

Typical diseases caused by defective or mutated genes are the hereditary ones, which are usually caused by a single gene defect in dominant genetic diseases or two defective alleles for a gene in recessive hereditary diseases. Tumors, immune disorders and metabolic disorders are all have genetic factors that make patients more susceptible to those diseases. Since multiple factors contribute to their pathogenesis, it becomes more difficult to elucidate the causes and processes of those diseases, so does it to prevent and cure such diseases.

Diseases Caused By Inherited Defective Genes or Mutated Genes

There are many diseases due to defective genes which are inherited from a patient's parents or mutations in patients. When the defective gene is located on the autosomes, it may cause an autosomal dominant disorder (if one defective allele can cause symptoms) or autosomal recessive disorder (if two alleles need to be defective to have symptoms). The common autosomal dominant disorders include polycystic kidney (Potter's I syndrome or Pearlman's syndrome) (Gabow, 1993), Huntington's disease (chorea) (Kostić, Dobričić, Stanković, Ralić, & Stefanova, 2014), familial hypercholesterolemia (Hobbs, Brown, & Goldstein, 1992), polydactyly (Radhakrishna, Wild, Grzeschik, & Antonarakis, 1997), syndactyly (Kozin, 2001), multiple familial polyposis of colon, multiple familial polyposis of colon, neurofibromatosis, achondroplasia, thalassemia, retinoblastoma, etc. The common autosomal recessive disorders include phenylketonuria (Blau, Van Spronsen, & Levy, 2010), alkaplonuria, albinism, congenital glucose galactose malabsorption (Wright, 1998), sickle cell anemia (Pauling, Itano, Singer, & Wells, 1949), postural proteinuria, glycogen storage disease, galactosemia, pyruvate kinase deficiency, Tay-Sachs' disease, Gaucher's disease, etc.

When the defective gene is located on the heterosomes, it may cause heterosomal dominant disorders or heterosomal recessive disorders. Common X chromosome-linked recessive inheritable diseases include red-green colorblindness (daltonism) (Wong, 2011), hemophilia (Hoyer, 1994), familial hereditary-optic atrophy, congenital night blindness (Rao, Cohen, & Oprian, 1994), angiomatosis, testicular feminization syndrome, and congenital agammaglobulinemia (Edwards, Magilavy, Cassidy, & Fox, 1978), etc. Common X chromosome-linked dominant inheritable diseases include hereditary nephritis (Grünfeld, 1985), pendulum nystagmus, orofaciodigital syndrome (Melnick & Shields, 1975), vitamin D-resistant rickets, and familial Alzheimer's disease (Sherrington et al., 1995), etc. Y chromosome-linked inheritable diseases include hypertrichosis of external auditory meatus, etc.

Before the advent of gene therapy, hereditary diseases can only be treated for preventing or alleviating symptoms. With the advances in molecular biology and gene therapy (Verma et al., 2000), all existing hereditary diseases will be readily diagnosed and those with only functional changes be cured with gene therapies in the near future. Once the disease is cured by gene therapy, the defective gene will not be passed on to the patient's offspring. For those diseases with structural changes, prenatal diagnosis and prevention are more important. Since the causes and pathogenesis of hereditary diseases will be clearer than those caused by biological pathogens, given the rapid progress in genetic engineering and gene therapy, most existing hereditary diseases are more likely to be eradicated in 15 to 20 years by gene therapy. However, defective genes need not to be inherited because mutation can occur in normal genes. As existing hereditary diseases are under control and eradicated, new ones may arise due to gene mutation from time to time.

Tumors

Tumor is a benign or malignant new growth of tissue that possesses no physiological function and arises from uncontrolled cellular proliferation. Malignant tumors, also called cancers, are one of the major killers especially in developed countries since the severe infectious diseases have been brought under control by mass vaccination, public hygiene measures and antibiotics. There are many different forms of cancer with diverse causes. Some cancers have a strong hereditary cause. Viral infection have been found to be an important factor for several cancers. Human papilloma virus (HPV) is related to cervical cancers (Li, Franceschi, Howell-Jones, Snijders, & Clifford, 2011), so is Epstein-Barr virus (EBV) to Burkitt's lymphoma and nasopharyngeal carcinoma (Zur Hausen et al., 1970). Hepatitis virus B (HBV) infection increases the probability of hepatocellular carcinoma (Anzola, 2004). Human T-cell leukemia/lymphoma virus I (HTLV-1) is among the causes of human T-cell leukemia/lymphoma (Jaffe et al., 1984). Exposure to carcinogenic substances, such as aflatoxin, arsenic trioxide, asbestos, hexavalent chromium, and dioxin, etc., and radiation may also cause cancers.

Cancer is in essence a genetic disease. It is caused by changes to genes that control how cells function, especially how they grow and divide. Genetic changes that cause cancer may arise during a person's lifetime as a result of replication errors that occur as cells divide. Viral infection and exposure to carcinogenic substances and radiation increase such errors by causing damage to DNA. Cancer-causing genetic changes can also be inherited. One function of the body's immune system is immune surveillance, which detects and destroys premalignant or malignant cells in the body. This monitoring process is performed by cells of the immune system, such as natural killer cells, cytotoxic T cells, or macrophages. When immune surveillance is weakened, it becomes more likely for a malignant cell to survive and multiply to form a tumor in the body. This is why cancer morbidity is high in elderly people.

Cancer-causing genetic changes tend to affect mainly three types of genes, proto-oncogenes, tumor suppressor genes, and DNA repair genes. Proto-oncogenes are involved in controlling normal cell growth and division. When they are altered to allow cells to grow and survive in ways not allowed under normal conditions, they may become oncogenes, i.e. cancer-causing genes (Slamon, 1987). Tumor suppressor genes are also involved in controlling cell growth and division, such that certain alterations in them may allow cells to divide in an uncontrolled manner (Weinberg, 1991). DNA repair genes are involved in fixing damaged DNA (Wood, Mitchell, Sgouros, & Lindahl, 2001). Since DNA damage is a main cause of mutation, cells with alterations in these genes tend to develop additional mutations in other genes such as proto-oncogene and tumor suppressor genes, which in turn may cause the cells to become cancerous.

Traditional cancer treatments include surgical removal of solid tumors as well as radiotherapy and chemotherapy to kill tumor cells. Later hormone therapy and immune therapy are also used to restrict tumor cell growth or kill tumor cells. Nanorobots in the future might be able to check health inside the body, identify cancer in situ, and remove cancers at the earliest time. They may also deliver anti-cancer drugs directly to the cancer tissue, so that a high drug concentration can be achieved with minimum damage to normal tissues. Progress in biotechnology will also improve the existing cancer therapies and make them safer and more effective, but gene therapy may provide the best approach to eradicate cancers. With the advances in gene therapy and better understanding of the genetic changes in various cancers, gene therapy will be able to correct the cancer-causing genetic changes and cure cancers by removing their causes in the near future. With gene therapy, it might be possible to strengthen the tumor suppression function and immune surveillance function in people who are more susceptible to cancers, whereby cancers can be prevented. All these approaches for treating cancer provide plenty business opportunities for entrepreneurs. According to the website Medical Startups (https://www.medicalstartups.org/), cancer is the No.1 disease target for medical startups. We may expect that by 2030 new cancer therapies can sharply decrease cancer mortality and by 2040 most cancers can be cured by novel as well as traditional therapies.

Immune Disorders

The body's immune system has three main functions: 1) immune surveillance, which we discussed earlier; 2) immune response, the defensive reaction of the body to microorganisms, cancer cells and antigenic substances or materials; and 3) immune memory, the function through which the immune system has a much stronger immune response to an antigen which was met before than met for the first time. In a certain sense, most diseases are an immune issue, which can be dysfunctional immune surveillance, weak immune response, unnecessarily strong immune reaction, or autoimmune reaction. Here immune disorders denote mainly three types of disease: 1) immunodeficiency, a weakened immune system; 2) allergy, hypersensitivity to certain antigens; and 3) autoimmune diseases, attacking the body's own tissue by its immune system.

Primary immunodeficiency disorders weaken the immune system and allow infections and other health problems to occur more easily (Ballow, 2002). They are usually caused by defective genes inherited from parents. They can be broadly classified into six groups based on the affected part of the immune system: 1) B cell (antibody) deficiencies; 2) T cell deficiencies; 3) combination B and T cell deficiencies; 4) defective phagocytes; 5) complement deficiencies; and 6) unknown (idiopathic). Up to now treatments are mainly to alleviate symptoms and to prevent infections. Gene therapy will be used to cure such diseases. The main obstacle for developing gene therapy to cure these diseases may be the costs, because they tend to be rare diseases and the cost per patient will be very high.

Secondary immunodeficiency disorders, which also weaken the immune system, are caused by an outside source like a toxic chemical or viral infection, or another disease. AIDS, leukemia and multiple myeloma are some examples. Chemotherapy, radiation, malnutrition, diabetes and severe burns, etc., may cause secondary immunodeficiency disorders. When the diseases causing them are cured, the secondary immunodeficiency will disappear. If genetic changes due to DNA damaged by chemicals or radiation cause immunodeficiency, gene therapy can be used to correct the defective gene.

An allergy is an immune response to a foreign substance that is not typically harmful to the body. These foreign substances are called allergens, which can be almost any substance. Some people are allergic to anti-allergy drugs. Red eyes, an itchy rash, sneezing, a runny nose, shortness of breath, or swelling are common symptoms of allergies, which include hay fever, food allergies, atopic dermatitis, allergic asthma, and anaphylaxis, etc. Allergies are common, with about 20% of people are affected by allergic rhinitis in the developed world (Wheatley & Togias, 2015), about 6% of people have at least one food allergy (Sicherer & Sampson, 2014), and about 20% have atopic dermatitis at some point in time (Nutten, 2015). Both environmental and genetic factors may underlie allergic reaction to typically innocuous substances (Kay, 2000). Anti-allergic drugs (usually anti-histamines or steroids) can prevent and relieve allergy, and vasoconstrictors can also alleviate a runny nose. If the responsible antigen is identified, it can be used for desensitization treatment (Rocklin, Sheffer, Greineder, & Melmon, 1980). When the genes responsible for allergy are identified, gene therapy may be used to make susceptible people less allergic, which requires thorough understanding of the genetic underpinning of allergies.

Autoimmune diseases are caused by attacks on auto-antigen in the body by auto-antibody produced by the immune system (Davidson & Diamond, 2001). It is still not very clear why the immune system attacks self-tissues, but autoimmune diseases may be caused by 1) loss of immune tolerance that prevents the immune system from attacking the body's self-tissues; 2) abnormal regulation of immune response, whereby low activity of suppressor T (Ts) cells and hyperactivity of helper T (Th) cells lead to a high level of autoantibody; 3) genetic factors, which may contribute to most autoimmune diseases; 4) viral or bacterial infection which induces antibody that attacks self-tissues; and 5) some chemicals.

Autoimmune diseases can be classified into organ-specific autoimmune diseases and systemic autoimmune diseases. The former includes chronic lymphocytic thyroiditis (Hashimoto's thyroiditis), hyperthyroidism, type 1 diabetes, myasthenia gravis, ulcerative colitis, pernicious anemia with chronic atrophic gastritis, Goodpasture syndrome, pemphigus vulgaris, pemphigoid, primary biliary cirrhosis, multiple sclerosis, acute idiopathic polyneuritis, etc. The latter includes systemic lupus erythematosus (SLE) (Mok & Lau, 2003), Sjögren syndrome, rheumatoid arthritis (McInnes & Schett, 2011), rheumatism, scleroderma (Gabrielli, Avvedimento, & Krieg, 2009), polyarteritis nodosa, granulomatosis with polyangiitis (GPA, also called Wegener's granulomatosis), etc. Treatments so far are to relieve symptoms and to use immunosuppressants to inhibit overactive immune system. With better understanding of the autoimmune system and its pathogenesis, more effective methods will be invented for treating and eradicating those diseases. In the near future, gene therapy may come as a method that cures common autoimmune disorders. We may expect that by 2040 mankind will have sufficient knowledge about immune disorders and gene therapy can be used to cure most of them.

Metabolic Disorders

Metabolism is the process the body uses to get or make energy from the food taken into the body. Metabolic disorders are diseases or disorders that disrupt normal metabolism, the process of converting food to energy on a cellular level. They affect the processing or transport of proteins (and amino acids), carbohydrates, or lipids. Many metabolic diseases are hereditary and discussed in the hereditary diseases section already, so this section will focus on those with less clear causes. Several common endocrine diseases cause metabolic disorders, such as diabetes, which affects the metabolism of carbohydrates, and hyperthyroidism, which increases the body's metabolic rate. Obesity may lead to type-2 diabetes (Kahn, Hull, & Utzschneider, 2006), while obesity, hypercholesterolemia or hypertriglyceridemia may cause heart and blood vessel disease, such as coronary artery disease, hypertension and atherosclerosis (Reaven, 2002). It is generally believed that high cholesterol and high blood pressure can contribute to the buildup of plaques in arteries. These plaques narrow and harden the arteries (atherosclerosis), which can lead to a heart attack or stroke.

Multiple factors contribute to the pathogenesis of metabolic disorders that are not considered congenital. Many have a genetic background of high susceptibility, but environmental factors often play an essential role in the pathogenesis. For example, obesity is the result of an interplay between genetic and environmental factors (Albuquerque, Nóbrega, Manco, & Padez, 2017). Type 1 diabetes is partly inherited with multiple genes, and its onset can be triggered by one or more environmental factors among which is viral infection (Butalia, Kaplan, Khokhar, & Rabi, 2016). Lifestyle factors play a big role in type 2 diabetes (Risérus, Willett, & Hu, 2009), but it also has genetic factors (Hanis et al., 1996). Autoimmune reaction may be responsible for some metabolic disorders. Latent autoimmune diabetes in adults (LADA) is a form of diabetes that exhibits clinical features similar to both type 1 diabetes and type 2 diabetes (Buzzetti, Zampetti, & Maddaloni, 2017). Graves' disease, which makes up about 50% to 80% of the cases of hyperthyroidism in the United States, is an autoimmune diseases (Brent, 2008).

Metabolic disorders generally have standardized therapies, but most treatments are to control or alleviate the symptoms. Many patients have to take the medicines for the rest of their life. Eradication of these diseases depends on better understanding of their pathogenesis. Improving lifestyle and diet may be important for controlling and eradicating obesity, type-2 diabetes, hypercholesterolemia, high blood pressure, cerebrovascular disease and coronary artery disease. Better medicines due to progress in biotechnology will help control and eradicate these diseases. Gene therapy may reduce the susceptibility of people with vulnerable genetic background. Coronary artery disease and cerebrovascular disease are major killers in the world (Mathers, Boerma, & Ma Fat, 2009). Biomedical engineering also produces treatments that physically remove blockages and open up the blocked artery such as heart stents (Javaid et al., 2007). Nanorobots might be able to clean plaques and blockages in blood vessels more precisely and more effectively (Cavalcanti, Shirinzadeh, Freitas Jr, & Hogg, 2007). Therapeutics for metabolic disorders present ample business opportunities for entrepreneurs. According to the website Medical Startups (https://www.medicalstartups.org/), diabetes alone is the No.3 disease target for medical startups, and heart diseases are the No.4 disease target. New therapies may enable mankind to eradicate most of the existing metabolic disorders by 2040.

MENTAL DISORDERS

A mental disorder or psychiatric disorder is a behavioral or mental pattern that causes significant distress or impairment of personal functioning (Bolton, 2008). Common mental disorders include depression, bipolar disorder, dementia, and schizophrenia and other psychoses. The causes of mental disorders are often unclear. A number of psychiatric disorders including depression, autism, schizophrenia and narcissistic personality disorder are linked to a family history or genetic background (Bienvenu & Ginsburg, 2007; Insel, 2009; Torgersen et al., 2000). Environmental influences are also important. Abuse, neglect, bullying, social stress, traumatic events, employment problems, socioeconomic inequality, and other negative or overwhelming life experiences are all risk factors (Hiday, 1995; Krabbendam & Van Os, 2005).

Mental disorders are usually defined by a combination of how a person behaves, feels, perceives, or thinks. Anxiety disorder denotes anxiety or fear that interferes with normal functioning, which is a form of mood disorder (Spitzer, Kroenke, Williams, & Löwe, 2006). Mood disorder involving unusually intense and sustained sadness, melancholia, or despair is known as major depression. Milder persistent depression is also called dysthymia. Other affective processes can also become disordered. Bipolar disorder involves abnormally "high" mood states known as mania, alternating with normal or depressed moods (Belmaker, 2004; Müller-Oerlinghausen, Berghöfer, & Bauer, 2002). Psychotic disorders such as schizophrenia are characterized by delusions, thought disorder and hallucinations, indicating that patterns of belief, language use and perception of reality can become dysregulated (Lieberman & First, 2018). Eating disorders, which include anorexia nervosa, bulimia nervosa, exercise bulimia or binge eating disorder, involves disproportionate concern in matters of food and weight (Polivy & Herman, 2002). Sleep disorders are associated with disruption to normal sleep patterns (Sateia, 2014).

Several approaches are currently in use for treating mental disorders. One option for many mental disorders is psychotherapy. Psychoanalysis, which addresses underlying psychic conflicts and defenses, has been a dominant school of psychotherapy. Psychiatric medication is a major option for many mental disorders. Antidepressants are used for the treatment of clinical depression and for anxiety and a range of other disorders (Frazer, 1997). Anxiolytics are used for anxiety disorders and related problems (Nemeroff, 2003). Mood stabilizers are used primarily in bipolar disorder (Rapoport, Basselin, Kim, & Rao, 2009). Antipsychotics are used for psychotic disorders, notably for positive symptoms in schizophrenia (Woods, 2003).

Although modern psychiatric drugs are more efficacious and safer than early drugs, to eradicate or to better control mental disorders requires a thorough understanding of their pathogenesis. Mental disorders are a diverse group of diseases, which also have diverse causes. Fully understanding their causes and pathophysiological processes will help develop effective treatments. For those underlain by genetic changes, gene therapy will be able to cure them by correcting the defective genes. With human society entering an age of affluence and robots replacing most human workers, social factors such as work pressure and low income that cause mental disorders will disappear, because people will no longer need to work for a living and society will become more harmonious. Then mental disorders caused by such social factors might be eradicated. By 2040, almost all jobs may be performed by robots and genetic factors that underlie mental disorders can also be corrected by gene therapy, morbidity of mental disorders will sharply decrease.

INJURIES AND POISONING

Injuries are damages caused by physical and chemical factors to the body. The most common ones are those caused by mechanical force, which can results cuts, bruises, laceration, contusion, sprain, bleeding, bone or skull fracture, severed or crushed digit or limb, severed or crushed nerve, internal bleeding, visceral rupture, pneumothorax, and brain concussion, etc. The impact of injury on the body depends on its modality (mechanical force by sharp or blunt objects, strain or sprain, heat, cold, radiation, and chemicals) location (skin, muscle, nerve, soft tissue, limb, head, brain or body) and magnitude (size and depth). A major injury has the potential to cause prolonged disability or death. Heat and corrosive chemicals can cause severe burns. High doses can cause visually dramatic radiation burns as well as rapid fatality through acute radiation syndrome (Bowers, 1951). Long-term exposure to low doses of radiation causes chronic radiation syndrome (Reeves & Ainsworth, 1995), which occurs at dose rates low enough to permit natural repair mechanisms to compete with the radiation damage during the exposure period. Radiation can induce cancer by damaging DNA as discussed earlier. Exposure to low temperature causes frostbites.

Poisoning is a condition or a process in which an organism becomes chemically harmed severely by a toxic substance or venom of an animal. Acute poisoning is exposure to a poison on one occasion or during a short period of time. Toxic substances in acute poisoning usually act through nervous system by blocking the neuromuscular junction or inhibiting the enzymes in mitochondria that provide energy for cellular function. Venoms of animals include four major classes: 1) necrotoxins, which kill the cells they encounter and destroy all types of tissue; 2) cytotoxins, which kill cells by acting at the level of individual cells; 3) neurotoxins, which affect nervous systems; and 4) myotoxins, which damage muscles. Venomous animals cause tens of thousands of human deaths per year (White, 2000). Chronic poisoning is long-term repeated or continuous exposure to a poison. Although symptoms do not occur immediately or after each exposure, depending on what toxic substance a person has been exposed to, it can cause severe damage to body functions.

Prevention is the most important measure to eradicate injury and poisoning. Most injuries and poisonings are caused by accidents and bites or stings from animals. Intentional injury and poisoning are social issues which need to be addressed by law and government policies. Treatment of injury depends on its severity and emergent surgical intervention is often the first step to save life for severe injuries caused by mechanical force. Nanorobots might be able to check internal injury and internal bleeding, stop internal bleeding inside the body, and deliver medicine directly to the internal injury. Infection and loss of fluid can be serious issues during wound healing. Supportive therapies are important in treating injuries. Some toxic substances have antidote, which can be used in combination with supportive therapies. Poisoning by venoms can be treated by anti-toxin as well as supportive therapies. With better understanding of regeneration and advances in gene therapy, it is hoped that damaged organs can be regenerated in the future (McCusker & Gardiner, 2011). New materials may also be used to help tissue and organ regeneration (Seal, Otero, & Panitch, 2001).

OVERUSE AND AGING

Aging is the process of becoming older. As a person grows older, most organs become less able to perform tasks than before. The decrease in organ functionality could be considered a normal process of aging, but it might arise due to overuse like many parts in a car need to be replaced after certain mileages because of wear-out. For younger people, repetitive use, vibrations, compression or long periods in a fixed position may cause an injury to part of the musculoskeletal or nervous system, i.e. a repetitive strain injury (Van Tulder, Malmivaara, & Koes, 2007). Most elderly people suffer from degenerative joints, ligaments, and tendons, which impair their mobility and ability to live independently. Some elderly have degenerative changes in the nervous system which affect their mental ability. The morbidity of dementia increases with age. With reduced function in the immune system especially immune surveillance, other diseases such as cancer and infection also increase with age.

The causes of ageing are uncertain and current theories think that the accumulation of damage may cause biological systems to fail, or that internal processes may cause aging. Caloric restriction substantially affects lifespan in many animals, including the ability to delay or prevent many age-related diseases (Guarente & Picard, 2005). There are many researches to develop anti-aging medicine (Mykytyn, 2006), but many researchers have doubts on them (Binstock, 2003; Butler et al., 2002). If the view of internal processes programming aging is correct, it would be possible to delay or stop the aging process by some medicine to switch off the process. However, there is still the issue of organ wear-out, such as degenerative joints, unless switching off the aging process will switch on the regeneration/repair process for the wear-out organs. Although the degenerative changes in organs are not acutely life-threatening, they are the main causes of the decreased quality of life for elderly people. They are also more resistant to treatment.

The objective for dealing with age-related medical issues should be to tackle degenerative changes in the nervous system and locomotor system among others. Hopefully with better understanding of degenerative pathogenesis, effective medicines for dementia and degenerative arthritis will be developed. Stem cell technology and regeneration technology may be used to rejuvenate the aged organs or to place them with new organs grown in vitro. Then elderly people will be rejuvenated in a way like an old car refurbished and equipped with a new engine. Biomedical engineering might also play a key role in addressing age-related medical conditions. Nanorobots may be used to clean cardiovascular blockage and plaques and repair damaged joints (Cavalcanti et al., 2007). Novel materials may be used to build artificial organs to replace aged and damaged organs. The ultimate goal is to modulate the aging process and maintain healthy organs for a high quality of life. Household service robots may entertain the elderly, help them carry heavy object, and provide personal care as well. Treatments and equipment for addressing age-related issues will provide plenty of business opportunities for entrepreneurs. According to the website Medical Startups (https://www.medicalstartups.org/), Alzheimer's disease alone is the

No.2 disease target for medical startups, and Parkinson's disease is the No.5 disease target for medical startups. We may expect that regeneration technologies become mature by 2040, and age-related diseases then can be treated with such therapies.

CONCLUSION

Mankind has made huge progresses in combatting diseases and pursuing a world without suffering. During this process, human beings have invented and produced more and more powerful weapons against biological pathogens as well as causes within their body. Mankind is much closer to the goal of eradicating all diseases than ever before. Although viral epidemics arise from time to time, they will no longer be as deadly as those in the history. Vaccines, quarantine and public hygiene measures, and anti-microbial drugs have enabled people to get bacterial and viral diseases under control. New DNA or RNA drugs are expected to strengthen human's toolkits in fighting against biological pathogens.

Mankind has no real treatment for hereditary diseases until the advent of genetic engineering and gene therapy. Now it is likely that in the near future hereditary diseases caused by genetic changes can be eradicated by gene therapy before those caused by biological pathogens. One reason for this is that genetic changes are relatively stable, while bacteria and viruses especially RNA viruses are prone to mutate. The immune system may no longer recognize a mutated virus, such that the mutated virus can avoid attacks by the immune system. Cancers, immune disorders and metabolic disorders all have some genetic background. Eradicating them needs not only better existing drugs, but also new approaches such as gene therapy to control the individual susceptibility.

Mental disorders need better understanding of their causes and pathophysiological processes for developing more effective therapies, especially for gene therapy which corrects defective genes that underlie some mental disorders. Injuries and poisoning need better regulations as well as better treatment. In a situation of acute injury or poisoning, time is very important for severe cases in order to rescue lives. Antidotes and antitoxins can be used to treat poisoning. Degenerative changes during aging may be normal physiological processes, but they severely decrease the quality of life for the elderly. Stem cell technology and regeneration technology may provide an opportunity to repair and rejuvenate the damaged organs of the elderly.

If we look forward to the future of medical progress, the major advances depend largely on how well we understand mechanisms of pathogenesis of diseases and how much we can intervene in their pathogenesis. It is very likely that by 2030 we will have a good understanding of the genetic underpinning of most diseases and genetic engineering will be refined sufficiently to carry out gene therapy for many diseases. By 2035, gene therapy will be widely used not only for treating hereditary diseases, but also for improving health conditions in people who are susceptible to cancer, cardiovascular diseases and metabolic disorders, etc. By 2040, with better understanding of regeneration mechanism, many age-related diseases can be treated with regeneration technologies. Although new virus and new genetic defects will arise from time to time, mankind is moving toward the goal of a disease-free world.

REFERENCES

Albuquerque, D., Nóbrega, C., Manco, L., & Padez, C. (2017). The contribution of genetics and environment to obesity. *British Medical Bulletin*, *123*(1), 159–173. doi:10.1093/bmb/ldx022 PMID:28910990

Anzola, M. (2004). Hepatocellular carcinoma: Role of hepatitis B and hepatitis C viruses proteins in hepatocarcinogenesis. *Journal of Viral Hepatitis*, *11*(5), 383–393. doi:10.1111/j.1365-2893.2004.00521.x PMID:15357643

Ballow, M. (2002). Primary immunodeficiency disorders: Antibody deficiency. *The Journal of Allergy and Clinical Immunology*, *109*(4), 581–591. doi:10.1067/mai.2002.122466 PMID:11941303

Belmaker, R. (2004). Bipolar disorder. *The New England Journal of Medicine*, *351*(5), 476–486. doi:10.1056/NEJMra035354 PMID:15282355

Bienvenu, O. J., & Ginsburg, G. S. (2007). Prevention of anxiety disorders. *International Review of Psychiatry (Abingdon, England)*, *19*(6), 647–654. doi:10.1080/09540260701797837 PMID:18092242

Binstock, R. H. (2003). The war on "anti-aging medicine". *The Gerontologist*, *43*(1), 4–14. doi:10.1093/geront/43.1.4 PMID:12604740

Blau, N., Van Spronsen, F. J., & Levy, H. L. (2010). Phenylketonuria. *Lancet*, *376*(9750), 1417–1427. doi:10.1016/S0140-6736(10)60961-0 PMID:20971365

Bolton, D. (2008). *What is mental disorder? An essay in philosophy, science, and values*. Oxford University Press. doi:10.1093/med/9780198565925.001.0001

Bowers, J. Z. (1951). Acute radiation syndrome: A problem for practitioners. *Journal of the American Medical Association*, *145*(2), 63–65. doi:10.1001/jama.1951.02920200003002 PMID:14794399

Brent, G. A. (2008). Graves' disease. *The New England Journal of Medicine*, *358*(24), 2594–2605. doi:10.1056/NEJMcp0801880 PMID:18550875

Butalia, S., Kaplan, G. G., Khokhar, B., & Rabi, D. M. (2016). Environmental risk factors and type 1 diabetes: Past, present, and future. *Canadian Journal of Diabetes*, *40*(6), 586–593. doi:10.1016/j.jcjd.2016.05.002 PMID:27545597

Butler, R. N., Fossel, M., Harman, S. M., Heward, C. B., Olshansky, S. J., Perls, T. T., ... West, M. D. (2002). Is there an antiaging medicine? *The Journals of Gerontology. Series A, Biological Sciences and Medical Sciences*, *57*(9), B333–B338. doi:10.1093/gerona/57.9.B333 PMID:12196485

Buzzetti, R., Zampetti, S., & Maddaloni, E. (2017). Adult-onset autoimmune diabetes: Current knowledge and implications for management. *Nature Reviews. Endocrinology*, *13*(11), 674–686. doi:10.1038/nrendo.2017.99 PMID:28885622

Carter, R. J., Sorenson, G., Heffernan, R., Kiehlbauch, J. A., Kornblum, J. S., Leggiadro, R. J., Nixon, L. J., Wertheim, W. A., Whitney, C. G., & Layton, M. (2005). Failure to Control an Outbreak of Multidrug-Resistant Streptococcus pneumoniae in a Long-Term–Care Facility Emergence and Ongoing Transmission of a Fluoroquinolone-Resistant Strain. *Infection Control and Hospital Epidemiology*, *26*(3), 248–255. doi:10.1086/502534 PMID:15796275

Cavalcanti, A., Shirinzadeh, B., Freitas, R. A. Jr, & Hogg, T. (2007). Nanorobot architecture for medical target identification. *Nanotechnology, 19*(1), 015103. doi:10.1088/0957-4484/19/01/015103

Davidson, A., & Diamond, B. (2001). Autoimmune diseases. *The New England Journal of Medicine, 345*(5), 340–350. doi:10.1056/NEJM200108023450506 PMID:11484692

De Clercq, E. (2009). Anti-HIV drugs: 25 compounds approved within 25 years after the discovery of HIV. *International Journal of Antimicrobial Agents, 33*(4), 307–320. doi:10.1016/j.ijantimicag.2008.10.010 PMID:19108994

DeLeo, F. R., & Hinnebusch, B. J. (2005). A plague upon the phagocytes. *Nature Medicine, 11*(9), 927–928. doi:10.1038/nm0905-927 PMID:16145573

Dondorp, A. M., Nosten, F., Yi, P., Das, D., Phyo, A. P., Tarning, J., ... Lee, S. J. (2009). Artemisinin resistance in Plasmodium falciparum malaria. *The New England Journal of Medicine, 361*(5), 455–467. doi:10.1056/NEJMoa0808859 PMID:19641202

Edmond, M. B., Ober, J. F., Weinbaum, D. L., Pfaller, M. A., Hwang, T., Sanford, M. D., & Wenzel, R. P. (1995). Vancomycin-resistant Enterococcus faecium bacteremia: Risk factors for infection. *Clinical Infectious Diseases, 20*(5), 1126–1133. doi:10.1093/clinids/20.5.1126 PMID:7619987

Edwards, N. L., Magilavy, D. B., Cassidy, J. T., & Fox, I. H. (1978). Lymphocyte ecto-5′-nucleotidase deficiency in agammaglobulinemia. *Science, 201*(4356), 628–630. doi:10.1126cience.27864 PMID:27864

Enright, M. C., Robinson, D. A., Randle, G., Feil, E. J., Grundmann, H., & Spratt, B. G. (2002). The evolutionary history of methicillin-resistant Staphylococcus aureus (MRSA). *Proceedings of the National Academy of Sciences of the United States of America, 99*(11), 7687–7692. doi:10.1073/pnas.122108599 PMID:12032344

Forterre, P. (2003). The great virus comeback—From an evolutionary perspective. *Research in Microbiology, 154*(4), 223–225. doi:10.1016/S0923-2508(03)00111-6 PMID:12901366

Frazer, A. (1997). Antidepressants. *The Journal of Clinical Psychiatry, 58*(SUPPL. 6), 9–25. PMID:9227669

Gabow, P. A. (1993). Autosomal dominant polycystic kidney disease. *The New England Journal of Medicine, 329*(5), 332–342. doi:10.1056/NEJM199307293290508 PMID:8321262

Gabrielli, A., Avvedimento, E. V., & Krieg, T. (2009). Scleroderma. *The New England Journal of Medicine, 360*(19), 1989–2003. doi:10.1056/NEJMra0806188 PMID:19420368

Gandhi, N. R., Nunn, P., Dheda, K., Schaaf, H. S., Zignol, M., Van Soolingen, D., Jensen, P., & Bayona, J. (2010). Multidrug-resistant and extensively drug-resistant tuberculosis: A threat to global control of tuberculosis. *Lancet, 375*(9728), 1830–1843. doi:10.1016/S0140-6736(10)60410-2 PMID:20488523

Grünfeld, J.-P. (1985). The clinical spectrum of hereditary nephritis. *Kidney International, 27*(1), 83–92. doi:10.1038/ki.1985.14 PMID:3981875

Guarente, L., & Picard, F. (2005). Calorie restriction—The SIR2 connection. *Cell, 120*(4), 473–482. doi:10.1016/j.cell.2005.01.029 PMID:15734680

Haensch, S., Bianucci, R., Signoli, M., Rajerison, M., Schultz, M., Kacki, S., ... Achtman, M. (2010). Distinct clones of Yersinia pestis caused the black death. *PLoS Pathogens*, *6*(10), e1001134. doi:10.1371/journal.ppat.1001134 PMID:20949072

Hanis, C., Boerwinkle, E., Chakraborty, R., Ellsworth, D., Concannon, P., Stirling, B., ... Gogolin-Ewens, K. (1996). A genome–wide search for human non–insulin–dependent (type 2) diabetes genes reveals a major susceptibility locus on chromosome 2. *Nature Genetics*, *13*(2), 161–166. doi:10.1038/ng0696-161 PMID:8640221

Hiday, V. A. (1995). The social context of mental illness and violence. *Journal of Health and Social Behavior*, *36*(2), 122–137. doi:10.2307/2137220 PMID:9113138

Hobbs, H. H., Brown, M. S., & Goldstein, J. L. (1992). Molecular genetics of the LDL receptor gene in familial hypercholesterolemia. *Human Mutation*, *1*(6), 445–466. doi:10.1002/humu.1380010602 PMID:1301956

Hoyer, L. W. (1994). Hemophilia a. *The New England Journal of Medicine*, *330*(1), 38–47. doi:10.1056/NEJM199401063300108 PMID:8259143

Insel, T. R. (2009). Disruptive insights in psychiatry: Transforming a clinical discipline. *The Journal of Clinical Investigation*, *119*(4), 700–705. doi:10.1172/JCI38832 PMID:19339761

Jaffe, E. S., Blattner, W. A., Blayney, D. W., Bunn, P. A. Jr, Cossman, J., Robert-Guroff, M., & Gallo, R. C. (1984). The pathologic spectrum of adult T-cell leukemia/lymphoma in the United States. Human T-cell leukemia/lymphoma virus-associated lymphoid malignancies. *The American Journal of Surgical Pathology*, *8*(4), 263–275. doi:10.1097/00000478-198404000-00003 PMID:6324600

Javaid, A., Steinberg, D. H., Buch, A. N., Corso, P. J., Boyce, S. W., Pinto Slottow, T. L., ... Torguson, R. (2007). Outcomes of coronary artery bypass grafting versus percutaneous coronary intervention with drug-eluting stents for patients with multivessel coronary artery disease. *Circulation*, *116*(11, supplement), I-200–I-206. doi:10.1161/CIRCULATIONAHA.106.681148 PMID:17846304

Johnson, N. P., & Mueller, J. (2002). Updating the accounts: Global mortality of the 1918-1920" Spanish" influenza pandemic. *Bulletin of the History of Medicine*, *76*(1), 105–115. doi:10.1353/bhm.2002.0022 PMID:11875246

Kahn, S. E., Hull, R. L., & Utzschneider, K. M. (2006). Mechanisms linking obesity to insulin resistance and type 2 diabetes. *Nature*, *444*(7121), 840–846. doi:10.1038/nature05482 PMID:17167471

Kay, A. (2000). Overview of 'allergy and allergic diseases: With a view to the future'. *British Medical Bulletin*, *56*(4), 843–864. doi:10.1258/0007142001903481 PMID:11359624

Koeleman, J., Parlevliet, G., Dijkshoorn, L., Savelkoul, P., & Vandenbroucke-Grauls, C. (1997). Nosocomial outbreak of multi-resistant Acinetobacter baumannii on a surgical ward: Epidemiology and risk factors for acquisition. *The Journal of Hospital Infection*, *37*(2), 113–123. doi:10.1016/S0195-6701(97)90181-X PMID:9364260

Kostić, V. S., Dobričić, V., Stanković, I., Ralić, V., & Stefanova, E. (2014). C9orf72 expansion as a possible genetic cause of Huntington disease phenocopy syndrome. *Journal of Neurology*, *261*(10), 1917–1921. doi:10.100700415-014-7430-8 PMID:25034271

Kozin, S. H. (2001). Syndactyly. *Journal of the American Society for Surgery of the Hand*, *1*(1), 1–13. doi:10.1053/jssh.2001.21778

Krabbendam, L., & Van Os, J. (2005). Schizophrenia and urbanicity: A major environmental influence—conditional on genetic risk. *Schizophrenia Bulletin*, *31*(4), 795–799. doi:10.1093chbulbi060 PMID:16150958

Li, N., Franceschi, S., Howell-Jones, R., Snijders, P. J., & Clifford, G. M. (2011). Human papillomavirus type distribution in 30,848 invasive cervical cancers worldwide: Variation by geographical region, histological type and year of publication. *International Journal of Cancer*, *128*(4), 927–935. doi:10.1002/ijc.25396 PMID:20473886

Lieberman, J. A., & First, M. B. (2018). Psychotic disorders. *The New England Journal of Medicine*, *379*(3), 270–280. doi:10.1056/NEJMra1801490 PMID:30021088

Ma, Q.-P. (2012). Government and individual strategy for dealing with the increasing post-retirement life expectancy. In A. Tavidze (Ed.), *Progress in Economics Research* (Vol. 26, pp. 1–29). Nova Science Publisher.

Ma, Q.-P. (2017). Pension provision in the robotic age. In A. Webb (Ed.), *Pensions: Global issues, perspectives and challenges* (pp. 1–38). Nova Science Publisher.

Mathers, C. D., Boerma, T., & Ma Fat, D. (2009). Global and regional causes of death. *British Medical Bulletin*, *92*(1), 7–32. doi:10.1093/bmb/ldp028 PMID:19776034

McCusker, C., & Gardiner, D. M. (2011). The axolotl model for regeneration and aging research: A mini-review. *Gerontology*, *57*(6), 565–571. doi:10.1159/000323761 PMID:21372551

McGowan, J. E. Jr. (2001). Economic impact of antimicrobial resistance. *Emerging Infectious Diseases*, *7*(2), 286–292. doi:10.3201/eid0702.010228 PMID:11294725

McInnes, I. B., & Schett, G. (2011). The pathogenesis of rheumatoid arthritis. *The New England Journal of Medicine*, *365*(23), 2205–2219. doi:10.1056/NEJMra1004965 PMID:22150039

Melnick, M., & Shields, E. D. (1975). Orofaciodigital syndrome, type I: A phenotypic and genetic analysis. *Oral Surgery, Oral Medicine, and Oral Pathology*, *40*(5), 599–610. doi:10.1016/0030-4220(75)90370-9 PMID:1059061

Mok, C., & Lau, C. (2003). Pathogenesis of systemic lupus erythematosus. *Journal of Clinical Pathology*, *56*(7), 481–490. doi:10.1136/jcp.56.7.481 PMID:12835292

Morris, A., Kellner, J. D., & Low, D. E. (1998). The superbugs: Evolution, dissemination and fitness. *Current Opinion in Microbiology*, *1*(5), 524–529. doi:10.1016/S1369-5274(98)80084-2 PMID:10066531

Müller-Oerlinghausen, B., Berghöfer, A., & Bauer, M. (2002). Bipolar disorder. *Lancet*, *359*(9302), 241–247. doi:10.1016/S0140-6736(02)07450-0 PMID:11812578

Mykytyn, C. E. (2006). Anti-aging medicine: A patient/practitioner movement to redefine aging. *Social Science & Medicine*, *62*(3), 643–653. doi:10.1016/j.socscimed.2005.06.021 PMID:16040177

National Academies of Sciences Engineering Medicine. (2017). Addressing Continuous Threats: HIV/AIDS, Tuberculosis, and Malaria. In *Global Health and the Future Role of the United States*. National Academies Press (US).

Nemeroff, C. B. (2003). Anxiolytics: Past, present, and future agents. *The Journal of Clinical Psychiatry*. PMID:12662127

Nutten, S. (2015). Atopic dermatitis: Global epidemiology and risk factors. *Annals of Nutrition & Metabolism*, *66*(Suppl. 1), 8–16. doi:10.1159/000370220 PMID:25925336

Oeppen, J., & Vaupel, J. W. (2002). *Broken limits to life expectancy*. American Association for the Advancement of Science.

Olshansky, S. J., Carnes, B. A., & Désesquelles, A. (2001). Prospects for human longevity. *Science*, *291*(5508), 1491–1492. doi:10.1126cience.291.5508.1491 PMID:11234076

Pas, J. F. (1995). *Visions of Sukhavati: Shan-tao's Commentary on the Kuan wu-liang-shou-Fo ching*. SUNY Press.

Patterson, K. B., & Runge, T. (2002). Smallpox and the native American. *The American Journal of the Medical Sciences*, *323*(4), 216–222. doi:10.1097/00000441-200204000-00009 PMID:12003378

Pauling, L., Itano, H. A., Singer, S. J., & Wells, I. C. (1949). Sickle cell anemia, a molecular disease. *Science*, *110*(2865), 543–548. doi:10.1126cience.110.2865.543 PMID:15395398

Polivy, J., & Herman, C. P. (2002). Causes of eating disorders. *Annual Review of Psychology*, *53*(1), 187–213. doi:10.1146/annurev.psych.53.100901.135103 PMID:11752484

Radhakrishna, U., Wild, A., Grzeschik, K.-H., & Antonarakis, S. E. (1997). Mutation in GLI3 in postaxial polydactyly type A. *Nature Genetics*, *17*(3), 269–271. doi:10.1038/ng1197-269 PMID:9354785

Rao, V. R., Cohen, G. B., & Oprian, D. D. (1994). Rhodopsin mutation G90D and a molecular mechanism for congenital night blindness. *Nature*, *367*(6464), 639–642. doi:10.1038/367639a0 PMID:8107847

Rapoport, S. I., Basselin, M., Kim, H.-W., & Rao, J. S. (2009). Bipolar disorder and mechanisms of action of mood stabilizers. *Brain Research. Brain Research Reviews*, *61*(2), 185–209. doi:10.1016/j.brainresrev.2009.06.003 PMID:19555719

Reaven, G. (2002). Metabolic syndrome: Pathophysiology and implications for management of cardiovascular disease. *Circulation*, *106*(3), 286–288. doi:10.1161/01.CIR.0000019884.36724.D9 PMID:12119239

Reeves, G., & Ainsworth, E. (1995). Description of the chronic radiation syndrome in humans irradiated in the former Soviet Union. *Radiation Research*, *142*(2), 242–243. doi:10.2307/3579035 PMID:7724741

Risérus, U., Willett, W. C., & Hu, F. B. (2009). Dietary fats and prevention of type 2 diabetes. *Progress in Lipid Research*, *48*(1), 44–51. doi:10.1016/j.plipres.2008.10.002 PMID:19032965

Rocklin, R. E., Sheffer, A. L., Greineder, D. K., & Melmon, K. L. (1980). Generation of antigen-specific suppressor cells during allergy desensitization. *The New England Journal of Medicine*, *302*(22), 1213–1219. doi:10.1056/NEJM198005293022201 PMID:6154241

Sateia, M. J. (2014). International classification of sleep disorders. *Chest*, *146*(5), 1387–1394. doi:10.1378/chest.14-0970 PMID:25367475

Seal, B., Otero, T., & Panitch, A. (2001). Polymeric biomaterials for tissue and organ regeneration. *Materials Science and Engineering R Reports*, *34*(4-5), 147–230. doi:10.1016/S0927-796X(01)00035-3

Selgelid, M. J. (2005). Ethics and infectious disease 1. *Bioethics*, *19*(3), 272–289. doi:10.1111/j.1467-8519.2005.00441.x PMID:16167406

Sherrington, R., Rogaev, E., Liang, Y., Rogaeva, E., Levesque, G., Ikeda, M., ... Holman, K. (1995). Cloning of a gene bearing missense mutations in early-onset familial Alzheimer's disease. *Nature*, *375*(6534), 754–760. doi:10.1038/375754a0 PMID:7596406

Shrestha, P., Cooper, B. S., Coast, J., Oppong, R., Thuy, N. D. T., Phodha, T., ... Lubell, Y. (2018). Enumerating the economic cost of antimicrobial resistance per antibiotic consumed to inform the evaluation of interventions affecting their use. *Antimicrobial Resistance and Infection Control*, *7*(1), 98. doi:10.118613756-018-0384-3 PMID:30116525

Sicherer, S. H., & Sampson, H. A. (2014). Food allergy: epidemiology, pathogenesis, diagnosis, and treatment. *Journal of Allergy and Clinical Immunology, 133*(2), 291-307.

Slamon, D. J. (1987). Proto-oncogenes and human cancers. Mass Medical Soc. doi:10.1056/NEJM198710083171509

Spitzer, R. L., Kroenke, K., Williams, J. B., & Löwe, B. (2006). A brief measure for assessing generalized anxiety disorder: The GAD-7. *Archives of Internal Medicine*, *166*(10), 1092–1097. doi:10.1001/archinte.166.10.1092 PMID:16717171

The Economist. (2016). Adding ages. *The Economist, 420*(9002), 14-16.

Torgersen, S., Lygren, S., Øien, P. A., Skre, I., Onstad, S., Edvardsen, J., Tambs, K., & Kringlen, E. (2000). A twin study of personality disorders. *Comprehensive Psychiatry*, *41*(6), 416–425. doi:10.1053/comp.2000.16560 PMID:11086146

Van Tulder, M., Malmivaara, A., & Koes, B. (2007). Repetitive strain injury. *Lancet*, *369*(9575), 1815–1822. doi:10.1016/S0140-6736(07)60820-4 PMID:17531890

Varo, R., Crowley, V. M., Sitoe, A., Madrid, L., Serghides, L., Kain, K. C., & Bassat, Q. (2018). Adjunctive therapy for severe malaria: A review and critical appraisal. *Malaria Journal*, *17*(1), 1–18. doi:10.118612936-018-2195-7 PMID:29361945

Verma, I. M., Naldini, L., Kafri, T., Miyoshi, H., Takahashi, M., Blömer, U., & Gage, F. (2000). Gene therapy: promises, problems and prospects. In *Genes and Resistance to Disease* (pp. 147–157). Springer. doi:10.1007/978-3-642-56947-0_13

Weinberg, R. A. (1991). Tumor suppressor genes. *Science*, *254*(5035), 1138–1146. doi:10.1126cience.1659741 PMID:1659741

Wheatley, L. M., & Togias, A. (2015). Allergic rhinitis. *The New England Journal of Medicine*, *372*(5), 456–463. doi:10.1056/NEJMcp1412282 PMID:25629743

White, J. (2000). Bites and stings from venomous animals: A global overview. *Therapeutic Drug Monitoring*, *22*(1), 65–68. doi:10.1097/00007691-200002000-00014 PMID:10688262

Wong, B. (2011). Color blindness. *Nature Methods, 8*(6), 441-442.

Wood, R. D., Mitchell, M., Sgouros, J., & Lindahl, T. (2001). Human DNA repair genes. *Science*, *291*(5507), 1284–1289. doi:10.1126cience.1056154 PMID:11181991

Woods, S. W. (2003). Chlorpromazine equivalent doses for the newer atypical antipsychotics. *The Journal of Clinical Psychiatry*, *64*(6), 663–667. doi:10.4088/JCP.v64n0607 PMID:12823080

Wright, E. M. (1998). I. Glucose galactose malabsorption. *American Journal of Physiology. Gastrointestinal and Liver Physiology*, *275*(5), G879–G882. doi:10.1152/ajpgi.1998.275.5.G879 PMID:9815014

Zhu, N., Zhang, D., Wang, W., Li, X., Yang, B., Song, J., . . . Lu, R. (2020). A novel coronavirus from patients with pneumonia in China, 2019. *New England Journal of Medicine*.

Zur Hausen, H., Schulte-Holthausen, H., & Klein, G. (1970). Epstein–Barr virus in Burkitt's lymphoma and nasopharyngeal carcinoma: EBV DNA in biopsies of Burkitt tumours and anaplastic carcinomas of the nasopharynx. *Nature*, *228*(5276), 1056–1058. doi:10.1038/2281056a0 PMID:4320657

KEY TERMS AND DEFINITIONS

Autosomal Dominant Disorder: A hereditary disease where one defective allele in an autosome can cause symptoms.

Autosomal Recessive Disorder: A hereditary disease where alleles in both paired autosomes need to be defective to cause symptoms.

Chlamydia: *Chlamydia trachomatis*, a gram-negative bacterium that can replicate only within a host cell.

Hereditary Disease: A disease caused by one or more abnormalities in the genome.

Immune Disorder: A dysfunction of the immune system.

Immunodeficiency: A state in which the immune system's ability to fight infectious disease and cancer is compromised or absent.

Life Expectancy at Birth: The number of years a newborn infant could expect to live if prevailing patterns of age-specific mortality rates at the time of birth stay the same throughout the infant's life.

Metabolic Disorder: A disease or disorder that disrupts normal metabolism, the process of converting food to energy on a cellular level.

MDR-TB: Multidrug-resistant tuberculosis.

MDRSP: Multi-drug resistant *Streptococcus pneumonia.*

MRAB: Multi-resistant *Acinetobacter baumannii.*

MRSA: Methicillin-resistant *Staphylococcus aureus*.

Mycoplasma: A genus of bacteria that lack a cell wall around their cell membranes.

Pathogen: Any organism that can produce disease.

Protozoan: A group of single-celled eukaryotes feeding on organic matter such as other microorganisms or organic tissues and debris, which used to be regarded as "one-celled animals."

Sukhavati: The western pure land of Amitābha in Mahayana Buddhism.

Superbug: Bacteria resistant to multiple antibiotics

VRE: Vancomycin-resistant *Enterococcus*.

X Chromosome-Linked Dominant Inheritable Diseases: Diseases in which symptoms are caused by mutations in genes on one X chromosome.

X Chromosome-Linked Recessive Inheritable Diseases: Symptoms do not appear if one X chromosome does not have the mutated genes.

XDR-TB: Extensively drug-resistant tuberculosis.

Chapter 15
Ageing:
Everybody's Doing It – Life-Affirming Technology and the Longevity Economy

Wadad Kathy Tannous
https://orcid.org/0000-0002-3153-5652
Western Sydney University, Australia

Kathleen Quilty
Western Sydney University, Australia

ABSTRACT

In 2019, the number of people in the world aged over 65 was 703 million. By 2050, this number is projected to be 1.5 billion. However, it is not only the number of older persons but the proportion that is changing. Italy was the first country to experience a demographic milestone known as the 'Historic Reversal', reporting in 1995 that the population of people aged 65 years and older outnumbered children under the age of 15 for the first time ever. By 2050, this number is expected to reach 90 countries. This social 'megatrend' collides with another shifting paradigm: technological advancement, accelerating every year at an extraordinary pace. While this trend of continuous technological evolution is not developing uniformly around the world, the scope of potential support that new and emergent technologies can bring to older adults is enormous. This chapter explores key areas of consideration to ensure that the promise of the digital future is realised for older people and that it is an enabling force, rather than a barrier to enjoying a longer, more supported later life.

INTRODUCTION

The technology that we should be looking for is not one device at a time, but a new story. The new frontier in which we are all the pioneers is longevity, to live longer and better. This is not just a social need, but an amazing business opportunity as well. This is the longevity economy.

- Dr. Joseph Coughlin[1]

DOI: 10.4018/978-1-7998-6772-2.ch015

Youth is a cultural obsession that drives the social and economic machinery of modern life. Our dominant narratives are steeped in the desire for, and the aspirational value of, youth. This is particularly potent in the context of Western consumerism, wherein 'old' is set up as the ultimate foil to 'young'. One simple way to structure something as 'good' or 'desirable' is to define and denigrate an opposite that is equally 'bad' and 'undesirable'. These polarities make a useful shorthand language for advertisers to convince or remind us to buy their goods and services: to get the good and avoid the bad. In this context, 'old' is the negative antithesis to the positive 'young': insofar as the latter marks out beauty, energy, vitality, strength, vigour and the luxury of time, the former is saddled with images of frailty, weakness, tiredness, sickness and mortality itself. But over time, these cease being convenient marketing symbols and instead become permeating features of our cultural landscape that are so ubiquitous as to seem like the truth of 'how things are'. In this youth-driven culture there are few stories more ripe for redefinition than the way we structure and relate to the idea of ageing. This is not just for reasons of factual accuracy or social justice but also for economic possibility. The need to communicate new stories about ageing and older people in general has never been greater because our flattened, one-dimensional ideas have never been more incorrect. As life expectancies increase and later life is stretched across multiple decades, the gap between these stories and reality grows. Rather than being understood as an opportunity for economic and social impact, the growing ageing population is viewed through a negative lens as a problem to be solved, or worse, avoided altogether. This chapter asks you to pull up a seat, listen and really *hear* a new story about life-affirming ageing in the era of technology.

A New Story: Life-Affirming Longevity As A Strategic Opportunity

Before diving into the specifics, let's dwell for a moment on the idea of a new story. As previously mentioned, this chapter invites readers to consider certain assumptions and associations that often go unnoticed. Many of these identifications operate on an unconscious level; few of us consciously set out to perpetuate outdated beliefs and practices around age. While many of us, including those living our later years, may resist the thought that we have internalised negative ideas around ageing and later life, these natural fears and avoidances around ageing have become enmeshed in the very fabric of our daily life. These unconscious associations can mask strategic opportunities for innovation, investment and development of technology-enabled goods and services for ageing populations. Perhaps the failings of the broader cultural imagination about the possibility of life-affirming longevity are reflected in our ability to imagine certain possibilities beyond the present; perhaps not. Wherever you are situated, the authors hope you learn something. After all, the research shows that when it comes to technology, you can most definitely teach an old dog new tricks – it might just take a different approach than with a young dog.

'Historic Reversal': The Population Ageing Megatrend

To understand the true extent of this strategic opportunity, one must first recognise the broader context in which it emerges. While people can die at any age or any stage of the lifespan, for millennia there has been a common shape to living populations. Relatively consistent rates of birth and death have ensured that a higher proportion of the population alive at any moment have been young rather than old. In essence, it has always been the case that there are more living grandchildren and great-grandchildren than grandparents and great-grandparents. For the first time in history, this population shape is fundamentally changing. Known as the 'Historic Reversal', this change has been dubbed a 'global megatrend' by the

United Nations (UN). Italy was the first country to experience this demographic milestone in 1995, reporting that the population of people aged 65 years and older outnumbered children under the age of 15. As of 2020, this phenomenon is being experienced in 34 developed countries and will likely hit 90 by 2050. By 2075, this phenomenon is expected to occur worldwide (Chamie, 2020).

While the global number of people over the age of 65 rises each year, the proportion also rises. Since 1990, the share of the global population aged over 65 has risen from 6 per cent to 9 per cent, and it is expected to reach 16 per cent, or one in six people, by 2050 (United Nations, 2019). This shift in the global demographic profile is a combination of many factors, the main being increasing longevity and declining fertility. Population ageing is also a testament to incredible advancements in public health and technology as well as a better understanding and management of social and economic factors that previously limited lifespans. The shape of life itself has been stretched to have a longer tail-end than at any other time in human history. As a species, we are now afforded the chance not just to live *longer* than our ancestors but to live *better*. Like any novel situation, these unprecedented developments require adaptation and learning, marking one of the most important challenges and opportunities facing the world in the 21st Century.

The Longevity Economy: Older Adults As An Economic Juggernaut

Although this trend is distributed differently both within and between populations and countries, the ageing population megatrend has enormous and far-reaching consequences both socially and economically. Socially, ideas about ageing and older people must be adjusted to reflect the new reality of later lives that can be both longer and healthier, with more widespread opportunities to meaningfully participate in the world in a way that honours the inherent value and dignity of people of all ages and abilities. Again, this is not *just* about what is factually accurate or socially inclusive but also what makes economic sense; and the economic consequences of population ageing are significant. Vanguard American gerontologist Robert Butler, MD coined the term "longevity revolution" to describe the incredible change that took place over his lifetime, wherein three decades were added to the average lifespan in America (Bengtson & Whittington, 2014). Yet, until recently, awareness and recognition of the attendant longevity economy has been relatively low. Dr Joseph Coughlin, founder of Massachusetts Institute of Technology's (MIT) AgeLab, describes the longevity economy as the world's fastest growing and most misunderstood market. Like a "new continent rising from the sea, populated with eager consumers", this market is growing (Coughlin, 2018). However, its consumers have not traditionally been well served by technology and, of course, have a strong opposition to being treated as 'old' where 'old' is historically negative, meaning 'a problem to be solved' (Coughlin, 2018).

The longevity economy is a testament to the immense power and potential that older adults represent as an economic force. The Economist Intelligence Unit estimated that, in 2018, the economic contribution from Americans aged 50 and over was $8.3 trillion dollars. For context, if these figures were for a country, it would only be behind the US and China in scale. This contribution is expected to grow to $28.2 trillion in 2050. In terms of direct spending, this group was responsible for 56% of every dollar spent, a figure anticipated to increase to 61% by 2050 (American Association of Retired Persons [AARP], 2019). Further, the economic impact to the US economy of age discrimination against older workers is estimated at $850 billion dollars (AARP, 2020). While the growth of the older adult market seems predictable and guaranteed, the same cannot be said for the success of the companies, products and services meant to meet the changing needs of these consumers, at least not without a radical updating of exactly what these needs are and how these specific consumers want them to be met.

Technology-Enabled Ageing: Beyond Barriers of 'Big, Beige and Boring'

The population ageing megatrend collides with another shifting paradigm as technological advancement accelerates at an extraordinary pace. As with population ageing, this is not developing uniformly around the world. The level of digital literacy, technological access and social license for using technology to support healthy ageing are highly variable by population group, country and region. This chapter focuses on the US context, however, the scope of potential support that new and emergent technologies can bring to this situation globally are enormous. The technology landscape is changing every day, and in order to ensure that the promise of this digital future is realised for older people as much as their younger counterparts, key areas of consideration must be understood and used to ensure that technology is an *enabling* rather than *disabling* force. Technology can and should be a key enabler of life-affirming ageing but this is only possible if those who design these goods and services do so in collaboration with older users and caregivers, integrating their input and experiences into every stage of the development process.

There are many compelling reasons to better understand these experiences, including the fact that, in America, the population over the age of 50 is expected to control 51% of technology spending by 2030 (AARP, n.d.). There is also a huge opportunity to support caregivers: a group whose technology needs have been largely unmet. Research indicates that by 2020, there will be 45 million unpaid caregivers in the US, providing support to over 117 million adults. Despite 71% of caregivers being interested in supportive technology, only 7% actually use it. This market opportunity is enormous, with direct spending estimated at $72 billion and opportunity costs estimated at $522 billion (AARP, 2016). Despite a range of technologies having been available, many older adults simply choose to go without supportive technologies when they have not designed with their actual needs and preferences in mind (AARP, n.d.).

Many of the traditional supportive technologies aimed at these markets are cumbersome and infantilising. Nicknamed "BBB: big, beige and boring", the proliferation of unappealing solutions to medical issues few wish to admit they experience can have serious health consequences (Coughlin & Yoquinto, 2018). For example, while one in four American adults aged over 65 fall each year at a medical cost of $50 billion dollars, the traditional personal emergency response systems that can detect falls are estimated to only be used by 4% of those in need (Orlov, 2020; Coughlin & Yoquinto, 2018). Another example is age-related hearing loss. Meaningful hearing loss is experienced by half of Americans over 60 and two-thirds of those over 70, and despite its high treatability in many cases, older adults typically delay diagnosis and treatment for years, mostly due to cost, accessibility factors, and social stigma (David et al., 2018). Recently, results from a large, ten year longitudinal study found independent associations between untreated hearing loss and a range of other adverse health outcomes in adults over 50, with 3.2 more dementia diagnoses, 3.6 more falls and 6.9 more depression diagnoses per 100 people (Deal et al., 2019). The use of hearing aids to treat hearing loss has also been found to slow cognitive decline and the onset of dementia by 75% (Maharani et al., 2018). Technology has a range of potential roles and use cases for sustainably supporting health and wellbeing in older populations but again, this needs to evolve beyond narratives that situate older people as a problem to be solved. As Dr Coughlin writes, "companies worthy of investment will not try to solve their consumers, but rather their consumers' problems—as those consumers define them. They will view older adults in a new light: as pioneers to be championed as they break new ground in our ever-evolving life span" (Coughlin, 2018).

Designing Technology: Solve Our Problems So We Can Get On With Living!

There is a vast array of age-supportive technology available today and many developments in this space have significant overlap with assistive technology for people with disabilities or those experiencing reduced independence or access to an enabling environment (Woodcock et al., 2020; Sapci & Sapci, 2019). Again, the value lies not in the ostensible 'coolness' of a product but in how and why a technology enables people to live the life that *they* want to live and continue to do more of the things *they* want to do the way *they* want to do them. While function and safety are important, so too are consumers' "goals, aspirations, aesthetic preferences, social needs and talents" (Coughlin, 2018). Unsurprisingly, people don't lose their desires, interests and preferences just because they get older or their circumstances change. There is a role for technology-enabled goods and services that support these goals as well as a whole field of study dedicated to providing evidence-based insights at the intersection of gerontology (the study of ageing and older people) and technology.

The emerging field of 'gerontechnology' is dedicated to exploring these issues through inter and multidisciplinary academic research and practice across a range of life domains. This field explores the complex matrix of social, biological, psychological, cultural and cognitive factors that contribute to the experience of ageing and asks how the technological environment can better match the mobility, housing, health, leisure, communication and work needs of older people around the world. Gerontechnology research and design uses a matrix framework to consider the role and application of technology across a range of domains of human activity, including: "Housing & Daily activities, Communication & Governance, Health & Self-esteem, Mobility & Transportation and Work & Leisure" (Kort et al., 2014, p. 328). Within these domains the goals of different technologies can be considered as "enhancement and satisfaction, prevention and engagement, compensation and assistance" as well as for care and organisation in each area of life (Kort et al., 2014, p. 328). To truly write a new story, innovations and breakthroughs in both thinking and technologies are needed across all domains. A 2019 National Science & Technology Council report outlined six core areas of function that are important to ageing populations and could benefit from support by emerging technologies: independent living, cognition, communication and social connectedness, personal mobility, transportation, and access to healthcare (Office of Science and Technology Policy, 2019). Each of these areas, the report stressed, require products and services that do not yet exist in order to solve problems that may not yet have been identified. For example, the generation of baby boomers (born 1945–65) are now ageing into later life in a rapidly changing world and have needs and demands that are different from the generations that came before them. To meet the technological innovations of the moment and the future needs of this generation is both an enormous challenge and an incredible opportunity. Success in this endeavour, first and foremost, requires a fundamental shift in ideas about what chronological ageing looks like, as so many of these ideas have traditionally been based in fear and denial.

Dr Robert Butler coined the phrase "the façade of chronological age" in order to communicate the highly variable relationship between years of life and functional capacity and to underpin his charge that ageism functions as a form of socio-cultural denial in which the projections of our collective fear about ageing, infirmity and death get transposed onto those who are marked out by time alone (Butler, 1963; Bengtson & Whittington, 2014). To succeed in the pursuit of a more clear-headed ageing agenda, technology and technology-enabled services must be understood, developed and presented in a non-condescending manner to the older person and/or their caregiver. The traditional approach, especially in the technology world, has made meeting the needs of these customers an afterthought. Indeed, fewer than half of companies globally even factor longevity and the ageing population into their strategic planning (Irving et al., 2014). Yet, recent world events, including the global coronavirus pandemic, have rapidly escalated the need to make telehealth solutions, remote monitoring and virtual opportunities for care, support and service delivery available for everyone, particularly those who are already sick or at greater risk of COVID-19 infection such as older adults. As the virus has presented a unique challenge to the global community, many more governments, industries and companies are prioritising alternatives to traditional models of providing services and support and looking to find novel ways to meet people's social, work and health needs.

Hitting Stride: Global Pandemic Shifts Norms and Policies on Technology-Enabled Care

The coronavirus pandemic has accelerated the timeline on many major changes to the surrounding policy and payment landscape for goods and services related to ageing populations (Mageit, 2020). While 2020 was already anticipated to be a year in which servicing this market became more mainstream, the pandemic has created an environment of urgency that has both upsides and downsides in relation to the realisation of this goal (Orlov, 2020; Irving et al., 2014). In the US, Medicare Advantage plans had already begun to cover some in-home technology to help older adults maintain independence, and other digital health services such as telehealth consultation and at-home monitoring have been enabled by recent government decisions brought about by the coronavirus pandemic (Centers for Medicare & Medicaid Services, 2019; U.S. Food and Drug Administration, 2020).

Another significant shift has been to bring the rate of reimbursement for telehealth services in line with that of in-person services (Donlan, 2020), a particularly important development as the demand for these services has understandably increased significantly, with telehealth claims increasing more than 4,000% in the past year (Lagasse, 2020). These changes support both supply and demand as people want to avoid unnecessary visits to hospitals and other medical settings as those settings are frequently stretched beyond their capacity (Sanger-Katz et al., 2020). In addition, as the impacts of physical distancing and reduced social contact measures are particularly acute for older people and other high-risk groups, there is an accelerated need for remote care technologies as well as those that support social connection and reduce feelings of isolation and loneliness. Loneliness is very common and impacts both physical and mental health through behavioural, psychological and physiological factors and increases the likelihood of mortality by 26 percent, making it comparable to other risk factors such as obesity or smoking (Holt-Lunstad et al., 2015).

While for some older adults technology has been lifesaving during COVID-19, in many ways the pandemic has highlighted the 'digital divide', or the gap between those who are readily included in the benefits of technology and those who are not. Unfortunately, many older adults have not been able to enjoy these supports for a range of reasons. For example, a survey of 215,000 plan-covered Californians conducted in the first few months of the pandemic showed that, despite the expanded coverage of telehealth in Medicare Advantage plans, about one-third reported that they lacked the technology required for a telehealth appointment, including the necessary devices or adequate Internet to support the calls (Graham, 2020). Even when the technology is available, a high number of older adults experience visual or hearing impairment, meaning that these technologies are not inaccessible to them. The major issue of technology accessibility is compounded during these times where there are fewer ways to access support, adaptation and training in their use. For example, a person with both visual and aural impairment who finds an app to quickly translate speech to captions may face difficulty actually reading these captions as they are oftentimes too small to be read by those with visual impairment.

The need to rapidly shift to new technologies with very little support means that those most vulnerable, including ageing populations, are most likely to be left behind. For people typically excluded by technology, the pandemic is an additional blow to the already challenging experience of navigating a world without access. However, there is an incredible range of technology-enabled goods and services that support older adults to write a new story of life-affirming ageing, pandemic or not. These successful offerings share some common elements in the approach to design, marketing and distribution that incorporate the needs and preferences of older people and show a strategic awareness of how to improve buy-in of these goods and services so that their benefits can be realised. Finally, in developing assistive technologies and solutions for ageing populations, it is vital to have an approach that treats older adults not just as a homogenous group of customers, consumers or patients but as whole persons, worthy of time, care and attention to their specific needs, desires, preferences and priorities for their life: whatever that might look like as they age.

Best-Practice: Technology-Enabled Products and Services for Life-Affirming Ageing

Supporting older adults to complete activities of daily living (ADLs) – such as eating, drinking, showering, maintaining personal hygiene, taking medications and maintaining communication and social engagement – are vital areas of focus. When these ostensibly simple and routine tasks are temporarily or permanently impacted, for people at any age, there can be a knock-on effect to physical and psychological well being. In many scenarios, the need for assistance with life-long habits and activities – or the need to adapt to changes in these domains of life – can produce feelings of reduced independence and associated shame and frustration. Yet, with appropriate support, there is an opportunity to re-orient these challenges and find solutions that enhance and enable the desired goals and behaviours of the person as well as provide caregivers with the single most important thing they want from technology: peace of mind (AARP, 2016). None of this, however, can come from a single product or service; instead, what is needed is an entire ecosystem of connected and communicative technology-enabled products and services that provide life-affirming support for later life across the whole continuum of life and care.

Thus, it is not individual technologies that are the great opportunity here but rather their potential to create a web of assistance and support that enables older adults to live their best and most fulfilled lives as they age. In this fragmented market, the value of these technologies must be communicated in the context of their use by trusted authorities and delivered through a range of channels and care-partners, with products linked to the services they enable and sold with ongoing training and support (Orlov, 2020). The following information is included to illustrate examples of age-friendly technology-enabled goods and services and demonstrate key opportunities, trends and considerations for players in this complex market space.

New Generation Wearables, Hearables, Neuroprostheses and Other Assistive Devices

Big and small technology companies are entering and disrupting the health and ageing market. In 2018, Apple released a new Apple Watch with fall detection, emergency response and ECG technologies built in and enabled by default for those over 55. This is an example of assistive technology that takes traditional emergency functions and packages them within a desirable consumer product, with recent patents indicating that more advanced sensors and functions are in development (Phelan, 2020). Other examples of this include smart jewellery products and new generation hearing aids.

As previously mentioned, the hearing technology market, while highly relevant, is underutilised by older adults. Driven by the ageing of the overall population profile, in the US alone it is estimated that the number of adults with hearing loss is expected to double from approximately 44 million in 2020 to 73.5 million by 2060 (Goman et al., 2017). Traditionally, audiology products are highly specialised with many expensive solutions that require ongoing expert management and are accessible only through medical gatekeepers. However, the new wave of consumer hearables feature more attractive design and lower prices and are sold via direct-to-consumer channels. This emerging market reflects technological breakthroughs with AI-enabled hearing aids featuring integrations with other personal devices, direct to device in-ear audio streaming, real-time gait monitoring, location-tracking and automatic fall-detection in addition to a smart noise cancellation capability that detects whose voice the listener is trying to focus on in crowded environments (Orlov, 2020).

Another new assistive technology for older people is the Tango Belt: a wearable, automatically-inflating personal air-bag to prevent broken hips from falls (Tango Technologies, 2019). This technology is a life-saving invention as falls and associated breaks are the leading cause of fatalities and other injuries for older Americans, causing one emergency room visit every eleven seconds and one death every nineteen minutes (National Council On Aging [NCOA], n.d.). Another innovation is a rechargeable neuroprosthesis that can replace lost foot pressure sensation from peripheral neuropathy (the most common complication of diabetes worldwide, with 30 million sufferers in the US alone) and help to restore balance and the ability to walk (RxFunction, Inc., 2020). Other potentially game-changing innovations are in development for managing kidney disease and failure, including lower-cost, environmentally-friendly portable home dialysis machines which can revolutionise the long-term, expensive and inconvenient treatment for kidney issues associated with many diseases experienced by older adults such as diabetes, hypertension and obesity (Ahmed, 2020; Huff, 2020). For telehealth, stimulation and social engagement during the COVID-19 pandemic and beyond, there are numerous devices that can be dispatched and used with little to no set-up or training. These include the GrandPads tablet and the iN2L (It's Never Too Late) technology which are specialised to cater for people aged 75+ and those with dementia and come pre-loaded with an accessible interface, engaging senior-friendly apps and content to help reduce social isolation as well as simple video chat, photo and video sharing and messaging functions (GrandPad, 2020; iN2L, 2020).

Robots As Workers, Helpers, Companions and Pets

A systematic review of the literature on robots for supporting ageing populations identified nine types of robot, "companion, manipulator service, telepresence, rehabilitation, health monitoring, reminder, entertainment, domestic and fall prevention/detection", that could support older adults with "social isolation, dependent living, physical or cognitive impairment, mobility problems, poor health monitoring, lack of recreation, memory problems and fall problems" (Shishehgar et al., 2018, p. 1). Aged care robots are particularly well-developed in Japan, in part due to an overall cultural acceptance of technology as a part of daily life and the fact that Japan, along with Italy, is one of the countries on the leading edge of the Historic Reversal trend. Due to workforce shortages, robots frequently support Japanese care staff in the delivery of culturally-relevant games, music and social activities, including natural language conversation with residents based on their personal history. Supportive exoskeletons can also help staff safely lift and transfer older patients without injury. Elderly residents are reported to enjoy bonding and companionship with robots such as an adorable plush seal pup and an infant-style robot that is popular with otherwise non-verbal residents (Lufkin, 2020). There are also robots that support people to manage their medications as well as companion 'pet' robotic plush dogs and cats that provide the comfort and companionship of live pets without the need for mess, care and maintenance (Orlov, 2020; Ageless Innovations LLC, 2018).

Smart Home and Textile Technology to Support Independence and Safety at Home

The last few decades has seen the institutional model shift in favour of an 'ageing in place' approach, increasing choice and control for older adults. Under this new framework, the priority is successful ageing in the preferred location of the individual, most often one's existing home. People aged 65 and over are statistically much more likely to live alone, which can make ageing in place a choice between independence and safety. In America, the majority of older adults live in their own homes, and whether due to personal preference or financial limitations, 76 percent would like to, or will have to, stay there as they age (Joint Center for Housing Studies of Harvard University, 2018). Technologies for 'smart' homes and appliances can improve safety by regulating water temperature in showers, managing fire risk in cooking and detecting wandering behaviours. For example, a current joint-initiative in Canada is a 'smart' apartment living lab built with non-invasive, integrated sensors throughout that detect a range of potential stimuli, such as a pot boiling dry on the stove, as well as pressure-sensitive mattress pads that monitor a person's health during sleep (Age-Well Nce, 2020). At Stanford University, a 'smart' toilet has been developed that can remotely monitor for health, hydration and diseases (Stanford Medicine, 2020). Other groundbreaking innovations include 'smart textile' technologies such as bed sheets embedded with biosensors that monitor pressure distribution to prevent bedsores; socks that monitor vital signs or sense patterns and changes in pressure and temperature to signal ulcers that can cause amputation; and bandages that can monitor wounds, detect infection and automatically distribute medication (Reyzelman et al., 2018; Russey, 2019).

One important feature for any 'smart home' technology is a voice-first interface. A voice-first interface ensures that the primary functions of 'smart home' technologies are accessible through speech, with these interfaces being used by many leading players such as Apple, Google, Samsung (Bajorek, 2018). For many older adults, interfaces that use voice as the primary mode of engagement can provide a way into technologies that might otherwise be inaccessible. Speaking is a natural behaviour and being able to speak to one's device can bypass issues of fine motor control, vision impairment and general unfamiliarity with gestures needed to navigate screens such as swiping, pinching, tapping and scrolling. In a pilot study, Amazon Alexa facilitated independence and wellbeing for residents aged 79-100 in an aged care community due to its voice-first interface, with residents able to initiate social engagement, hear about the weather and control their environment using only their voice (Front Porch Communities and Services, 2018). Smart technologies that are developed with an emphasis on independence and autonomy help to support older populations as they age and ensure they are included rather than left behind when it comes to technological engagement.

VR & AI Offer Monitoring, Training, Recreation and Therapeutic Applications

Virtual reality (VR) technologies can be used for a range of applications and have particular, unique benefits for older people. For example, when given the opportunity to visit bucket-list destinations together through VR headsets, groups of aged care residents saw significant benefits to their psychological and social well-being (Rendever, Inc., 2020). Similarly, there are incredible opportunities to support cognition, memory and mood issues related to Alzheimers and other dementias through VR-driven reminiscence therapy, in which people can visit locations of personal significance or soothing locations in nature without mobility limitations or safety issues (Rendever, Inc., 2020; Rehman, 2019). VR can also be useful for training caregivers to better understand and empathise with the experiences of older adults. One example of this is Embodied Labs, which uses "immersive, first-person VR experiences" to provide training and insight into the experiences of common medical conditions (Embodied Labs, 2020).

Another beneficial innovation is artificial Intelligence (AI), which can take a range of forms but includes the use of data-driven technologies, natural language processing approaches, machine learning and neural networks. One novel use of these technologies is in early detection of dementia by tracking changes in memory, cognition and speech patterns from established baselines to recognise early symptoms (Orimaye et al., 2017). The pattern recognition and analytics enabled by these technologies can "extend and augment simple monitoring" by using dashboards and predictive algorithms for flagging issues for caregiver follow-up (Orlov, 2020). Alzheimers and dementia are anticipated to impact over 150 million people globally by 2050 and, in 2015, had an estimated total global societal cost of USD $818 billion, equivalent to 1.1% of global GDP (World Health Organization, 2019). Expanding the use of VR and AI technologies to assist older populations, specifically those with Alzheimers and dementia, is an incredible opportunity both for companies and consumers.

CONCLUSION

The role of technology for life-affirming ageing is less about fragmented, individual instances of technology products and technological breakthroughs and more about the need for an entirely new and more affirming story about older people and ageing. Technology in and of itself is not and never will be the catch-all solution to the problems and challenges of the present. This is in part because all technology is a direct product of the humans who design, develop and distribute it, and is therefore laden with their assumptions, biases and beliefs, not just in the technology itself but also in the processes of conceptualisation, design and creation. Technological developments reflect both our capacity for breakthroughs and the limitations of our current thinking. As the reach of technology and the length of lifespans stretch, there is a crucial need to continue questioning, observing and refining what it is that goes into the making of these products and services. This helps ensure that the technology produced is not just inventor-centric but actually functional, fit-for-purpose, accessible and useful as a tool and enabler for longer, better lives and more affirming life experiences. The real story is not just case-by-case change but instead a wholesale shift to the dominant narrative of who technology is by and for and what age has to do with it. As an enabler for life-affirming ageing, the possibility and potential for technology is infinite, limited only by our individual and cultural imagination. Indeed, the new story needed is a larger one in which the default is to always question: how can *every single example* of a technology-enabled product or service be adjusted or packaged to better support the needs of people as they age: as they navigate temporary or permanent changes in their abilities or circumstances and as they experience illness or disability? Indeed, there are incredible and important real-world examples of this in action, but what if the core of important technological breakthroughs resides in us, in each individual? In the technology of our hearts and minds, our ideas, our imaginations, our stories? In our ability to see beyond the present moment and collaborate with those who have come before us, for the benefit of those that will come after? After all, it is so easy to forget when talking about technology that it is, first and foremost, human beings from which everything else flows.

REFERENCES

Age-Well Nce. (2020, August 4). *Canada, we need to talk about COVID-19, pandemics, technology and the Way forward.* YouAreUNLTD. Retrieved from https://www.youareunltd.com/2020/08/04/canada-we-need-to-talk-about-covid-19-pande mics-technology-and-moving-forward/

Ageless Innovations, L. L. C. (2018). *Joy for All: Companion pets.* Retrieved from https://joyforall.com/

Ahmed, N. (2020, March 18). *Portable dialysis machines, artificial kidneys to revolutionize treatment.* HealthUnits. Retrieved from https://healthunits.com/news/portable-dialysis-machines-artificial-kidneys-to-revolutioniz e-treatment/

American Association of Retired Persons. (2016). *Caregivers & technology: What they want and need.* https://www.aarp.org/content/dam/aarp/research/surveys_statistics/ltc/2018/caregivers-t echnology-needs.doi.10.26419-2Fres.00191.002.pdf

American Association of Retired Persons. (2019). *The Longevity Economy® outlook: How people age 50 and older are fueling economic growth, stimulating jobs, and creating opportunities for all.* Retrieved from https://www.aarp.org/content/dam/aarp/research/surveys_statistics/econ/2019/longevity-economy-outlook.doi.10.26419-2Fint.00042.001.pdf

American Association of Retired Persons. (2020). *The economic impact of age discrimination: How discriminating against older workers could cost the U.S. economy $850 billion.* Retrieved from https://www.aarp.org/content/dam/aarp/research/surveys_statistics/econ/2020/impac t-of-age-discrimination.doi.10.26419-2Fint.00042.003.pdf

American Association of Retired Persons. (n.d.). *The Longevity Economy® outlook.* AARP. Retrieved from https://longevityeconomy.aarp.org/?CMP=RDRCT-PRI-OTHER-LONGECON-111319

Bajorek, J. P. (2018, October 22). *Voice first verses the multimodal user interfaces of the future.* UXMatters. Retrieved from https://www.uxmatters.com/mt/archives/2018/10/voice-first-versus-the-multimodal-us er-interfaces-of-the-future.php

Bengtson, V. L., & Whittington, F. J. (2014). From ageism to the longevity revolution: Robert Butler, pioneer. *The Gerontologist, 54*(6), 1064–1069. doi:10.1093/geront/gnu100

Butler, R. N. (1963). The façade of chronological age: An interpretative summary. *The American Journal of Psychiatry, 199*(8), 19–28. doi:10.1176/ajp.119.8.721 PMID:14017385

Centers for Medicare & Medicaid Services. (2019, April 5). *CMS finalizes policies to bring innovative telehealth benefit to Medicare Advantage.* CMS. Retrieved from https://www.cms.gov/newsroom/press-releases/cms-finalizes-policies-bring-innovative-te

Chamie, J. (2020, February 11). World Population: 2020 overview. *YaleGlobal Online*. Retrieved from https://yaleglobal.yale.edu/content/world-population-2020-overview

Coughlin, J. (2018, December 7). *The longevity economy: Why seniors are a fast-growing emerging market.* Barron's. Retrieved from https://www.barrons.com/articles/seniors-have-spending-power-especially-as-life-spans-lengthen-1544226021

Coughlin, J. F., & Yoquinto, L. (2018, October 15). Technology for older people doesn't have to be ugly. *Wall Street Journal.*

David, D., Zoizner, G., & Werner, P. (2018). Self-stigma and age-related hearing loss: A qualitative study of stigma formation and dimensions. *American Journal of Audiology*, *27*(1), 126–136.

Deal, J. A., Reed, N. S., Kravetz, A. D., Weinreich, H., Yeh, C., Lin, F. R., & Altan, A. (2019). Incident hearing loss and comorbidity: A longitudinal administrative claims study. *JAMA Otolaryngology-Head & Neck Surgery*, *145*(1), 36–43.

Donlan, A. (2020, March 17). *Trump administration announces more relaxed restrictions on telehealth services during COVID-19 emergency.* Home Health Care News. Retrieved from https://homehealthcarenews.com/2020/03/trump-administration-announces-more-relaxed-restrictions-on-telehealth-services-during-covid-19-emergency/

Embodied Labs. (2020). *Content library.* Embodied Labs. Retrieved from https://embodiedlabs.com/content-library/

Front Porch Communities and Services. (2018, July 23). *Front Porch Center for Innovation and Wellbeing releases results of Amazon Alexa pilot project in San Diego retirement community.* Front Porch Center for Innovation and Wellbeing. Retrieved from https://fpciw.org/news/front-porch-center-for-innovation-and-wellbeing-releases-results-of-amazon-alexa-pilot-project-in-san-diego-retirement-community/

Goman, A. M., Reed, N. S., & Lin, F. R. (2017). Addressing estimated hearing loss in adults in 2060. *JAMA Otolaryngology-Head & Neck Surgery*, *143*(7), 733–734.

Graham, J. (2020, July 23). *Seniors who struggle with technology face telehealth challenges and social isolation.* CNN. Retrieved from https://edition.cnn.com/2020/07/23/health/seniors-technology-telehealth-wellness-partner/index.html

GrandPad. (2020). *GrandPad.* https://www.grandpad.net/

Holt-Lunstad, J., Smith, T. B., Baker, M., Harris, T., & Stephenson, D. (2015). Loneliness and social isolation as risk factors for mortality: A meta-analytic review. *Perspectives on Psychological Science*, *10*(2), 227–237.

Huff, C. (2020, March 11). How artificial kidneys and miniaturized dialysis could save millions of lives. *Nature*, *579*, 186–188. https://www.nature.com/articles/d41586-020-00671-8

iN2L. (2020). *iN2L Tablet.* Retrieved from https://in2l.com/our-solution/in2l-tablet/

Irving, P., Beamish, R., & Burstein, A. (2014). *Silver to Gold: The business of aging.* Milken Institute. Retrieved from https://milkeninstitute.org/sites/default/files/reports-pdf/FINAL-Silver-to-Gold-0226.pdf

Joint Center for Housing Studies of Harvard University. (2018). *Housing America's older adults, 2018.* Retrieved from https://www.jchs.harvard.edu/sites/default/files/Harvard_JCHS_Housing_Americas_Older_Adults_2018_1.pdf

Kort, H. S. M., Woolrych, R., & van Bronswijk, J. E. M. H. (2014). Applying the gerontechnology matrix for research involving ageing adults. In L. Pecchia, L.L. Chen, C. Nugent, & J. Bravo (Eds.), *Lecture Notes in Computer Science: Vol. 8868. Ambient assisted living and daily activities: 6th international work-conference, IWAAL 2014, Belfast, UK, December 2-5, 2014, proceedings.* Springer.

Lagasse, J. (2020, June 3). *Telehealth claim lines increased more than 4,000% in the past year.* Healthcare Finance. Retrieved from https://www.healthcarefinancenews.com/news/telehealth-claim-lines-increased-more-4000-past-year

Lufkin, B. (2020, February 7). What the world can learn from Japan's robots. *BBC.* Retrieved from https://www.bbc.com/worklife/article/20200205-what-the-world-can-learn-from-japans-robots

Mageit, S. (2020, July 23). *Technology to the rescue - COVID-19 as an effective accelerator of digital health adoption.* Healthcare IT News. Retrieved from https://www.healthcareitnews.com/news/europe/technology-rescue-covid-19-effective-accelerator-digital-health-adoption

Maharani, A., Dawes, P., Nazroo, J., Tampubolon, G., & Pendleton, N. (2018). Longitudinal relationship between hearing aid use and cognitive function in older Americans. *Journal of the American Geriatrics Society*, *66*(6), 1130–1136.

National Council On Aging. (n.d.). *Fall management facts.* National Council on Aging. Retrieved from https://www.ncoa.org/news/resources-for-reporters/get-the-facts/falls-prevention-facts/

Office of Science and Technology Policy. (2019, March). *Emerging technologies to support an aging population.* https://www.whitehouse.gov/wp-content/uploads/2019/03/Emerging-Tech-to-Support-Aging-2019.pdf

Orimaye, S. O., Wong, J. S., Golden, K. J., Wong, C. P., & Soyiri, I. N. (2017). Predicting probable Alzheimer's disease using linguistic deficits and biomarkers. *BMC Bioinformatics*, *18*(34), 1–13.

Orlov, L. (2020, March). *Technology for aging: 2020 market overview.* Aging and Health Technology Watch. Retrieved from https://www.ageinplacetech.com/files/aip/Market%20Overview%202020%20Final-March-2020.pdf

Phelan, D. (2020, August 11). Apple Watch Series 6: Patents reveal game-changing hardware upgrades. *Forbes.* Retrieved from https://www.forbes.com/sites/davidphelan/2020/08/11/apple-watch-series-6-patents-reveals-game-changing-hardware-upgrades/#5a364f5354e6

Rehman, A. (2019, July 26). *How virtual reality can transform dementia care.* The Academy of International Extended Reality. Retrieved from https://aixr.org/insights/how-virtual-reality-can-transform-dementia-care/

Rendever, Inc. (2020). *Rendever.* https://rendever.com/

Reyzelman, A., Koelewyn, K., Murphy, M., Shen, X., Pillai, R., Fu, J., Scholten, H. J., & Ma, R. (2018). Continuous temperature-monitoring socks for home use in patients with diabetes: Observational study. *Journal of Medical Internet Research, 20*(12), 1–14.

Russey, C. (2019, July 30). *How smart bandages are transforming medicine and healthcare.* Wearable Technologies. Retrieved from https://www.wearable-technologies.com/2019/07/how-smart-bandages-are-transforming-medicine-and-healthcare/

RxFunction, Inc. (2020). *RxFunction.* https://rxfunction.com/

Sanger-Katz, M., Kliff, S., & Parlapiano, A. (2020, March 17). These places could run out of hospital beds as coronavirus spreads. *The New York Times.* Retrieved from https://www.nytimes.com/interactive/2020/03/17/upshot/hospital-bed-shortages-coronavirus.html

Sapci, A. H., & Sapci, H. A. (2019). Innovative assisted living tools, remote monitoring technologies, artificial intelligence-driven solutions, and robotic systems for aging societies: Systematic review. *JMIR Aging, 2*(2), 1–16.

Shishehgar, M., Kerr, D., & Blake, J. (2018). A systematic review of research into how robotic technology can help older people. *Smart Health, 7-8*, 1–18.

Stanford Medicine. (2020). *'Smart toilet' monitors for signs of disease.* Stanford Medicine News Center. Retrieved from http://med.stanford.edu/news/all-news/2020/04/smart-toilet-monitors-for-signs-of-disease.html

Tango Technologies. (2019). *Tango: Life in Motion.* https://www.tangobelt.com/

United Nations. (2019). *World population ageing 2019: Highlights.* (Report No. ST/ESA/SER.A/430). New York, NY: United Nations. Retrieved from https://www.un.org/en/development/desa/population/publications/pdf/ageing/WorldPopulationAgeing2019-Highlights.pdf

U.S. Food and Drug Administration. (2020, June). *Enforcement policy for non-invasive remote monitoring devices used to support patient monitoring during the coronavirus disease 2019 (COVID-19) public health emergency (revised): Guidance for Food and Drug Administration staff.* Retrieved from https://www.fda.gov/media/136290/download

Woodcock, A., Moody, L., McDonagh, D., Jain, A., & Jain, L. C. (Eds.). (2020). *Design of assistive technology for ageing populations.* Springer.

World Health Organization. (2019, September 19). *Dementia.* World Health Organization. https://www.who.int/en/news-room/fact-sheets/detail/dementia

ENDNOTE

1 Coughlin, J. (2018, December 7). *The longevity economy: Why seniors are a fast-growing emerging market.* Barron's. Retrieved from https://www.barrons.com/articles/seniors-have-spending-power-especially-as-life-spans-lengthen-1544226021.

Chapter 16
Never Say Die:
The Techno-Politics of Radical Life Extension

Amy Lynn Fletcher
Futurist, A Fletcher Strategies LLC, USA

ABSTRACT

This chapter evaluates the longevity industry and the claims being made about the viability of radical life extension. It distinguishes between the established science of human aging and the emerging promissory economy being built upon visions of extending life well beyond the theoretical optimal human lifespan of approximately 115 years. With reference to examples such as cryo-preservation and cellular interventions, this wide-ranging exploration of the contemporary Western obsession with prolonging vibrant life focuses on the techno-political dimensions of "disrupting death" and the complex relationship between hope and hype that sustains the immortality imaginary.

INTRODUCTION

The Greek poet Homer (c. 800 BCE—700 BCE) often referred to old age as a condition that is hateful, miserable, and harsh. In the Western tradition, this fear and resentment of aging and death threads through the culture in guises ranging from Mary Shelley's *Frankenstein, or The Modern Prometheus* (1818) to inventor Ray Kurzweil's professed belief that digital immortality might one day be achieved by uploading the brain to networked computers.

As venture capital begins to move into the volatile sector of radical age extension, it is important to consider the claims being made about the viability of anti-aging start-ups and initiatives. This chapter introduces the emerging techno-politics of "disrupting death" and focuses on the complex relationship between bio-hope and bio-hype that is shaping the future of aging. The next section begins with a brief discussion of the major scientific and ethical objections to the pursuit of radical aging and immortality. The chapter then considers prominent stakeholders in the anti-aging sector, the major research pathways being developed and supported, and the claims being made about what is possible with respect

DOI: 10.4018/978-1-7998-6772-2.ch016

to extending the human lifespan. The main argument of this chapter is that the pursuit of anti-aging constitutes a socio-technical imaginary of immortality that envisions human ingenuity breaking through the ultimate boundary—death—and emerging into health and life on the other side through the power of science and technology.

THE LAST EXIT: SCIENCE AND IMMORTALITY

The anti-aging sector can be divided into four distinct subsectors: 1) established researchers who want to make the normal lifespan of approximately eighty-five years healthier; 2) the moderate life extensionists, who hope to push the average human lifespan to one hundred and twenty-five years; 3) the radical age extensionists, led by figures such as Aubrey de Grey, who envision future lifespans of at least three hundred years; and 4) the immortalists (Stambler 2014), who see in nanotechnology, biotechnology, and/or digital technology the skeleton keys that will eventually unlock the secret of life itself. Key to the growth of the anti-aging industry is a rhetorical and scientific shift from seeing aging and death as intrinsic to the human condition to the notion that "death is a curable disease" (Istvan 2019) and thus vulnerable to advances in science and technology. This distinction is not simply semantic; rather, aging or death as a curable disease metaphor unmoors these conditions from the realms of theology and philosophy and places them squarely within an engineering framework. From this vantage point, aging becomes amenable to human ingenuity and biotechnological interventions that reverse time and deterioration. Via the systematic reverse engineering of natural processes, the hope is that we can potentially unlock the secrets of aging and thus find a way to forestall a fate that once seemed inexorable.

Synthetic biology explicitly conceptualizes biology as a subfield of engineering, but the basic idea has a long lineage in the Western tradition. Since the late nineteenth century, many scientists have sought to elucidate the mechanistic basis of life and thus render natural processes amenable to human intervention. In *Artificial Parthenogenesis and Fertilization* (Loeb 1913), for example, biologist and physiologist Jacques Loeb documents experiments in which he manipulated the eggs of sea urchins *in vitro* to develop embryonically without sperm. Throughout his career Loeb focused on the possibilities of controlling biological processes in the lab, writing to his colleague, physicist Ernst Mach, in 1890, "the idea is hovering before me that man himself can act as a creator, even in living Nature, forming it eventually according to his will. Man can at least succeed in a technology of living substance" (Pauly pp. 50-51). Professor Victor de Lorenzo's assertion that "what you want [in biology] is to create or recreate systems that have some properties of life from engineering principles" (Briethaupt 2006, p. 21), or entrepreneur J. Craig Venter's belief that "biology in the twenty-first century has become an information science" (Venter p. 5), represent the contemporary ascendance of a paradigm that began to take shape at least a century earlier.

Radical age extension and immortality research remain quite controversial within science and it is likely that some serious biological engineers would be reluctant to entertain these notions publicly. Nevertheless, the essential idea threads through the history of biology as a counterpoint to less ambitious claims and, as synthetic biology advances, more stakeholders are emerging who purport to believe that transcending aging and possibly even death will become possible. In the words of Michael Solana, Vice President of the Founders Fund, "we're quickly approaching a world of biological malleability, in which the natural world and the organisms within it, including ourselves, can be whatever we want them to be" (You et al. 2016). This optimism persists despite both physics and biology arguably putting a hard stop to more extreme visions of life extension. To date, Jeanne Louise Calment, born in Arles, France in 1875, is believed to have lived the longest known human lifespan, reaching the advance age of one hundred and twenty-two years (and one hundred and sixty-four days), though this record is shrouded in controversy. The question of whether it is feasible to push the human life span beyond approximately one hundred and twenty-five years is an ongoing dispute within several disciplines (Scharping 2016).

Perhaps the bedrock scientific objection to physical immortality and even radical age extension is the second law of thermodynamics, which holds that entropy always increases in any natural process, including aging. Consequently, as Jonathan Keeney puts it, "the components of our system want to tend towards the lowest energy state, just as the components of a rocket want to tend towards the surface of the Earth. In order to resist that, both require energy use, but energy use wears down those very components. If we want to stay alive, we have to use the components in our system, which means we must age" (Keeney 2015). The second law thermodynamics also predicts that the universe itself must eventually end, though, as this process is measured in billions of years, this fact does not deter the most ardent proponents of human immortality from speculation. From a biological standpoint, we also need to acknowledge the salience of the Hayflick limit. The Hayflick limit (named by Frank Macfarlane Burnet in 1974, referring to influential work on cellular aging done by Leonard Hayflick in the early 1960s) holds that a normal human cell can replicate a finite number of times. For humans, the limit is estimated between forty and sixty cell divisions, at which point apoptosis—programmed cell death—occurs. Hayflick's work also informs much contemporary research on whether and how cancer results from abnormal cell replication and is the underlying basis for subsequent work on the role of telomere attrition, for which Elizabeth Blackburn, Carol Grieder, and John Szostak jointly won the 2009 Nobel Prize in Physiology or Medicine. In 1974, Hayflick co-founded the National Institute on Aging (part of the National Institutes of Health) and in 1982 became the president of the Gerontological Society of America.

Despite these objections, science proceeds in sometimes surprising ways and age extension enthusiasts often point to the fact that the mammalian cloning procedure that produced Dolly the Sheep, as just one example, was once considered impossible in legitimate scientific circles. This ongoing tension between what we know and what we might know someday underwrites the speculative economy of age extension and blurs the demarcation between science and science fiction. For example, a Salk Institute study suggests it is possible to slow or even reverse aging, at least in mice, by undoing changes in gene activity (Weintraub 2016). Scientific findings such as this provide crucial information for those seeking to improve the normal human lifespan through pharmaceuticals or other medical interventions, but also tantalize those who want to push the boundaries of the field into radical age extension. Scientists themselves must walk a thin line between extrapolating from their findings in a way that engages the mass media and general public, while also respecting the limits and complexities of the actual science. Per the Salk research cited above, for example, senior author and expert in gene expression Juan Carlos Izpisua Belmonte contends that "aging is something plastic that we can manipulate," (Weintraub 2016), an observation that may be amenable to multiple different readings but is technically correct in a scientific sense per that specific study. Yet the non-specialist is also confronted with David Sinclair, a Harvard biologist of some eminence, stating that the extension of the human life span will be "the biggest thing that is going to happen in the 21st century. It's going to make what Elon Musk is doing look fairly pedestrian" (Regalado 2018). At this juncture, the line between hope and hype becomes quite blurry and the construction of a promissory economy built upon visions of near limitless human life becomes viable.

Betting on Life

The field of anti-aging and life extension is vast, ergo this chapter can only introduce a small sample of the ideas being touted as potential longevity elixirs. Some of the more intriguing proposals include the prospect of microscopic nano-bots that will be able to travel through the bloodstream and sweep away disease, brain-machine uploading, interventions in the cellular process of aging, and the replacement of human organs on a routine basis via 3-D printing. These visions converge into a nascent transhumanist philosophy that exerts an outsized influence on markets and investment patterns due to the prominence of figures, such as Elon Musk or Peter Thiel, who either tacitly or explicitly seek the pathway to a utopian future in socio-technical innovation. Point one of the *Transhumanist Declaration* states "humanity stands to be profoundly affected by science and technology in the future. We envision the possibility of broadening human potential by overcoming aging, cognitive shortcomings, involuntary suffering, and our confinement to planet Earth" (humanity+ 2009). Moreover, just as synthetic biology's roots can be tracked to the nineteenth century, transhumanism has a crucial predecessor in the Russian cosmism of Nikolai Fedorov (1829 – 1903). Drawing upon a unique philosophical mixture of Russian Orthodox Christianity and mysticism, Fedorov envisioned a future in which mankind would eventually travel into space, there to reunite with the resurrected ancestors of the entire human race. Though Stalin banned discussion of Fedorov by the 1930s, due to his anti-materialistic and religious views, his ideas survived within a small circle of acolytes that included the prominent Russian rocket scientist Konstantin Tsiolkovsky (1857-1935). As the twenty-first century hurtles into the technological future, Fedorov and his work are being excavated for a new generation of thinkers who want to push the boundaries of the possible well beyond Earth and its mortal coil. Indeed, Fedorov even found his way into the *National Geographic* (Stone 2018), thus assuring a mass audience for this previously secret history.

Fedorov is also considered one of the earliest proponents of cryonics, the controversial field that purports to freeze deceased human bodies or severed heads at sub-zero temperatures in the hopes of resurrecting them in the future. Speculative visions of immortality often converge with a focus on space exploration, whether in the era of Russian cosmism, the Cold War, or the early twenty-first century (see chapter 18 in this volume). In 1962, Evan Cooper (pseudonym Nathan Duhring) published *Immortality: Scientifically, Physically*, and founded the original Life Extension Society in 1964 to further research and promote cryonics. His work was superseded in terms of visibility by the publication of Robert Ettinger's best-selling *The Prospect of Immortality*, which was released privately in 1962 and then re-released to commercial success by Doubleday in 1964. Both Cooper and Ettinger caught the zeitgeist in the sense that rapid technological advances in computing, artificial intelligence, and biotechnology dovetailed in the popular imagination with the drama of the space race between the United States and the Soviet Union and the imminent moment when man would step onto the Moon. Even a B-movie such as *The Brain that Wouldn't Die* (1962) suggests the extent to which the freezing temperatures of space and cryonics blended and blurred throughout the culture in a decade both obsessed with and fearful of runaway technological power.

Today, in an era in which "vitality can now be decomposed, stabilized, frozen, banked, stored, commoditized, accumulated, exchanged, traded across time, across space, across organs and species, across diverse contexts and enterprises in the service of both health and wealth" (Rose 2007), cryonics is experiencing a revival of interest and continues to occupy that liminal space between science fiction and fact. Newspaper headlines such as "Cryogenics [*sic*] chilling prospect of return of the frozen dead" (Donnelly 2017) and "Cryonics, brain preservation, and the weird science of cheating death" (Reilly

2020) demonstrate the unstable platform upon which cryonics rests: is it serious science or fantasy or perhaps even a joke? Nevertheless, as the subject of cryonics resurfaces in this tech-obsessed age, David Crippen argues that "cryonics deserves open-minded discussion, as do mainstream efforts to understand the nature of consciousness, preserve human tissue and organs for life-saving transplants, and rescue critically injured patients by understanding the boundaries between biological life and death" (Crippen et al. 2015). Indeed, cryonics has been joined by a slew of research proposals and investment opportunities in anti-aging and immortality research such that in the twenty-first century it is appropriate to speak of the immortality imaginary.

Immortality Imaginary

In *Dreamscapes of Modernity: Sociotechnical Imaginaries and the Fabrication of Power*, the editors define socio-technical imaginaries as "collectively held, institutionally stabilized, and publicly performed visions of desirable futures, animated by shared understandings of forms of social life and social order attainable through and supportive of advances in science and technology" (Jasanoff and Sang-hun 2015, p. 4). These visions can be seeded by a small group of enthusiasts but eventually exert an outsized influence both on the direction of technological development and on collective understandings not only of how society should be lived but what constitutes "good" and "evil" within the culture (Jasanoff and Sang-hun 2015). In the United States, the origin of what I refer to as the immortality imaginary has its locus in the visions and resources of Silicon Valley, that real-world technology corridor in Northern California that is also a mythical place of seemingly limitless technological and financial possibility. Here, a small group of billionaires and the visionaries that they choose to fund are constructing a powerful discourse within which aging, and possibly even death, become evils that must be vanquished through the systematic application of human ingenuity.

For example, Peter Thiel, a co-founder of PayPal and the first outside investor in Facebook, jointly manages the Founders' Fund, based in San Francisco, California, a venture capital firm that invests in "companies building revolutionary technologies" (Founders Fund 2020). A prominent figure in the anti-aging sector, Thiel told the Washington Post in 2015 "that evolution is a true account of nature, but I think we should try to transcend it in our society . . .I think it is against our human nature not to fight death [and] believe if we could enable people to live forever, we should do that. I think that is absolute" (Cha 2015). Both Thiel and Jeff Bezos, the CEO of Amazon Corporation, have invested in Unity Biotechnology, which focuses on "developing therapeutics to slow, halt, or reverse diseases of aging" (Unity Biotechnology 2020). For its part, Google spun-off the affiliated (and secretive) company Calico, which *Time* magazine greeted with a cover article asking, "Can Google solve death?" (McCracken and Grossman 2013). These significant investments seem to be driven by several factors including the focus and expertise that Silicon Valley CEOs can bring to bear on an issue, as well as a sense that disrupting death could be the ultimate technological coup. More prosaic but no less important is the staggering amount of money that stands to be made by products that could successfully forestall aging in wealthy societies even if death proves a bridge too far even for Amazon. As Michael West, gerontologist and CEO of AgeX Therapeutics acknowledges, "whenever you meet a fundamental human need, there's a market. And in this case, the market for age-related disease and aging is a trillion-dollar market" (CNBC 2020).

The luxury spending of a small group of billionaires alone would not constitute a socio-technical imaginary. What lifts these investments to that level is both the promissory nature of much of the research that undergirds these investments and the proliferation of affiliated non-profits and advocacy organizations to promote the new discourse of anti-aging and death disruption. For example, the Coalition for Radical Life Extension references the plurality of views within the anti-aging epistemic community about what radical life extension means and how it could be achieved but argues that "what we all agree on is that the deathist paradigm has to go; it's time to look beyond the past of dying to a future of unlimited living" (https://www.rlecoalition.com). The construct of a "deathist paradigm" indicates that the project of radical age extension extends beyond a few mega-wealthy gamblers in Palo Alto towards a culture that rejects death not only because it can be painful and frightening, but because it simply should not be possible at all. The Maximum Life Foundation, which seeks to "reverse aging" by 2033, does not go as far as the Coalition for Radical Age Extension, but it does exult that "you are put here to relish life. Its excitement. Its miracles. Its rewards. MaxLife is here to help you realize those treasures NOW" (https://www.maxlife.org/). It is becoming increasingly difficult, within this complex matrix of funders, press releases, and non-profit organizations, for the media and laypeople to know where to draw a line between exuberant but normal corporate spin and fantastical claims that ascend far beyond what the science can bear.

There is even a Church of Perpetual Life, which acknowledges Nikolai Fedorov as an inspiration. The Church website notes "we refer to ourselves as Immortalists, not because we have defeated death, but because we believe that future technology will conquer disease and aging, as well as death itself. While we fully understand that this technology does not appear to be currently available, the impressive history of human problem-solving and technological advancement gives us faith in its inevitability" (https://www.churchofperpetuallife.org). The Church of Perpetual Life elides the traditional Christian belief in spiritual immortality through acceptance of Jesus Christ as Savior with a thoroughly modern and secular faith in technology such that the pursuit of techno-physical (as distinct from spiritual) immortality becomes the apotheosis of a socio-technical imaginary. Fittingly, the headquarters of this faith is in Hollywood, California, a place that embodies the convergence of money, fantasy, technology, and a relentless drive towards the future.

CONCLUSION

Never Say Die

Paul Martin argues that ours is an era of "the development of knowledge economies in which scientific research and technical innovation are seen as the motor for growth and competitive advantage. Nowhere is this more apparent than in the biosciences, where the emerging bioeconomy is surrounded by high hopes, but remains an area with few working technologies entering routine use" (Martin 2015, p. 422). Despite the attention and money currently being lavished on anti-aging research and radical age extension, the longevity industry has thus far fallen far short of its promises though it remains an alluring prospect and one that seems to captivate the mass media and Silicon Valley. The promissory nature of longevity research and investment, as well as the fragile credibility of many of the scientific claims used to bolster radical age extension, are key reasons that many ethicists find the prospect of immortality highly suspect. As Jill Lepore argues, with reference to one highly touted pathway to longer life:

Neuropreservation has a scientific attitude, too, but that doesn't make it a science; it's more like extremely optimistic cosmetic surgery. It promises to cure hair loss, wrinkles, senescence, impotence, and death, all at once. Ah, yes, but will it work? It would be going too far to say that stranger things have happened because they haven't. Reanimating and rejuvenating the dead would be several orders of magnitude stranger than, say, landing on the moon (Lepore 2015).

Similarly, Hendrix, in a cogent critique of cryonics, argues "while it might be theoretically possible to preserve these features in dead tissue, that certainly is not happening now. The technology to do so, let alone the ability to read this information back out of such a specimen, does not yet exist even in principle. It is this purposeful conflation of what is *theoretically conceivable* with what is *ever practically possible* that exploits people's vulnerability" (Hendrix 2015).

This human vulnerability brings us full circle to considering important objections to the radical age extension project, though these are not scientific but rather ethical and theological concerns. In the article "Radical life extension, cybernetic immortality, and techno-salvation," Ted Peters asks, "what does it take for us to greet the end of life with contentment" (Peters 2018, p. 250)? Across religious and ethical belief systems and through time, one of the most enduring objections to pursuing immortality or even radical age extension is the sense that life has meaning precisely because it is finite. In Franz Kafka's words, the meaning of life is that it stops. Just as the Ancient Greeks worried about man's hubris and the divine punishment to ensue for such arrogance, science fiction of the modern era (drawing as well from ancient myths) produced such figures as Dracula the Vampire and Frankenstein's Monster, both of whom suggest quite strongly that the pursuit of life after death results not in exuberance but rather in dread and loneliness.

In *The Immortalization Commission*, John Gray argues that "during the nineteenth and early twentieth century science became the vehicle for an assault on death. The power of knowledge was summoned to free humans of their mortality. Science. . .became a channel for magic" (Gray 2011, p. 1). Yet it also became a channel for carnage beyond human reckoning, as seen in the depredation of World War I, and we might productively pause in this supremely technological age– which will, it should be acknowledged, advance the frontiers of human health in positive ways—to ask if we are so very different from those who considered these questions in earlier times. Early in the twentieth century, the combined trauma and death toll of World War I and the 1918 Spanish flu precipitated a surge in popular interest in spiritualism. Before obvious fakery brought the practice into disrepute, prominent Victorians such as Sir Arthur Conan Doyle and Sir Oliver Lodge paid serious attention to spiritualism, as did the general public, many of whom hoped to assuage catastrophic loss by reconnecting with loved ones in the afterlife (Daugherty 2020).

In this era, confronted by climate change, geopolitical tension, and a pandemic that within a year, to late December 2020, had caused the deaths of an estimated 1.7 million people worldwide, we seem to be turning to biotechnology, artificial intelligence, cryo-preservation, and neuroscience, just as the Victorians turned to spiritualism, to find a way to contain and perhaps even conquer death (and, in the end, all anti-aging research is about life but really about death). Yet the immortality imaginary, while seductive, carries within it the shadow realization that while there may be a purpose in trying to escape the inevitable, the house in this case always wins. The more fruitful path might be to accept this fact and focus on enabling the healthiest normal human lifespan for as many people as possible, rather than on extreme life extension for the few. As one imminent Victorian reminds us, "if death is your enemy, life is not your friend" (Gould 1891, p. 372).

REFERENCES

Breithaupt, H. (2006). The engineer's approach to biology. *EMBO Reports*, *7*(1), 21–23. doi:10.1038j.embor.7400607 PMID:16391532

Cha, A. E. (2015, April 3). *Peter Thiel's quest to find the key to eternal life.* https://www.washington-post.com/business/on-leadership/peter-thiels-life-goal-to-extend-our-time-on-this-earth/2015/04/03/b7a1779c-4814-11e4-891d-713f052086a0_story.html

CNBC. (2020, February 23). *The ultra-rich are investing in companies trying to reverse aging. Is it going to work?* https://www.cnbc.com/2020/02/19/the-ultra-rich-are-investing-in-companies-trying-to-reverse-aging.html

Crippen, D. (2015, October 19). The science surrounding cryonics. *MIT Technology Review.* https://www.technologyreview.com/2015/10/19/109714/the-science-surrounding-cryonics/

Daugherty, G. (2020, April 21). *Talking to the dead: How the 1918 pandemic spurred a spiritualism craze.* History. https://www.history.com/news/flu-pandemic-wwi-ouija-boards-spiritualism

Donnelly, M. (2017, January 11). Cryogenics' chilling prospect of return of the frozen dead. *The Daily Telegraph.* https://www.dailytelegraph.com.au/news/cryogenics-chilling-prospect-of-return-of-the-frozen-dead/news-story/afd172f56e6d10decbda80e01afac75c

Gould, G. M. (1891). Immorality. *The Monist*, *1*(3), 372–392. doi:10.5840/monist18911339

Gray, J. (2011). *The immortalization commission: Science and the strange quest to cheat death.* Farrar, Strauss, & Giroux.

Hendrix, M. (2015, September 15). The false science of cryonics. *MIT Technology Review.* https://www.technologyreview.com/2015/09/15/109906/the-false-science-of-cryonics/

Humanity+ (2009). *Transhumanist declaration.* https://humanityplus.org/philosophy/transhumanist-declaration/

Istvan, Z. (2019, May 14). Death could soon become a curable disease. *Metro.* https://metro.co.uk/2019/05/14/death-could-soon-become-a-curable-disease-9191393/

Jasanoff, S., & Kim, S.-h. (2015). *Dreamscapes of modernity: Sociotechnical imaginaries and the fabrication of power.* The University of Chicago Press.

Keeney, J. (2015). *To beat death and become immortals, we first must defeat entropy.* Inverse. https://www.inverse.com/article/8867-to-beat-death-and-become-immortals-we-first-must-defeat-entropy

Lepore, J. (2015, January 25). The iceman. *The New Yorker.* https://www.newyorker.com/magazine/2010/01/25/the-iceman

Loeb, J. (1913). *Artificial parthenogenesis and fertilization.* University of Chicago Press.

Martin, P. (2015). Commercialising neurofutures: Promissory economies, value creation and the making of a new industry. *Biosocieties*, *10*(4), 422–443.

Pauly, P. J. (1987). *Controlling life: Jacques Loeb and the engineering ideal in biology*. Oxford University Press.

Peters, T. (2018). Radical life extension, cybernetic immortality, and techno-salvation. Really? *Dialog*, *57*(4), 250–256. https://doi.org/10.1111/dial.12432

Regalado, A. (2018, May 9). A stealthy Harvard startup wants to reverse aging in dogs, and humans could be next. *MIT Technology Review*. https://www.technologyreview.com/2018/05/09/142971/a-stealthy-harvard-startup-wants-to-reverse-aging-in-dogs-and-humans-could-be-next/

Reilly, C. (2020, July 9). *Cryonics, brain preservation, and the weird science of cheating death*. Cnet. https://www.cnet.com/features/cryonics-brain-preservation-and-the-weird-science-of-cheating-death-alcor/

Rose, N. (2007). Molecular biopolitics, somatic ethics, and the spirit of biocapital. *Social Theory & Health*, *5*(1), 3–29.

Scharping, N. (2016, October 6). *Humans may not be able to live past 125*. Discover. https://www.discovermagazine.com/health/humans-may-not-be-able-to-live-past-125

Stambler, I. (2014). *A history of life-extensionism in the twentieth century*. CreateSpace Independent Publishing Platform.

Stone, D. (2018, February 21). These people believe death is only temporary. *National Geographic*. https://www.nationalgeographic.com/photography/proof/2018/february/nucci-transhumanists-technology/

Venter, J. C. (2013). *Life at the speed of light: From the double helix to the dawn of digital life*. Penguin Books.

Weintraub, K. (2016, December 15). Aging is reversible—at least in human cells and live mice. *Scientific American*. https://www.scientificamerican.com/article/aging-is-reversible-at-least-in-human-cells-and-live-mice/

You, E. (2016). *Anatomy of next: Utopia/Bio*. Founders Fund. https://foundersfund.com/2016/06/anatomy-next-utopia-episode-2-bio/

Chapter 17
Enhancing Mental Health:
A Role for Technology

Parvind Gambhir
Independent Researcher, Australia

ABSTRACT

Mental health is one of the health priorities in the 21st century along with cardiovascular, cancer, and diabetes. Technology nowadays is most sought for getting any information. There are numerous online platforms that have been developed to assist in assessment, provide feedback, suggest management, and thereafter monitor mental disorders in target patients with a view to enhancing the quality of mental health care provided by traditional face-to-face services. In the current situation where the world is facing a challenging situation in COVID-19, technology has become highly sought after for managing the health of patients.

INTRODUCTION

Mental health is one of the health priorities in the 21st century along with cardiovascular, cancer and diabetes. With enhancing mental ability, it can help a person to make a right decision at correct time which can be handy in maintaining health of the person. This mental alertness and ability to interpret the symptoms based on the available data can be handy. Technology nowadays is the most sought for getting any information. There are numerous online platforms has been developed to assist in assessment, provide feedback, suggest management and thereafter monitoring with mental disorders in target patients with a view to enhancing the quality of mental health care provided by traditional face-to-face services. Like in current situation where the world is facing a challenging situation for Covid-19, technology has become a highly sought after for managing health of patients. Many trials were done for comparing the impact of an enhanced community care intervention that uses interactive SMS communication tools with treatment as usual. Service users with a diagnosis of Severe Mental Illness who received mental health care in Newham / London were recruited from community mental health teams (CMHTs).

DOI: 10.4018/978-1-7998-6772-2.ch017

Main Developments

Doctors are using telehealth for managing their patients which is basically technology chatting for consultation. Nowadays not only consultation is done by chatting over the phone but also video conference is becoming quite common. Even emergencies surgeries can be done in remote area over the video telehealth. Similarly, technology is a key saviour for those having mental health problems. Internet is available in rural and remote areas and it is hard for people living there to travel hundreds of kilometres for getting mental medical help. Most of them rely on chatting via their website. There are many projects running to enhances the mental ability by co-designing technology-enabled solutions for Australian mental health. The technology mainly supports the person using the technology to decide the suitable options care options. It does not help in the medical advice or clinical assessment or treatment. But digital tools, combined with new personal sensors and monitoring technologies, can underpin clinical assessment, objective testing, initiation of treatments or collection of additional information from family or supportive others.

Mental health has affected over hundreds of million people worldwide. We can say that there are over 300 million people suffering from depression, 250 plus million suffer from anxiety. We can say that there are many drugs but there is huge challenge in researching for any new drugs. On the other hand, new technological innovations are helping some patients connect with a therapist, get diagnosed, track moods, manage or mitigate symptoms, and stick to treatments.

Technology is also extremely helpful for many males in general as it is hard for them to discuss the mental health issues and getting initial help from technology regarding their symptoms. Like there is an app for every mental health problem such as depression, post-traumatic stress disorder, schizophrenia to addiction. Some apps are designed to assuage and manage symptoms, track moods and thoughts, or help individuals stick with treatments. But experts caution prospective users against using them as a stand-in for professional treatment, as per National Institute of Mental Health (NIMH). Most aren't backed by peer-reviewed research or clinical trials, partially because tech development moves faster than traditional scientific testing.

The other benefit of technology could relate to staff shortage as then video or text chatting can be done for any guidance. Such technology is getting more relevant in current covid /pandemic situation where most of the consultation is done online. Technology is like a double edge sword for mental health work during such situation as most of the people are working from home. This has led to researching as how the existing studies gets relevance to technology driven impact at workplace. Suggestions as regular break, exercise, making notification easier to get across can be helpful in such pandemic situation. Even the private health insurance companies are jumping in to provide mental health through technology to their customers.

While some mental health apps are designed to deliver outcomes like improved mood or lessened anxiety, many researchers are working to harness mobile technology to detect suicidal inclinations, burgeoning manic episodes, or depressive episodes before they manifest. There is also software that can locate at risk patients. Even google teamed up with US researchers to upload questionnaire for depression on mobile phone if anyone searched for the same.

With virtual reality (VR), computer-generated interactive environments, individuals can repeatedly experience their problematic situations and be taught, via evidence-based psychological treatments, how to overcome difficulties (Freeman, 2017). VR has the potential to transform the assessment, understanding and treatment of mental health problems. The treatment possibilities will only be realized if - with the user experience at the heart of design - the best immersive VR technology is combined with targeted translational interventions. The capability of VR to simulate reality could greatly increase access to psychological therapies, while treatment outcomes could be enhanced by the technology's ability to create new realities. VR may merit the level of attention given to neuroimaging.

Like for the trail, SMS were used for Severe Mental Illness (Schizophrenia, Schizoaffective Disorder, Bipolar Disorder) ; Mental and Behavioural Disorders ; Schizophrenia, Schizoaffective Disorder, Bipolar Disorder. For thisthe service user satisfaction scores (subjective quality of life measured using the DIALOG scale (patient-reported outcome measure) at baseline and after six months.

Now a day's smartphone is the necessity. There are many apps on such phones that can help people cope with depression and anxiety. Yes, there are more apps that care for your physical health. The apps targeting mental health help the users to share their stories, how to be calm as stress relieving exercise, how to cope with symptoms. These apps not only help the patients, but they also provide useful data to physicians, therapists, and other health care informatics expert.

The discussion on use of technology in any sector say medicine, education, transport etc is never ending and keeps on improving day by day. We have new and improved technology evolving currently and in future. It is hard to imagine life without technology now. Earlier it was hard for any company to approve work from home and this covid has made it possible that nearly every company has its employee where possible working from home, and this has become possible within days. With the lockdown there has been drastic increase in the number of people needing mental health support. Government had to increase the budget allocation for mental health in particular. Males are more vulnerable to suffer mental health problem as it is hard for them to open up having any mental health issues.

Apart from alcoholism or addiction, there is mental health app covers many different conditions. There is alcohol addiction recovery app that supports, give strength and positivity to those who need it but not to forget that it is not a substitute for treatment.

Technology is so much used now that even if you are going to see the doctor next day, people would do some searches on internet and discuss their findings as differential diagnosis. Doctors do say that little knowledge is not good for arriving at any conclusion, but technology can be of some help in some circumstances as in mental health. Many people are so busy now with meetings etc that it is hard for them to see concerned specialist or they are not able to attend support groups. Some app allows users to remain anonymous, people can feel comfortable revealing their struggles and engaging with other participants.

One of the features of technology is telehealth where consultant can do the consultation via phone call or video chatting. Many time it is hard for people to ask medical questions in more anonymous or easy access setting, but virtual clinician makes it much easier. There are many apps on mobile that can help you in discussing your medical problems or be a friend to help you maintain your health or remind you of your medication in time. There are apps that track daily medication dosing, provide reminders, track physical activity or anxiety symptoms are all available with many more on the way.

Technology has advanced a lot that it has there are some public health initiatives trying to add onto many social media accounts in which ways they can reach greater number of population of whom there could be many targeted audience for whom such messages on online groups can discuss depression, alcoholism or even suicide thoughts. Many anonymous stories can be extremely helpful in enhancing peoples understanding of what happens in such illness and thereby reducing the stigma attached with these illnesses.

Various ways that technology can be helpful in mental health as having apps targeting specific users with conditions like depression, schizophrenia, anxiety where they trach the symptoms or moods, thoughts and help them in sticking to the treatment.

Overall technology has opened new frontier in mental health support and data collection. Mobile devices like cell phones, smartphones, and tablets are giving the public, doctors, and researchers new ways to access help, monitor progress, and increase understanding of mental wellbeing. The researcher complied the data and said that such support is simple but amazingly effective. Nearly everyone now a days has smartphone and it is easy to send text message to the crisis support unit. The new technology has been packaged into very sophisticated apps for all the smartphones, which has in built sensors that can collect relevant information on how the user will be using that app. Down the track, if there are some changes in the behaviours of the person, the app can then detect such change of behaviour and provide signal that help is required. This can prevent any crisis and relevant help can be provided in time. In such apps there are some programs like games etc that can promise to improve memory or thinking skills which can then result in good cognitive score. Other apps can help in connecting to some counsellor or other relevant health care professionals. Overall net benefit is crisis prevention resulting in better health of the person, less domestic violence. They can perform their daily activity at home or job.

There are some pros and cons for using the technology for mental health section. Yes, it is very convenient for the patients as they can access it in privacy, can look for various treatment options available for their symptoms and then choose accordingly and can read experiences from others in similar scenario. Apart from providing service to more people, also particularly important is that many aps are free of cost. The other benefit is that it can be accessed any time and any day, so it is available 24/7. There is consistency of the treatment on all the apps and many technology assists in motivating clients to continue therapy for longer duration. As per the current software's it is also possible to track the phone of the clients, their movements and other relevant information's.

Some of the cons can be that it is technology and there can be problems with technology. Most of the apps work on traditional way of management and that there are apps that work for all the mental health symptoms. Other question is that who is managing these apps and the major concern can be the privacy issue. Also some of the apps many involve cost.

When we talk of apps for mental health which is a necessity for current situation, it has also emerged a cool way to delivering psychotherapy. As for example mental health app will keep a tab on your behaviour activity that is interrelated to problem like depression as the app will then deliver some cognitive behaviour therapy/interventions to the person. In managing any mental health issue, it requires frequent visit to counsellor, psychologist or to mental health physician. If the patient is able to self-monitoring the symptoms which goes a long way in managing their problem and new technology with various apps can do that as guiding alongside the in-person therapy sessions with ultimate aim to improv behaviour change. Such type of quantified accredited self-mobile applications allows the individuals to easily track their behaviours and advice on specific cognitions therapy as needed for the management. The collected data in the app can then be conveniently shared with the therapists to review and chart out adequate treatment.

Since the monitoring is done regularly through these apps, it becomes easier to gauge the patient's progress and effectiveness of the treatment to achieve the desired patient outcome.

When a patient discusses their circumstances/condition with the health care professional, they might miss out discussing all the symptoms but with the app under the health behaviour intervention category solve this very problem. Moreover, many app deliver clinical education along with behaviour feedback.

The apps also assist in increasing health-promoting behaviours like exercising in individuals. Problems, such as stress management, are better addressed with such mobile apps.

Getting the correct and accurate information is highly crucial in mental health conditions, both for the patients and mental healthcare providers. With healthcare mobile apps, therapists or doctors can obtain useful, precise, and real-time patient data to craft an effective, individualised treatment plan.

Various studies were conducted as can apps be helpful. One such study had mentioned that around 50% of doctors reported that apps facilitated diagnosis during on-call work whereas 43% reported apps helped them interpret laboratory values. The American Psychiatric Association has produced a DSM-5 app that contains detailed information on the DSM-5 diagnostic categories as well as ICD-10 coding equivalents. The app includes many illustrative videos that help in the diagnosis. Such can be viewed by patients also that helps them in recognising and corelating the symptoms and keep a track of the management. This type of illustrative information helps building up patients' confidence and reliability for the application.

There are apps also that helps in phobias. Such apps can help in guiding way out of phobia by gradual exposure as with some games or getting into real world. On similar track there are apps that can track your mood and guide you out to better manage the situation.

Recent advances in smartphone technology have revealed new opportunities for the diagnosis, monitoring and provision of mental health care. Such digital approaches have the potential to enhance our understanding of the course of mental disorder, enable early diagnosis and generate new treatment targets. However, issues relating to mobile phone ownership, misuse of clinical data, commercial devices and the ethics of such approaches need to be overcome if smartphones are to be routinely used in mental health care (Patel, 2018).

Discussion

Technology has opened up many spheres for communication, there is plenty of data in academic or public health domain for discussion and understanding but there also needs to be ongoing ethical negotiation of how to balance these initiatives with privacy and profit motives moving forward in this brave new world.

People can find help/guidance through internet-based support groups. They can rely on telehealth. But there are not that many apps for mental health as compared to apps devoted for physical health support.

The advanced mobile apps with innovative technologies have certainly made availing mental health care simpler for everyone. But considering a broader picture, the role of a recognised mental healthcare professional cannot be ignored on the road to recovery. Hence, it is advisable to combine the use of such apps with the assisted treatment by an authorised professional.

Mental health professionals across a range of disciplines are increasingly developing and implementing Internet-based treatments for people with a variety of mental health conditions. Many therapeutic and economic benefits are associated with technology-enabled treatments for a range of mental health disorders. Despite this, the role of human computer intraction (HCI) and associated design elements remains poorly understood in regard to the impact on patient safety, effectiveness, and to adherence of treatment for computer users who engage with e-mental health interventions. The main finding of this review is that Internet-based e-mental health interventions are routinely implemented without sufficiently describing the relevant HCI design features applied. This is a limitation that in turn jeopardizes the assessment validity of e-mental interventions generally, leaving those who administer the interventions with incomplete evidence to support the safe, reliable, dependable, credible, and trustworthy implementation of the interventions. The recommendation arising from this review is that human computer interaction should be carefully considered when mental health nurses and other practitioners adopt e-mental health interventions for therapeutic purposes to assure the quality and safety of e-mental health interventions on offer to patients (Neilsen, 2019).

REFERENCES

Freeman, D., Reeve, S., Robinson, A., Ehlers, A., Clark, D., Spanlang, B., & Slater, M. (2017). Virtual reality in the assessment, understanding, and treatment of mental health disorders. *Psychological Medicine*, *47*(14), 2393–2400. doi:10.1017/S003329171700040X PMID:28325167

Neilsen, A., & Wilson, R. (2019). Combining e-mental health intervention development with human computer interaction (HCI) design to enhance technology-facilitated recovery for people with depression and/or anxiety conditions: An integrative literature review. *International Journal of Mental Health Nursing*, *28*(1), 22–39. doi:10.1111/inm.12527 PMID:30133096

Patel, S., & Saunders, K. (2018). Apps and wearables in the monitoring of mental health disorders. *British Journal of Hospital Medicine (London, England)*, *79*(12), 672–675. doi:10.12968/hmed.2018.79.12.672 PMID:30526097

Chapter 18
The Coming Space Age:
Forecasts From the TechCast Project

William Halal
George Washington University, USA

ABSTRACT

This chapter draws on forecasts from The TechCast Project to map out the beginning of the Space Age about 2050. The author's work on the Life Cycle of Evolution shows that the world is moving beyond the Knowledge Age, which began about AD 2000, and is now entering an Age of Consciousness about 2020. This seems to mark the culmination of civilization on Earth, when the planet reaches a stage of maturity needed to resolve historic threats such as climate change. If this passage to a unified global order is successful, it should mark the beginning of space exploration beyond the solar system. The commercialization of space will likely be well underway, and colonies established on the Moon and possibly Mars, solar satellites will likely be functioning, and space tourism will become normal. With Earth a stable civilization, attention should then turn to this final frontier of space. The intellectual resources of roughly 10 billion educated people will be drawn on to make the breakthroughs in our understanding of physics needed to travel to star systems.

The idea of commercial space tourism seemed silly at first. But when Dennis Tito traveled into space as a private astronaut in 2001, it soon became clear that more would follow. Now, commercial ventures are flying to near-Earth orbit, and many are planning space hotels, tours to lunar orbit and expeditions around the Moon. Driven by our need to explore, we are about to leave our planet *en masse.* Peter Diamandis said, "We're on the verge of the golden age in space. It should soon be possible to tour space on a routine basis." (Diamandis, 2020).

This article draws on the foresight research conducted by The TechCast Project to chart the path into a Space Age. TechCast draws on a state-of-the-art collective intelligence system to produce some of the most authoritative forecasts available. Results suggest the commercialization of space is poised to reach the critical take-off point in the next several years, about 2025 +/- 3 years, growing into a $1 trillion industry that promises to transform our world. A Moon colony should be established in about 15 years, solar satellites are likely to power earth systems about the same time, and humans are likely to land on

DOI: 10.4018/978-1-7998-6772-2.ch018

Mars a bit later, roughly 2040. Travel to neighboring star systems is suggested to occur about 2050, and first contact with alien civilizations in roughly 2060 or so.

However, the theoretical model guiding this research suggests this will require a higher level of global consciousness able to support a Space Age. Drawing on a theory describing the Life Cycle of Evolution, we see that the existential transition to space is only possible if the world is able to resolve the constellation of crises (climate change, etc.) that threaten civilization itself.

Evidence shows the world is now moving beyond the Knowledge Age into an Age of Consciousness and reach global maturity (Halal, forthcoming). If this can be done, the world should become almost fully industrialized, well-educated and integrated into a sustainable global order about 2050. Automation and artificial intelligence will eliminate routine knowledge work, turning resources and attention to the exploration of deep space.

The seven forecasts that follow outline our push into this final frontier. They are only rough estimates, of course, with accuracy of plus or minus roughly five years. But they capture the timing and broad themes likely to mark this historic passage.

Theoretical Framework: The Life Cycle of Evolution

The opening of a Space Age can be best understood as a historic shift in social evolution. This is illustrated by the graph shown in Figure 1 below. I have struggled with this problem for years, and the result is this trendline showing what I call the "Life Cycle of Evolution (LCE)." Similar graphs have been sketched in general terms, but this is the first to plot the long-term evolutionary trend using real scales and real data (Halal, 2004). The logarithmic time scale is needed to encompass the billions of years at the start of life, as well as mere decades today. Without a log scale, the shape of the LCE would not be recognizable; the trendline would run flat and make a sharp 90 degree turn straight up.

Above the fray, there is a rather precise direction and logic to this evolutionary process. Four **million** years of nomadic human development were needed to found Agrarian Civilizations. Eight **thousand** years to invent Industrial Society. Two **hundred** years for the Post-Industrial Economy. Five **decades** to a Knowledge Age. And roughly 20 **years** to an Age of Consciousness that is starting today about 2020. The completion of this mature phase in the LCE could then lead to the serious start of a Space Age about 2050.

Beneath this tectonic shift in consciousness is the driving force of artificial intelligence (AI), the most powerful agent of change today. Sundar Pichai, CEO of Google, said "AI is probably the most important thing humanity has ever worked on … more profound than fire or electricity." The advance of AI is automating knowledge work, threatening to eliminate roughly half of all jobs and posing one of the most perplexing questions of our time: *What lies beyond knowledge?*

It follows logically that everything beyond knowledge is consciousness. Civilization is at the cusp of transformation from a society based on knowledge to one guided by consciousness. The great Jesuit anthropologist, Pierre Teilhard de Chardin, has long fascinated us with his vision that the world would evolve into a "noosphere," a great web of consciousness enveloping the Earth (de Chardin, 1955).

Not too long ago, we relied on telephones and newspapers to communicate. We now use two billion personal computers, 14 billion cell phones and laptops, and two billion TVs. The information flows

Figure 1.

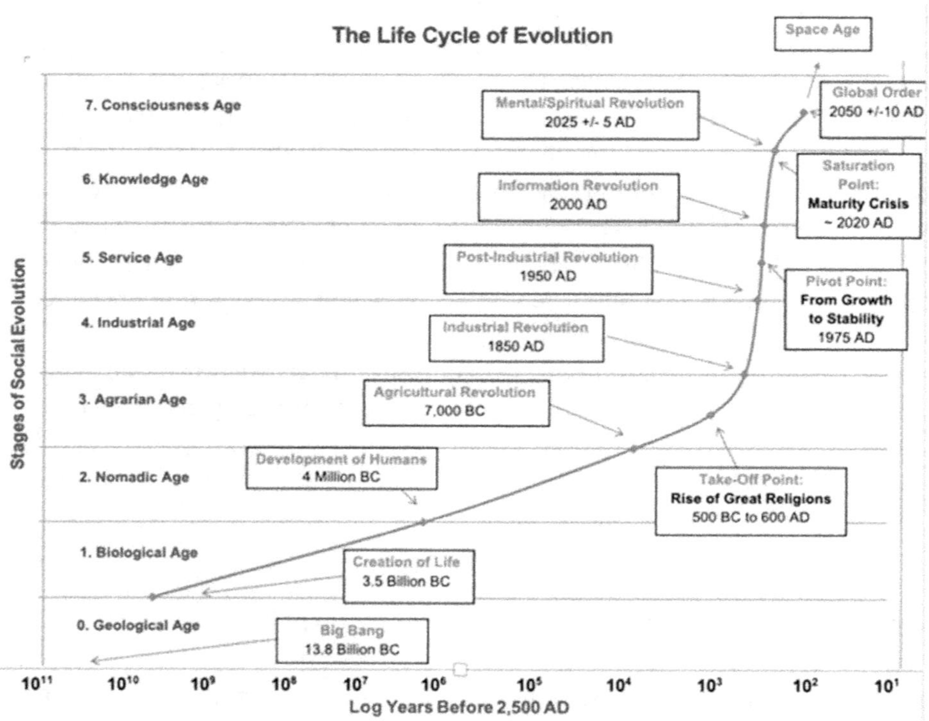

through 30 million Internet servers, 3,000 space satellites and almost one million miles of undersea cables. Kevin Kelly, founder of *Wired* magazine, thinks the Internet is subsuming the knowledge of all humans into a "global central nervous system." He observed, "There is only one time in the history of each planet when its inhabitants first wire up its innumerable parts to make one large machine." (Kelly, 2016). This planetary layer of digital connections knits eight billion people into a living overlay of thought – the noosphere is here.

This historic transition also poses enormous threats that seem almost impossible. Climate change and the entire constellation of end-of-the-world challenges comprise what I call the "Global MegaCrisis," or the "Crisis of Global Maturity." A survey I conducted estimates that roughly 70 percent of the public thinks the present world trajectory will lead to disaster. Ask anyone off the street and you will probably get the same answer. People have deep fears over today's failures in governance, and they attribute it to a lack of leadership, vision and cooperation. The late Stephen Hawking worried about "widening inequality, climate change, food, decimation of species, epidemic disease, acidification of the oceans. This is the most dangerous moment in the development of humanity, and our species must work together." (Hawking, 2016)

Today's Age of Consciousness faces the enormous challenge of creating a unified world able to resolve the MegaCrisis that limits our abilities today. We would then be able to harness its collective material and intellectual resources for the exodus of life into the universe.

The demands of deep space exploration outside our solar system are so beyond current scientific knowledge that a Space Age is inconceivable with today's technologies. The nearest star system, Alpha

Centauri, is four light years away from Earth. Travelling at 10 percent the speed of light (a generous assumption) would take 40 years. And that's the nearest star system.

Any serious travel in the larger universe would require fundamental breakthroughs in physics. The forecasts here suggest this will require solar sails, warp drives, anti-matter propulsion, artificial ecosystems to support astronauts, hibernation technologies, radiation shielding and other unknown intellectual breakthroughs that only a unified world can provide. Astronaut Edgar Mitchell thinks it can only be accomplished by a "combined global effort" (Fowler, 1985).

TechCast's Method of Collective Intelligence

The TechCast Project is an academic think tank that developed out of coursework at George Washington University. TechCast forecasts roughly 100 emerging technologies, social trends, and wild cards to cover the entire strategic landscape for planners and decision-makers. (Halal, 2008) The research method uses collective intelligence to pool background information and the knowledge of roughly 150 high-tech CEOs, scientists and engineers, academics, consultants, futurists and other experts to forecast breakthroughs in all fields. Experts are selected based on their experience, educational degrees, publications and inventions. The experts are not all world-renowned, but they are all thought leaders representing the cutting edge of knowledge.

Unlike ordinary Delphi forecasts, this work is science-based (Linstone, 1975). It draws on empirical adoption data, research underway, leading-edge ventures and other relevant knowledge to guide expert judgments. We make a point of including opposing trends that hinder development, such as political obstacles, high costs, or social resistance, to ensure that the analysis is balanced.

Experts look over this analysis and they use their judgment to estimate when each technology is most likely to reach its next adoption level, the potential size of the economic market, and their confidence in the forecast. The experts' estimates are aggregated automatically to update results in real time. More than snapshots in time, this is a continual tracking process that improves as comments and results help experts learn and as new data updates the analyses. The results show that an avalanche of innovative change should sweep over society during the next 20 years.

There is no lack of strong and extreme opinion on technology breakthroughs, ranging from "That's already here" to "It'll never happen." This confusion is exactly what our research method of collective intelligence is designed to resolve. By integrating a wide variety of data sources, we rise above the details to provide a higher-order analysis that is more authoritative. Experts may have bias, naturally, but these biases are usually distributed normally, canceling out in the aggregate results. In fact, collective human intelligence has the potential to surpass machine intelligence, giving us the edge in the coming conflict between AI and humanity.

It is often thought that forecasting methods like this are subjective, whereas quantitative methods are precise. However, quantitative methods also involve uncertainty because they require underlying assumptions that often are doubtful, so results vary widely. In fact, we use available quantitative forecasts in our background data, thereby subsuming this information and allowing experts to resolve the uncertainty that remains.

The results are compelling when considering the fact that the expert panel changed over this time, as did the prospects for various technologies. Prediction markets have demonstrated sound accuracy using a similar form of collective intelligence (Wolfers, 2004). This work also holds up well in our work for corporations and governments. On one consulting assignment, we conducted two parallel studies to forecast the maturing of energy technologies, one using a group of energy experts and the other using a group of general experts. The forecasts compared almost exactly, usually within one to two years (Laitner, 2004).

This method of collective intelligence integrates all available knowledge and expert judgment to approach a scientific consensus that is remarkably prescient. If the present level of uncertainty about future technologies is defined as 100 percent, we have found that this process reduces uncertainty to about 20 to 30 percent. Validation studies comparing previous forecasts with actual arrival dates find that the average error of all forecasts is roughly +1/-3 years at ten-year future horizons. The -3 error reflects the well-known tendency to be optimistic.

Forecasting The Space Age

With the theoretical framework of the LCE in mind, we now draw on the work of the TechCast Project to forecast the key breakthroughs needed for a Space Age able to travel through deep space. There is a plethora of forecasts that make similar claims. Many other forecasts are subsumed in the TechCast method of collective intelligence, which comprises a complete data base of prominent technological, economic, political and social factors affecting these 100 forecasts. The following seven forecasts form what is thought to be a fairly complete and authoritative guide to the evolution of space.

Commercialization To Take Off About 2025

After a long government monopoly on space flight, the private sector is commercializing space with a flourish of exciting new ventures to explore the final frontier. NASA's commercial space program supports 20 or more companies like Space X, Orbital Sciences, Blue Origin and Boeing in developing spacecraft for shuttling people and material to low Earth orbit. Commercial space is thought to decrease costs by a factor of ten.

DARPA, Bigelow, Rocket Lab, and other companies are also developing commercial space projects. Lockheed's Orion is replacing the Apollo vehicle. Space X, Blue Origin, and the Air Force all are successfully developing reusable rockets, while Russians are planning a private space station. Spaceports are being built around the globe and plans to mine the asteroids for minerals are underway.

Elon Musk's Space X is leading the field in many ways. Its heavy booster, the Falcon 9 rocket, would have cost NASA $4 billion to develop, while Space X did it for $443 million – a little more than 10 percent. The company made history in 2012 when its Dragon spacecraft became the first commercial vehicle to dock with the International Space Station, and it landed successfully the first reusable booster, reducing launch costs 100-fold. Space X did it again in 2018 when the Falcon Heavy became the most powerful rocket sent into space and by a private company. One US politician called it "the dawn of a new era."

Amazon founder Jeff Bezos has spent more than US$500 million to make his company, Blue Origin, the standard in low-cost space travel, using suborbital tourist flights to practice for more ambitious goals. Bezos wants to turn Earth into a residential planet, moving heavy industry beyond the atmosphere.

Projects like these merely hint at the possibility that private enterprise could ignite a boom in commercial space development. Humanity is gingerly stepping off the planet into a vast frontier, much as it did after the discovery of a New World in the Americas. In a few decades, this could be similar to the land rush that followed the opening of the Western United States.

Our experts are quite confident these trends should drive commercial space development to reach that critical take-off point at about 2025 +/- 3-5 years where rapid growth will soon produce a mature industry. They also estimate this could create a worldwide space industry with revenues of about US$1 trillion/year when the field saturates roughly about 2050.

Space Tourism Should Follow

The feasibility of private space flights was demonstrated when Virgin Galactic's SpaceShip One made its second flight to win the Ansari prize of US$10 million. Interest picked up when the crew tested their first passenger flights into near-Earth orbit, and they have tested the mother ship that will carry the spacecraft to launch. The company has built a US$300 million Spaceport America launch facility in New Mexico and had 500 clients ready to take suborbital flights for US$200,000. The crash of SpaceShip Two in 2014 set plans back, but the company and its competitors continue to move ahead.

Since then, a variety of space tourism ventures has started to explore the business opportunities to carry passengers into space, fly around the moon and explore the solar system. Some claim space tourism could in time become roughly comparable to air travel in safety, comfort and cost, extending eventually to the moon and other locations. An entrepreneur said, "Business is starting to look hard at space tourism; they don't want to miss an opportunity."

Bezos' space venture, Blue Origin, is conducting suborbital test flights with passengers and could start selling tickets soon. China is developing a large rocket ship to carry tourists into near-Earth sub orbit. Virgin Galactic, Hilton Hotels, British Airways, and billionaire Robert Bigelo are all working on space hotels, and some are planning tourists' trips around the moon.

We expect a flurry of space tourism activity to join the economic boom of commercial space about 2025 as entrepreneurs and adventurers rush to explore this final frontier. Thousands of tourists could be traveling into space each year, paying big money for a space flight and producing total revenues in the billions.

Moon Colony Built About 2035

The moon has long been recognized as a convenient launching pad for future space missions. It is relatively close, its mild gravity is easy to escape, and it has almost all materials needed for space travel, including water. Attracted by those assets, several private ventures are planning to revisit the moon. The US, Russia, and China are all developing programs to build lunar colonies in the next decade. The LiftPort Group plans to build a Lunar Space Elevator to facilitate ferrying materials to and from the moon's surface, similar to the space elevator they are developing for Earth.

NASA's plan for a moon colony recommends partnering with private rocket companies to put humans on the moon for about US$10 billion. This base camp would produce 200 tons of water/year to be broken down into hydrogen and oxygen rocket fuel, creating a gas station for missions to Mars or elsewhere later that decade. Total costs would be one-tenth of the present Mars-only plan. NASA also is planning an outpost at the Lagrange point on the far side of the moon to facilitate trips to the moon, Mars, and other locations.

The challenges of building a permanent base on the moon are not much more daunting than those of the International Space Station, which has been inhabited since 2000. The biggest obstacle is the cost, estimated at US$100 billion, at a time of nagging budget deficits. This is a serious concern, but a moon colony will be part of the ambitious attempt to land humans on Mars and travel to other planets.

A permanent lunar colony would allow extended research on space living, working, manufacturing, and other valuable activities. The moon's mineral resources could boost the development of medical and industrial research. Nations that miss the race to the moon are likely to set back their space programs and may impair their national security. A moon colony would make it easier to explore the solar system, develop commercial ventures, and become a space-faring civilization.

Solar Satellites Deployed About 2035

The US, Japan, China, and other nations are planning geosynchronous solar power satellites (SPS) with huge fields of photovoltaic cells to generate power in space and beam it to Earth. These systems are the equivalent of large nuclear power plants, cost between $US 20-30 billion and are expected to be operational about 2030-2040. The US Military and DARPA are planning solar satellites to give military units access to power anywhere. Some plan to put a ring of solar power stations directly on the moon, beaming energy to Earth.

With no nighttime, no atmosphere, and no weather, SPS should be far more efficient than collecting the sun's energy on Earth. There are dangers to beaming energy, but they are manageable, and the economic benefits are vast. "Whoever develops clean and renewable energy in space will be the world leader," said Wang Xiji, Chinese Academy of Sciences.

Solar Satellites have big potential for meeting energy in an environmentally safe way. It may take decades and many billions of dollars to realize the potential, but SPS could solve the problems of global warming and environmental deterioration due to carbon fuels. And they can be seven times more energy-intensive than Earth-based solar systems because there is no atmosphere in space, and they operate 24/7 without interference from weather.

SPS could also open up entirely new industries, spur economic growth, and provide opportunities for employment in high-tech jobs. A large-scale SPS could employ 5 million people annually and prove as important as railroads in the 19th century and automobiles in the 20th. Ralph Nansen of US-based Solar High Company said, "SPS is the ultimate energy source. I don't think there's any doubt that within the next century we will be getting the majority of our power from space."

Humans Land on Mars About 2040

Mars is so far away that a round trip would take a full year of space travel. The journey requires sophisticated life support systems beyond the state of the art, advanced propulsion (For instance nuclear and ion engines) to make the long journey, and protection from radiation belts that would harm the astronauts. And these are only the most obvious breakthroughs needed.

Still, the symbolic allure of colonizing the Red Planet is hard to resist, and a manned landing on Mars would be historic. NASA, Netherlands-based Mars One, Russia and other programs plan to use a window of 2030-35 for a likely trip. Most notably is Elon Musk's plan using Space X to land humans on Mars by 2025, which is slipping, and Boeing's program which is racing to beat Space X. Efforts are also underway with the Mars One Settlement projects, which has invited private astronauts to participate in a Mars landing. Interest is so strong that the project attracted more than 78,000 volunteers, even though it would be a one-way trip with no hope of return. Some are already thinking of terraforming Mars to make it habitable.

Advanced propulsion systems are needed because of the huge distances involved. NASA is developing a nuclear engine, as well as ion and plasma propulsion engines. This would increase efficiency so greatly that they could take a manned craft to Mars in 39 days instead of the three months required by chemical rockets. "Using existing rocket fuels, it's nearly impossible for humans to explore much beyond Earth," said the lead researcher. NASA is also planning to establish fuel depots in low Earth orbit to avoid carrying booster fuel on long missions.

This would be a six-month journey fraught with life-threatening risks. Astronauts would suffer loss of bone density from weightlessness, as much as 40-percent loss of brain cells from cosmic radiation and weakened immune systems. Estimates require five feet of water to protect from radiation, although scientists have proposed creating an electromagnetic shield for this purpose.

TechCast's data show considerable uncertainty, but our experts think humans are likely to land on Mars about 2040, marking the beginning of deep-space travel.

Star Travel Begins About 2050

Despite TV shows like Star Trek, human travel outside of our solar system requires deep breakthroughs in physics to become feasible. The distances are too vast, the time required too long, the technology too sophisticated, the risks too great, and the need arguably nonexistent. A 10,000-fold gain in speed is required to go from walking to space travel within the solar system. The challenge of deep space requires another 10,000-fold increase, demanding far more powerful and efficient propulsion systems than are now thought possible.

DARPA and NASA are conducting a "100 Year Starship" program to help send humans on a one-way mission to the stars within a century. Scientist Jim Bedford noted at a recent meeting of the British Interplanetary Society: "A few years ago, here was only one organization working on interstellar travel. Now there are five."

Possible propulsion sources need to achieve the needed speeds include nuclear fission and fusion engines and combining matter and anti-matter to create extraordinary power. There is also the prospect of using an "e-sail" relying on the solar wind to accelerate the craft to the outer planets in a few years. The e-sail relies on charged particles in the solar wind to generate an electric field around the spacecraft. The field deflects these ionized particles and generates a force to accelerate the craft through its journey. This should produce a velocity of about 20 km/s, reaching Uranus in six years and half that time to much closer Galileo.

NASA advanced propulsion program is developing the theory of warp drives that would create distortions in spacetime to carry a spacecraft across the universe. This would avoid the laws of physics that prohibit faster-than-light travel. Theoretical studies at the Johnson Space Center in Houston suggest that warp drives could travel at 10 times the speed of light.

Still, astronauts will need to hibernate for the decades-long journeys. The European Space Agency (ESA) is studying hibernation states that would allow astronauts to sleep through the decades-long journey to distant star systems. Recent breakthroughs have put dogs into hibernation for months with no ill effects.

TechCast has studied this possibility over many years, and the consensus is that travel outside our solar system is likely about 2050.

First Contact

The search for extraterrestrial intelligence (SETI) once was considered quack science, but the National Academy of Science endorsed it years ago: "SETI is an important national resource in astrobiology," the group pronounced.

Today, planets are routinely discovered orbiting other stars, evidence supports the existence of life throughout the universe, telescopes and communications are becoming exponentially more powerful, and many scientists think the odds strongly favor the existence of other civilizations. One estimates: "There are probably 10,000 to 100,000 advanced civilizations in our galaxy alone." NASA recently discovered seven Earth-sized planets at a system 40 light-years away, with three of the planets firmly located in the so-called habitable zone. The concept of "panspermia"—the theory that the seeds of life are prevalent throughout the universe—has been supported by many studies that find organic molecules exist in deep space.

A NASA mission is being planned to send a spacecraft on an interstellar mission to explore the Alpha Centauri system, 4 light-years away from Earth. Traveling at a tenth the speed of light, the trip would take 44 years to reach the star around 2113, and the data wouldn't get back to Earth until 4.4 years later.

Despite 50 years of fruitless search, the SETI program could detect its first alien civilization anytime. Two-way communications are impossible using known technologies because of the vast distances involved. Yet, we are fairly certain now that there is lots of intelligent life out there, and the challenge is to find it. TechCast experts estimate contact is most likely after the Space Age arrives in earnest, about 2060 or so.

Global Maturity Is Needed To Enter The Space Age

These forecasts cannot be accurate precisely, of course. Yet, they capture the essential timing and trends anticipated by the background data and the knowledge of our thought leaders who make up TechCast's international brain trust of experts.

With NASA's encouragement of private launch services, the late 2020s are likely to see the growth of a vibrant space industry. It should then expand with space tourism and eventually move deep into the solar system about 2050 to start the Space Age. In the process, a private space industry will relieve the planet of much polluting mining and manufacturing, provide an endless supply of clean and cheap energy, contribute to global prosperity, and in the long run give humanity a way to survive the next extinction-level asteroid impact. Next to all those benefits, the cost of these pioneering space programs is clearly a bargain.

All these breakthroughs are pioneering efforts to explore and map the new terrain of space, much as the first pioneers explored the American West centuries ago. But that's only half of the journey. Then comes the hard part of settling these new lands, just as ordinary citizens became settlers taming the Wild West. Judging from the mess the world is still in, we still have a long way to go here on Earth, much less to settle space.

The greatest obstacle to the coming Space Age is the confluence of threats making up the Crisis of Global Maturity. Absent some unexpected stroke of luck, the world is following a worn-out path that is no longer sustainable. On our present trajectory, climate change is starting to flood coastal zones, scorch temperate lands, foster disease and famine, spawn more tornadoes and hurricanes, and provoke other extreme weather. Extreme temperatures increase energy use for heating and cooling, perversely making climate change worse and causing even more extreme temperatures. UN Secretary General Antonio Guterres warned, "We are in deep trouble." (Dennis, 2018).

With nations deeply in debt and trillions of dollars zipping around the globe at lightning speeds, leaders in international banking are fearful it is only a matter of time to the next financial meltdown. The yawning gap between the wealthy and the middle class shows no signs of diminishing, and we may see more class conflict between rich and poor. With autocrats running large parts of the world, violence and war are on the rise.

The technology revolution will add even greater threats. Smart cars, for example, will follow a similar path as smartphones. "A car is like a cell phone, and that makes it vulnerable to attack," said Jonathan Brossard, a security engineer. Many are horrified at the prospect of AI guided weapons turning on people. Now, ponder what could happen when billions of intelligent devices are wired into the Internet of Things?

While these reactionary forces are pushing back, we are also being pulled forward by the technology revolution, creative entrepreneurs, the rise of women to power and a new breed of young people who seem united across cultures. This transition will require a different worldview among billions of people, transformations in leadership and reformed institutions.

Keep in mind that the LCE is maturing rapidly. The great S-curve formed by the LCE is the universal symbol of life. All living systems pass through this same process of birth (beginning of the S-curve), growth (upward phase), and maturity (saturation) along their life cycle – a culture of bacteria, a growing child or the life of a planet. Today, global civilization is poised to transform into a mature society able to enter space.

This transition is an infinitely larger version of the same crisis of maturity that forces teenagers to transform into adulthood. Anyone who has raised children knows that teens may be fully grown physically and "know everything," but they have not learned to control their impulses and get along in society. The typical teenager struggles with inner conflicts and doubts, is well-educated but unwise and unsure how to move ahead in a confusing life.

That is roughly the state of our world today. Most industrialized nations are fully developed physically, awash in information and with enough armaments to destroy us all. Yet they lack the will and wisdom to address climate change, regulate economies safely, curb terrorism and a flood of other crises. The world struggles with conflicts between capitalism and society, Islam and the West, growth and the environment and endless other deeply rooted problems that are ideological. People think we are heading toward a disaster of catastrophic proportions, and they have little faith in their leaders' ability to avert this collapse.

At some point, the stress becomes so severe that most teenagers eventually find the courage to act more wisely, grow up and become responsible people. In a roughly similar way, the global passage to maturity is humanity's challenge to grow into a sustainable civilization. The human race is being forced to grow up, to develop a mature global order – or perish. Progress may arise out of noble motives, but also out of necessity.

The superior power of consciousness provides the key to resolving the multiple crises of today. Each stage in social evolution has been propelled by revolutions – the Agrarian Revolution, the Industrial Revolution, Post-Industrial Revolution and, most recently, the Information Revolution. As the LCE shows, we are now in the beginning throes of what I call a "Mental/Spiritual Revolution" to kick start the Age of Consciousness. I think the world is heading toward some type of historic shift in consciousness, a collective epiphany, a new mindset, code of global ethics or spiritual revolution. This would fulfill Teilhard de Chardin's insight that planetary consciousness is a natural end-state of evolution. It is also what has been called a Type I Civilization, when a unified world harnesses its collective energy (Kardashev, 1984).

Such heroic change may seem daunting, especially at a time when hostilities seem endless and environmental disaster looms ahead. That is often the case before upheavals. Nobody thought the USSR would collapse up until it actually did. I suspect that this tumultuous transition is a normal but traumatic process in social evolution.

The main reason this view seems optimistic, and even foolhardy, is because we have no experience in global consciousness. Huddled in our small section of a limitless universe, humans have little conception of planetary evolution, much less the transition to a unified world. Our understanding is roughly similar to a naïve person who first witnesses the agony of a human birth or a teen struggling to adulthood. Without previous experience, these painful transitions would seem awful, too hard to bear. Yet they are entirely normal and usually successful.

So too is our passage to global maturity likely to appear in years to come. It should be a moment of truth for civilization, much like that crucial point when a teenager faces the need to become an adult. This may sound too good to be true, yet I think we could see the beginnings of a unified planet over the next few decades. Then it's on to the Space Age.

REFERENCES

De Chardin, P. (1955). *The Phenomenon of Man*. Harper.

Dennis, B., & Mooney, C. (2018, Dec. 6). Global Carbon Reaches Record. *Washington Post*.

Diamandis, P., & Kotler, S. (2020). *The Future is Faster Than You Think*. Simon & Schuster.

Fowler, L. (1985). *From Outer Space to Inner Odyssey*. New Age.Journal.

Halal, W. (2008). *Technology's Promise: Expert Knowledge on the Transformation of Business and Society*. Palgrave Macmillan. doi:10.1057/9780230582538

Halal, W. (2004). The Life Cycle of Evolution: A Meta-Technological Analysis of Civilization's Progress. *Journal of Future Studies, 9*(1).

Hawking, S. (2016, Dec. 1). This Is the Most Dangerous Time for Our Planet. *The Guardian*.

Kardashev, N. (1984). On the Inevitability and the Possible Structures of Supercivilizations. *The Search for Extraterrestrial Life: Recent Developments; Proceedings of the Symposium*.

Kelly, K. (2016). *The Inevitable: Understanding the 12 Technological Forces That Will Shape Our Future*. Viking.

Laitner, J. (2004). *Energy Impact of Emerging Technologies*. Environmental Protection Agency.

Linstone, H., & Turoff, M. (1975). *The Delphi Method*. Addison-Wesley.

Wolfers, J., & Zitezwitz, E. (2004). Prediction Markets. *The Journal of Economic Perspectives, 18*(2), 107–126. doi:10.1257/0895330041371321

Chapter 19
Space Tourism and Commercial Deep Space:
Humans Going to and Beyond Low Earth Orbit

Dennis Meyer Bushnell
NASA Langley Research Center, USA

ABSTRACT

After some 60 years of highly trained astronauts going into space, the related technologies and costs have altered to the point where increasing numbers of private citizens can become space tourists, initially suborbital for minimal times and Earth orbital for up to the order of two weeks. There has also developed a rapidly improving digital reality/immersive virtual presence technology providing space tourism experiences at minimal cost and available essentially to everyone. The safety aspects of physical space tourism need further development, but those that relate to the space environment are tolerable for a few weeks from the 60 years of manned space flight experience. As space tourism over the years expands beyond earth orbit to moon, Mars, asteroids, other planets, etc., the safety issues will need to be seriously worked. Tourism is only a portion of what will become major opportunities and expansion of commercial space beyond earth utilities into deep space, enabled by the ongoing major reductions in the costs of space access.

INTRODUCTION

Since the late 1950's, humans in space (defined herein as Low Earth Orbit (LEO) and beyond) have, with very few exceptions, been restricted to highly trained astronauts. Going forward, there is an increasing expectation that technologies will enable the public to visit and vacation in space. With the ever-increasing capabilities of what is now termed digital reality (DR) or immersive presence, there are two approaches to space vacations: virtual and physical. Both will be discussed herein (Bushnell, 2020). The potential space tourism experiences include destinations such as space stations, moons, planets, and asteroids.

DOI: 10.4018/978-1-7998-6772-2.ch019

Also, virtually only with currently known or projected technology, planets/moons around other stars. Discussed herein are the space tourism issues that technology needs to address to enable space tourism, resultant space tourism experiences, and developing commercial deep space.

Space is dark, cold, a nearly perfect vacuum, with microgravity, GEV level galactic space radiation, unimaginable distances, where solid matter is a trace species, but which supplies the energy that enables human existence. Space is often referred to as the final frontier, and the general environmental conditions, as stated, are very different than those under which humans have evolved on Earth. Therefore, considerable technology is required to enable humans in space. In fact, even physics appears to change at cosmological scales including dark matter/energy, huge disagreements between quantum theory and the cosmological constant, and the mystery of what happened to the anti-matter. There is also the increasing interest in, and search for, life on other worlds, which may be silicon- or sulfur-based instead of carbon. Overall, there is a lot to learn. Solar system destinations beyond the Moon, Mars, and nearby asteroids require greatly increased trip time (years to decade(s)), costs, distances, and improved health and safety technologies.

The Space Tourism Issues and Options

The basic issues which must be addressed to enable space tourism are safety/reliability and cost/price. Of these, the first is the more difficult and includes many aspects such as dust, gravity, radiation, micro/other meteorites, and reliability. There are very encouraging approaches to reducing cost/price.

Safety. Currently, rockets malfunction-to-explode on the order of one every 100 launches. That accident rate is 100 to 1000 times the accident rate of scheduled airlines (Bushnell, 2020).

Sampling of safety and reliability related issues in aerospace. These can involve human error, which is a major source of safety problems, as well as crashes, even after many decades of Research and Development (R&D) and technologies to reduce such.

- Equipment failures. These may be due to design, installation, operational employment, maintenance, aging, environmental effects, and tipping points for cascading system failures.
- Electron/photon related failures including cyber/software issues, Electromagnetic Pulse (EMP)/space weather/radiation, jamming, failure of equipment essential for navigation, controls, sensing, propulsion, and communications.
- Inadequate margins, especially regarding a cascading system of systems failures where reduced capability does not fail the parent part but adversely impacts in serious ways the functionalities/piece parts that the overall system depends on for robust performance.
- Operation in extreme environments/in the presence of discrete extreme environmental and operational conditions.
- Bio becoming more virulent in space (e.g., pathogenesis, corrosives).
- Potential impacts of reusability upon safety and reliability and certification for such.
- Human health issues (e.g., micro g, radiation, psychological, accidents, and illness).
- Cost/profit/schedule exigencies, “corner cutting”.
- Oxidative chemistry, causing equipment failure.
- Electrostatics, causing electronic failures, fires, and explosions.
- Planetary and moon dust, abrasive, corrosive, oxidative.
- Fatigue and fracture, a prime design metric for much of aerospace equipage.

- Unknown unknowns; it is usually not possible to anticipate all combinational situations, conditions which will result in a reliability issue; we can become surprised reliability-wise.
- Inadequate analysis and testing, due usually to either cost or inadequate knowledge or study

In aerospace in general, some 80% of accidents are due to human factors. Therefore, the major first order approach to safety is to utilize autonomous Artificial Intelligence (AI) robotics to both manufacture and operate rockets/space access. Then there is a long list of safety-related design/operational precepts that can be applied to greatly improve rocket safety as well as the safety and reliability of the spacecraft, habitats, on-planet transportation, and the systems that keep humans healthy in what, externally, is a terminally unhealthy space environment.

On the Moon, dust was found to be a problem during Apollo, and Mars, with its rarefied CO_2 atmosphere, has huge, long lasting dust storms. Therefore, equipment must be designed to keep the inhabited areas as dust free as possible. Also, spacecraft/habitats or habs, must be designed to protect against micrometeorites or larger space debris, ensure structural integrity, and be leak tight.

Radiation in space, outside the protection of Earth's magnetic fields and dense atmosphere, is orders of magnitude worse than the radiation produced by nuclear weapons. The most worrisome is Galactic Cosmic Rays or particle radiation, such as fully ionized iron at tens of GEV energy levels (Moses, 2018). Protection from this requires four meters of material, either regolith (living underground) or for in-space habs the same thickness of materials with a high hydrogen composition to reduce the major health concerns about secondary neutron production. This radiation is carcinogenic and affects nearly all physiological aspects, including the cardiovascular and central nervous systems. An alternative radiation protection approach is fast transits, i.e., 200-day round trips to Mars instead of three years using usual chemical rockets. There are at least two approaches to provide such fast transits. For destinations beyond Mars, well beyond what the human body can withstand regarding radiation received during in-space trip time, some form of serious radiation protection will be required.

The in-space micro g, and probably even the 0.38g on Mars (or the $1/6^{th}$ g on the Moon), have serious physiological effects on the human body, including wasting bones and muscles and eye problems (Moses, 2018). Rotational artificial gravity incorporated into the design and functionality of the deep space habs would be efficacious.

It is useful to contrast the space tourism issues faced by physical tourism with the increasingly popular alternative, immersive presence/digital reality (Bushnell, 2020). The development of digital reality (DR) or virtual/non-physical presence began with the telegraph and progressed to the telephone, then telephones with screens, and now to computers. The nearer-term developments of DR include augmented reality (AR), advanced virtual reality (VR) heading toward five senses VR, and holographic projection off flat screens. Going forward with the availability of ever greater bandwidth and direct machine to brain communications, instead of through the senses, DR is projected to be as good as, or better than, physical reality. Even the current DR technology provides the critical body language aspects of human communication.

Compared to physical travel, DR provides the following benefits:

- Massive cost reductions, and not just for transportation; total trip costs are reduced by factors of many millions.
- Large time savings, again, not just transportation time but total trip time.
- Much improved use of time - can visit many planets/moons in a day.

- Do not have to be absent from family.
- Travel/entertainment experiences for all, affordably, on their personal schedule, including the infirm and the young.
- Enables expansion of the huge ongoing societal shift, begun some 25 years ago with the Web, into the virtual age with tele-everything such as tele-commuting, work, shopping, education, medicine, socialization, commerce, and politics.
- Avoidance of health problems with regards to diet, sleep, exposure to space radiation, gravity, and other hazards of space

Even before COVID-19, a virtual leisure travel industry was developing (including a wholly virtual "airline"). This industry has greatly expanded during COVID-19. The industry provides virtual tourism as an alternative to physical tourism and along with the many benefits of DR already discussed, includes virtual trips to destinations without the increased crowding often associated with physical travel and virtual trips, and traveling adventures which are not physically possible. This includes travel to historical places as they were in previous ages and to inaccessible places in space. The virtual tourism industry has shown strong growth/success, especially during COVID. Similar success has been observed with virtual business conferences, with many more able to attend, no venue costs/scheduling issues and excellent networking opportunities.

Immersive presence space tourism to the Moon/Mars is a developing business. What could be put together for children is a Martian experience where, as they round a boulder a Martian of the parents' choice in terms of configuration is there to conduct a tour of their planet. The data to inform and enable building these improving virtual experiences are provided by many countries employing the many sensors, crawlers, etc. conducting scientific exploration of planets, asteroids, moons and now, for the Earth's Moon, industrial development. This increasingly realistic alternative for space tourism is in the context of digital reality and increasingly immersive with five-senses haptics and holographic projection under development.

The Current Status of Space Tourism

For physical space tourism there has been a near term emphasis upon suborbital flights going forward. Orbital space tourism was begun by the Russians, who in eight trips took seven suitably trained tourists to the International Space Station (ISS) from 2001 to 2009 on Soyuz. The nominal cost of such trips is now in the range of $20M to $25M, via Blue Origin, Virgin Galactic, Boeing, and Space X. There were several startups to provide orbital space tourism during and after the Russian space tourist trips. These were not successful, perhaps partially due to the economic conditions in that era. The U.S. is planning for Space X Crew Dragon and Boeing Starliner to transport tourists (sometimes termed spacecraft participants) to the ISS, with an ISS accommodation cost of $35K/day each. NASA is largely responsible for the reinvigorated emphasis upon private space travel. The NASA commercial crew program with Boeing and Space X has as one of its objectives the fostering of commercial/private human space access. With that program thus far a success, there are now serious plans to exploit that space access capability for space tourism (Wattles, 2020). Bigelow Aerospace and Orion Span have nascent luxury orbital hotel designs. Space X and Space Adventures plan Moon loop flights for $70 to $100 M. A survey indicated that there is significant public interest in spending up to two weeks on the ISS with a spacewalk, with some desiring a space hotel (Purdy, 2018). Space Adventures plans to fly two tourists to the ISS in 2023 on Soyuz. Axion/Space X plan to fly three tourists to the ISS in 2021 for 10 days. Space Adventures and Space X plan a four-to-five-day space trip, no ISS visit, in 2021. Axiom Space is planning a private module addition to ISS as a hotel.

There appears to be accelerating serious near-term interest in orbital space tourism. The current costs are daunting, but the activity, interest, and now emerging capabilities to do these trips plus the large Space X LEO access cost reductions bode well for the future of space tourism, providing that there are no accidents and loss of life (Matignon, 2019). The projected mean number of missions for loss of crew on space shuttle was a factor of order two more than what occurred. However, the shuttle was designed in the 1970's and built for half the projected cost due to budget cuts. We know more now and have more experience and larger budgets. Therefore, space tourism will probably experience at least the historical average of about 100 flights between failures. The extent to which rocket safety can be improved and receives public acceptance is not yet known. Serious space tourism in Earth orbit will start over the next few years. The two weeks or less in space should not produce many of the adverse health effects that characterize longer stays. ISS is within the planetary magnetic fields that offer some protection from radiation (even though it is above the protective atmosphere). In fact, the radiation exposure is half of what would be experienced in deep space (e.g., during trips to the Moon, Mars, and other destinations). The typical ISS duty period for highly trained ISS astronauts is six months. A year on ISS for Astronaut Mark Kelly was difficult. In deep space, travelers are exposed to full radiation, and serious radiation protection and probably artificial gravity will be needed for trips to Mars and beyond.

With respect to the present status of virtual/digital space tourism, NASA has several immersive/virtual tours of destinations such as the ISS and the planets (NASA, 2019). Companies offer virtual tours of the Moon, Mars, the planets, and the solar system. Google offers a tour of the ISS, space tours are on YouTube, and the VR companies are planning to offer live virtual tours of Earth from space. Digital reality will provide some 80% of the space tourism experience at some million times less expense and some thousand times greater safety.

The Outlook For Space Tourism

Due to the high costs of rocket production and operations including launch, those technologies that provide extreme reductions in the costs of space access are the fundamental enablers for space tourism (Bushnell, 2018). These costs are mainly due to rocket manufacture and operations including those associated with launch. The bulk of the costs are human labor. Space X has been extremely innovative and successful in reducing these launch costs via reusable rockets, cutting manufacture cost per launch, and in increasing automation, printing and robotization for manufacture and operations, thereby cutting human labor costs. The cost of fuel is very minor (less than 1%) and constitutes the limit to which LEO access costs can be reduced. Space X has produced a factor of six reduction in space access costs and is well along in providing a factor of 14 reduction. Such reductions are the basic enabler for affordable space travel. Increased use of autonomy and robotics for manufacture of the other equipment required for space tourism will also provide serious cost reductions with an overall resulting trip cost that will be increasingly manageable by the public. Therefore, of the two major issues/metrics for physical space tourism stated previously, cost and safety, requisite cost reductions are on the horizon. Unfortunately, the many aspects and issues surrounding safety are works-in-progress. For physical space tourism, short jaunts for the usual week(s) vacation period (e.g., to nearer destinations such as space stations, the Moon, near Earth objects) would be relatively safe regarding the major health safety issues (E.g., micro g and radiation).

However, launch safety and overall reliability probably still require significant improvements for public tourism utilization. Trips to Mars and beyond entail much longer trip times and probably artificial gravity and more effective radiation protection (Moses, 2018). Some have posited that psychological issues might occur during such long trips, which might be alleviated by including holographic members in the tourist group with characteristics updated periodically.

Radiation Protection Options in Decreasing Order of Technology Readiness Level (TRL)

- For on Moon, planet, etc., ~ under three to four meters of regolith.
- For in-space:
  - Fast transits (e.g., 200-day round trips to Mars) via inexpensive chemical fuel enabled by far lower LEO access costs.
  - Three to four meters of reusable polyethylene overcoat, the weight/cost of which is enabled by inexpensive chemical fuel and much less expensive space access.
  - Biological/medical countermeasures, a partial solution in space thus far,
  - Fast transits (e.g., 200-day Mars round trip) via 6,000 sec. Isp VASIMR high thrust MHD propulsion powered by an alpha of order one nuclear battery.
  - Magnetic redirection of Galactic Cosmic Ray (GCR) particles via superconductive (S C) magnets located extended distances from the spacecraft.
  - Silicon crystal reflection of GCR particles plus shielding for Gamma secondaries. This approach may be applicable to space suits, and in that event, an exoskeleton may be required to handle the additional inertial loads.

Prospective Commercial Space Activities Related to Space Tourism

- Major LEO constellations of small satellites for high-speed internet and Earth observation, expanding the number of satellites from the order of 1,000 now to some 10,000 plus by 2025. The associated Earth observation capabilities could enable "staring" anywhere 24/7/365.
- "Utilities" for beyond Geosynchronous Equatorial Orbit (GEO) to service both public and private customers, including communications, energy/fuel, transportation, maintenance/repair, and life support.
- Mining of the Moon, Mars, and asteroids for anything commercially viable or needed for colonization such as water, minerals, He3, rare earths, volatiles, and "mass.". There are purportedly 850 near-Earth asteroids larger than a kilometer in diameter and many smaller ones at lunar distances from Earth or less.
- Entertainment including virtual reality (VR), videos, virtual presence (e.g., to enable spending an evening exploring Mars from your living room)
- Asteroid defense including detection, tracking, and diversion of threats deemed capable of causing grievous harm.
- Space solar power for utilization on planets, moons, asteroids, in space, delivered via energy-beaming using microwave or lasers.

- "Space Beach Combing," the identification, collection, destruction, repurposing, and/or remanufacturing of space debris. Of special interest is boosting ISS, in due course, into a parking orbit and scavenging its parts by the piece.
- Space as a trash dump involves putting "trash" in parking orbits for "safe storage," including possibly some components of nuclear waste depending upon if it could be certified "launch indestructible."
- Space manufacturing in orbit, in-space, on other "bodies" or enroute. This could include initially products that are much improved by production in micro g such as pharmaceuticals, fiber-optics, ball bearings, LEDs, solar panels, organs, hearts, and protein crystals. There are also products that benefit from the near absolute vacuum of space, and the manufacture of fuels and on planet/body or in-space human or robotic equipage.
- Space hospitals if microgravity or other in-space conditions prove to be efficacious for specific human ills.
- Quantum technologies and quantum computing, utilizing the "quiet" conditions in space, vacuum, low temperature, etc. to delay de-coherence and stabilize quantum states.
- Enhanced positional Earth utilities including telecom, internet, navigation, weather, imagery/ Earth observation, and resource monitoring at lower cost and with larger apertures and greater resolution.
- Space weather forecasting for prediction of potential space condition impacts upon electronics both in space and on bodies including Earth.
- Communications and navigation, other satellite functionalities for the Moon, Mars, etc. to replicate or improve as is useful the extant "positional Earth utilities" that constitute most of current commercial space.
- Refueling depots. The fuel to supply such depots could be sourced from Earth, the Moon, Mars, asteroids, etc., anywhere that provides the fuel at lowest cost. These depots could be in Earth orbit/ Earth environs, anywhere suitable/convenient/required in-space or on "bodies." The fuel could be chemical of various flavors including CH_4, H_2/O_2, nuclear reactor, or nuclear battery "fuel." Where propulsive mass and energy have been and can be separated, such as utilization of solar or beamed energy using the various flavors of nuclear energy including positrons, propulsive mass for propulsion utilizing external energy addition could be supplied. The total refueling depot system architecture includes fuel sourcing, production, transportation, storage, and disbursement.
- Repair/servicing/maintenance functionalities-could be either itinerant or fixed, probably both, could connect to fuel depots in some cases, and ultimately be capable, possibly via printing, etc., of repair/maintenance of nearly everything. Reusability implies possible refurbishment, repair, maintenance, which are not an operational issue with a one-time use approach.
- To service humans, a combination lifeboat, hotel, search, and rescue functionality, which could also be sited at fuel depots, utilized for survivability, transit, and space tourism.
- Energy beamers, nuclear or solar powered, including by positrons, the cheap anti-matter. These could be located on bodies, in orbit or in space, and provide an alternative energy/powering source, perhaps involving less cost than producing, handling fuel per se. This would still require propulsive mass for utilization of beamed energy for propulsion instead of the many other uses of beamed energy including asteroid defense, mass drivers, on body/spacecraft power and energy, space manufacturing, and other industrial/commercial activities.

- Momentum tethers, space elevators, utilized as alternatives for rocket transportation. Some have envisaged these as Earth/Moon transport or for even longer distances and they are rather major putative reusable infrastructures, capital projects.
- Propulsive Ground Assist-propulsive infrastructures for space access including the slingatron, blast wave accelerators and mass drivers-alternatives to rockets or tethers or as a partial launch assist.
- Space manufacturing facilities and space “Business Parks”. These could be located in space and/or on bodies, all types of bodies, depending upon the business to be conducted, resource availability, transportation costs and products produced.
- Utilities for space colonization including sourcing, transfer, storage, disbursement of food, water, energy, “supplies”, including from In-Situ Resource Utilization (ISRU) activities.
- Space tugs, in-space transfer, including cyclers, to move raw materials and products. In the absence of time criticality, these could employ sails-solar sails, laser sails, magnetic sails, or particulate sails, aka “sloboats”.
- On-body transportation, nuclear powered or “fueled” otherwise, for “freight” and humans. Due to the typically rough terrain of bodies, in-atmosphere, near-body flying is probably the most efficacious transportation approach for other than very localized transportation.
- On-body infrastructures-habs, mines, manufacturing plants, energy sources, life support, etc. especially for colonization.

Potential Closed Business Cases for Commercial Deep Space

These include:

- Commercial space utilities beyond GEO including communications, energy/fuel/transportation, maintenance/repair and life support. In various forms and flavors, these will be needed for both government and commercial deep space activities, and especially for colonization. Development has already begun on several of these including communication.
- Space mining-asteroid water is particularly interesting, especially if the quantity of Moon water proves to be significantly less and/or extraction costs prove significantly more than anticipated. Given the extant competitive ocean (accessed via the increasingly low cost of renewable terrestrial energy)/other Earth resources, space mining may be more applicable to deep space utilization(s) than for use on Earth.
- Space beach combing/cleaning up space debris. Given the current situation with respect to space debris and the plans to loft many more satellites, we will probably have to move on from avoidance of space debris/decommissioned satellites to removal. The legal issues and costs have held that in abeyance. The costs could be addressed via use of E-M tethers powered by NTAC or solar, fuel-less transportation to collect space debris for space manufacture repurposing/remanufacturing.
- Space manufacturing. With the space access cost reduction in the offing from reusable rockets, the major impediments to space manufacturing, which are cost and schedule, are greatly mitigated. This capability would also enable in-space manufacture of equipage not suitable for launch (e.g., equipment that is too large or too fragile) even in piece parts. Also, there are many products that are either enabled or much improved by in-space micro g and/or deep vacuum conditions.

- Space tourism. When human health/safety issues including safety, reliability, radiation, and microgravity are addressed, which are major mission design issues, the inexpensive space access enabled by reusable rockets should greatly accelerate space tourism.
- Quantum computing/technologies in space. This is a "new-bee," with viability and realism yet to be determined. However, for delaying decoherence and maintaining quantum conditions, the temperature, vacuum, and "quiet" conditions of space appear to be of interest (initial experiments are in progress).
- Space solar power, especially for other bodies and in-space utilization initially, and eventually with the potential for Earth use, depending upon the business case vis-à-vis the ever-reducing cost of terrestrial renewables/storage with their massive potential capacity.
- Earth positional utilities. This is the current, very successful, commercial space. Inexpensive space access will make it even more so and several additional areas/functionalities will be enabled.
- Colonization of bodies such as Mars with all of its resources, would create a wholly new economy that could be essentially self-contained. Mars could, with its resources, become the department and building supply store for the solar system as human colonization moves beyond Mars.

Concluding Remarks

The bottom line on space tourism is that after some 60 years of highly trained astronauts going into space, space-related technologies and their costs have altered to the point where private citizens can become space tourists. Initially, space travel will be suborbital for minimal times and Earth orbital for up to two weeks, with near-term plans by Space X to do lunar flybys. There has also developed a now rapidly improving digital reality/immersive virtual presence technology providing space tourism experiences at minimal cost and available essentially to everyone. The safety aspects of physical space tourism need further development, but the 60 years of manned space flight experience suggest that those relating to the space environment are tolerable for a few weeks. As space tourism expands beyond Earth orbit to the Moon, Mars, asteroids, and other planets, the safety issues will need to be seriously worked, and that is doable. The apparent breakthrough in the resurgence of space tourism was the NASA commercial crew program, which was partially targeted on space tourism and has led to the transportation necessary to enable the nearer term, very few years out now, plans for a goodly expansion of physical space tourism. This is only a portion of what will become major opportunities and expansion of commercial space beyond Earth utilities/up to GEO into deep space, initially the Moon, Mars, asteroids, and near-Earth objects. All of this is greatly enhanced by the ongoing major reductions in the costs of space access. Given the wonderous ongoing progress in AI-enhanced autonomous robotics, for goodly cost and safety reasons, much to most of this commercial deep space will probably not include onsite humans. There are projections that travelers that colonize Mars will, over time, due to the reduced g and radiation exposure, evolve into Martians.

REFERENCES

Bushnell, D. M. (2020). The Coming Digital Reality. In *Aerospace America*. AIAA.

Bushnell, D. M., & Moses, R. W. (2019). *Reliability, Safety and Performance for Two Aerospace Revolutions – UAS/ODM And Commercial Deep Space*. NASA/TM.

Bushnell, D. M., & Moses, R. W. (2018). *Commercial Space in The Age of New Space, Reusable Rockets and The Ongoing Tech Revolutions*. NASA/TM.

Bushnell, D. M., & Moses, R. W. (2020). *Prospectives in Deep Space Infrastructures, Development and Colonization*. NASA/TM.

Matignon, L. d G. (2019). *Space Tourism Legal Aspects*. Space Legal Issues. https://www.spacelegalissues.com/space-law-space-tourism-legal-aspects/

Moses, R. W. (2018). *Maintaining Human Health for Humans on Mars*. AIAA.

NASA. (2019). *NASA Opens International Space Station To New Commercial Opportunities, Private Astronauts*. https://www.nasa.gov/press-release/nasa-opens-international-space-station-to-new-commercial-opportunities-private

Purdy, C. (2018). *Welcome to the first luxury space hotel*. Quartz. https://qz.com/1247091/orion-span-claims-its-luxury-space-hotel-will-open-to-its-first-guests-in-2022/

Wattles, J. (2020). *SpaceX Teams Up With Space Tourism Agency To Sell Rides Aboard Its' Spacecraft*. CNN Business. https://www.cnn.com/2020/02/18/tech/spacex-tourism-space-adventures/index.html

Chapter 20
Asteroid Futures:
The United States of America, Space Mining, and Sovereignty in Space 2.0

Nicholas Baldwin
Independent Researcher, New Zealand

Amy Lynn Fletcher
Futurist, A Fletcher Strategies LLC, USA

ABSTRACT

This chapter evaluates the emerging industry of asteroid mining and the pivotal role of the United States in shaping the new rules for an extra-terrestrial economy. The Outer Space Treaty 1967 (OST) governs the use of space, with over 100 signatories, including the United States and China. However, as space exploration expands to encompass both public and private stakeholders, there is a growing international debate about whether the OST's provisions prohibit the assertion of sovereignty and, hence, property rights, in outer space. With the Commercial Space Launch Competitiveness Act (2015), the United States has pursued a legal framework that facilitates commercial asteroid mining and a political strategy that focuses on bilateral space exploration agreements with countries such as Luxembourg, Italy, and the United Arab Emirates. Due to its dominant position in the space sector, the United States will strongly influence the regulatory roadmap for the era of Space 2.0.

INTRODUCTION

Space 2.0 refers to the emerging era of twenty-first century space exploration in which not only national governments but commercial entities, such as Blue Origin (Jeff Bezos, founder) and SpaceX (Elon Musk, founder) hope to establish a viable space economy that encompasses activities ranging from space tourism to rocket launches and even astro-burials, asteroid mining, and the colonization of Mars. Other major players seeking to monetize outer space include China, which plans to launch a space-mining robot into space via Origin Space, a Beijing-based private company, and India, which in 2019 established a private institution, Newspace India Limited (NIL), to institutionalize the country's focus on the commercial

DOI: 10.4018/978-1-7998-6772-2.ch020

space sector. Space 2.0 is, essentially, a rebooted space race that retains the geopolitical features of the initial Cold War era contest—particularly in the national security concerns and positioning of major players such as the United States, Russia, India, and China—but which has expanded to include private companies, new niche players such as Luxembourg and New Zealand, public-private partnerships, and venture capitalists. Yet as space technology and the celestial futures it underwrites advance rapidly into orbit, one issue, national sovereignty, and its relationship to property rights in outer space, remains vexingly Earth-bound and intractable.

This chapter evaluates the issue of sovereignty in space as it pertains to both the emerging industry of asteroid mining and the pivotal role of the United States in shaping the new rules of the road for an extra-terrestrial economy. Multiple sources suggest that the ability to mine the asteroid belt will be a necessary component of both geopolitical advantage and economic competitiveness in this century. The investment firm Morgan Stanley estimates that by 2040 the global space economy could be worth one trillion USD (Morgan Stanley 2020). The nascent asteroid mining industry, worth an estimated $720 million USD in 2017, could reach $3.8 billion USD by 2025 (Allied Market Research 2019). The scope and scale of the profits to be made in space portends a new gold rush in which the risk of loss will be very high but significant rewards are likely to accrue to the first movers, the innovative, and the lucky.

Here on Earth, the fundamental legal and ethical problem is how to reconcile rapid expansion of the extra-terrestrial economy with the established international norms of the *Outer Space Treaty-1967*. The OST remains the key treaty governing the use of space, with over a hundred signatories, including the United States, Russia, and China. It emerged at a unique moment in the twentieth century, only two years before Neil Armstrong would step onto the Moon from the Apollo 11 spacecraft, and at a time wherein an emerging global environmental consciousness dovetailed with both the new politics of an enlarged United Nations (as previous colonies gained independence and then membership status) and international concern to prevent outer space from becoming another front in the nuclearized stand-off between the United States and the Soviet Union. The OST, while concise, contains several textual gaps that allow for skillful parsing of its actual intent when applied to an era of commercial space exploration. Proponents of asteroid mining often argue that the extraction of space resources by a commercial entity does not constitute a claim of national sovereignty any more than removing fish from the deep sea, with due respect for environmental law and catch quotas, would violate the Law of the Sea Convention (1958). Opponents of asteroid mining, and those who urge that the United Nations, not nation-states, continue to be the site for resolving space law and disputes, counter with serious reservations about the potential rapid depletion of space resources (Nature Astronomy 2019) and/or the economic inequities that could confront countries that are not technologically or economically capable of outer space exploration and mining. With both space technology and space entrepreneurialism rapidly advancing, the window of opportunity for regulatory bodies to establish a multinational framework for asteroid mining is closing.

This chapter analyses the role of the United States as a key stakeholder in the burgeoning space economy. It covers the period from the signing by President Obama of the *U.S. Commercial Space Launch Competitiveness Act* in November 2015 to the Artemis Accords, a Trump-era initiative which has resulted in bilateral agreements between the United States and Australia, Canada, Italy, Japan, Luxembourg, the United Arab Emirates, and the United Kingdom (with additional expected signatory countries in future). Drawing upon close reading of the relevant media and policy documents, this chapter identifies emergent legal and public policy issues and evaluates the framework for space mining that is being constructed via U.S. actions and in its interactions with other nation-states. Due to its dominant position as a major stakeholder in space exploration, the United States will strongly influence the regulatory roadmap being set down for the era of Space 2.0.[1]

The fundamental research question driving this chapter asks, *how is the United States using legislation and policies such as the Artemis Accords to articulate and organize the extra-terrestrial economy?* The theoretical framework for the chapter draws upon the sociology of expectations and its emphasis on how the *anticipation* of future returns from highly disruptive and nascent technological sectors such as artificial intelligence or space exploration makes narratives and legislative actions instrumental to bringing these scientific fields into the domain of finance and commercial enterprise. As multiple analysts have noted, "space has been transformed into a site of speculative financial investment" (Kearnes and van Dooren 2017, p. 179), drawing in venture capitalists and entrepreneurs eager to wager now in the hopes of securing a place at the table of the next great fortunes. We contend that the overarching space economy remains poised at this juncture between the "real" and the "promissory" (Martin 2015) and must therefore be constructed through anticipatory stakeholder discourses and policy decisions. In the initial phase of creating a space economy, competing narratives will be contested, both within countries and between countries, often vociferously, and then begin to stabilize as a dominant framework emerges. To argue that the space economy is promissory is thus to foreshadow the role of expectations, powerful actors, and initial decisions in shaping the pathway to what finally becomes business-as-usual. Yet before venturing into twenty-first century space, it is important to begin briefly with the OST and the ways in which this Cold War artefact continues to structure contemporary debates.

From Cold War To Lunar Gas Stations: The Outer Space Treaty Confronts Space 2.0

The question of sovereignty in space, and hence the *Outer Space Treaty-1967*, is again visible at both national and global levels due to the ongoing development of new extra-terrestrial mining technology and plans for space tourism. On the legislative front, space tourism may require little more than regulations for health and safety and some particularly adventurous and deep-pocketed customers. However, space-based mining and the rapid growth in technology will drive the need for the establishment of property rights if conflict is to be avoided (Lieb 2015, p. 2). The establishment of property rights would allow both public and private actors to secure ownership of the minerals extracted from celestial objects and thus provide opportunities for new industries to arise. As speculation about the potential profits from mining the asteroid belt escalates, the pressure on governments to approve mining missions will increase and, should the regulatory environment lag behind economic and technological possibilities, some commercial actors might be inclined to ask for forgiveness rather than permission as private mining expeditions become feasible.

Asteroids are the "solid, rocky and irregular bodies that are the rocky remnants of the protoplanetary disk of dust and gas that formed around our young Sun over 4.5 billion years ago" (Atkinson 2015). These consolidated spheres of ancient space dust contain a spectacular amount of precious minerals, including elements that are typically rare on Earth such as gold, platinum, nickel, iridium, palladium, osmium, ruthenium, rhodium, and many others. This potential motherlode of mineral wealth has tantalized entrepreneurs for decades. The disruptive interim development—which moves the space economy further towards the real from the promissory—is the discovery of frozen water on the far side of the Moon. Treated lunar water could readily be used as a rocket propellant fuel. Though this sounds esoteric, the water-to-propellent variable is the chemical foundation for building refuelling stations on the Moon that could, in turn, support missions into deep space that would otherwise be physically impossible from the Earth. In the words of Bob Richards, co-founder and CEO of Moon Express, "The recent discovery of water on the moon is an economic game changer for humanity's future. Water is the oil of the solar system, and the moon has become a gas station in the sky" (McMullan 2016).

Where there is oil, there will be prospectors, and increasingly those who seek to mine space are eager to establish outposts. This, in turn, requires regulations that can support commercial activity in space and provide a stable risk and investment framework. As noted by Weaver, "effective control of airspace in the Roman Empire… extended as high as an arrow could be shot [but] only after the development of heavier-than-air flight…did states attempt to assert control over the airspace above their territories" (Weaver 1992, p. 213). As space mining moves from science fiction to Silicon Valley, advancing via rapid technological developments, effective commercialization, if not control, of outer space will require similar new thinking. The *Outer Space Treaty* remains the fundamental global agreement governing space exploration and its associated activities. Yet, though property rights are key to avoiding conflict in a space economy, the OST remains unclear on the distinction between sovereignty and property rights, reflecting instead rhetorical language more attuned to an era in which human space exploration was new and geopolitics seemed to be moving inexorably in the direction of multilateralism. As early as 1961, the *Harvard Law Review* noted that "since national sovereignty is the fundamental ordering principle of international relations, such rules must deal with the scope and limits of sovereignty in outer space." In particular, the Association held that defining the exclusive authority of the state was essential in order to establish the line between legitimate actions and provocative behavior that could incur the resistance of other states (Harvard Law Review Association 1961, p. 1154). Throughout the 1960s, understandably, the Janus face of space exploration split between optimism for a new era of global cooperation and the tangible reality of nuclear proliferation and mutual assured destruction, hence the focus on the questions of sovereignty and appropriation had more to do with avoiding outright territorial claims and military zones (in this the *Outer Space Treaty* has much in common with the *Antarctica Treaty* of 1959). Yet as Space 2.0 expands to encompass significant commercial stakes, a major regulatory and legal challenge confronting stakeholders today is how to adjudicate different interpretations regarding the OST and its implications for property rights. As Salter and Leeson highlight, "without some means of enforcing claims to mine and thine, individuals have little incentive to risk investing in and growing commercial enterprises" (Salter and Leeson 2014, p. 583).

The prohibitive legal approach/interpretation of the OST, which seeks to rule out commercial exploitation of space, derives primarily from article two. This provision explicitly states that "outer space, including the moon and other celestial bodies, is not subject to national appropriation by claim of sovereignty, by means of use or occupation, or by any other means" (NASA n.d.). As noted, those who drafted the OST were primarily concerned to remove outer space from the militarized logic of the Cold War. Yet in a contemporary context, article two could also be construed as putting commercial development of space outside prevailing international law, thus rendering such activity illegal and unethical. Furthermore, article one of the OST states that "the exploration and use of outer space, including the moon and other celestial bodies, shall be carried out for the benefit and in the interests of all countries, irrespective of their degree of economic or scientific development, and shall be the province of all mankind"(NASA n.d.). This provision raises difficult legal questions regarding eventual resource distribution and/or compensatory payments to countries that may be unable to participate directly in space exploration and mining. A literal interpretation of the text could, for example, lead to a presumption of reparations or entitlements for non-space faring countries if codified in international law.

In summary, both the province of all mankind clause and the non-appropriation clause, could form a basis for rejecting appeals to exercise sovereignty in space, thus weakening the economic rationale for space exploration and mining. In this case, outcomes for actual space exploration could range from "celestial anarchy" (Salter and Leeson 2014, p.528), which would likely ratchet up both uncertainty and conflict, to the under-utilization of the rich reserves of resources available in outer space. Failing to tap the extraordinary resources of space would also arguably constitute a violation of the spirit and intent of the OST and points to a potential contradiction within the text. As Jack Heise notes, "the potential of asteroid mining to reduce the cost of spaceflight, moreover, could facilitate the growth of the space economy. Asteroid mining thus aligns with another stated purpose of the OST in the sense that an expanded space economy could provide substantial benefits to all mankind" (Heise 2018, p. 195).

However, the prohibitive approach is not the only way in which to interpret the OST and its relevance to contemporary debates about space mining. The language of the OST permits alternative interpretations and legal arguments regarding the intent of the draftees with respect to potential commercialization of outer space. Jefferson Weaver, for example, suggests that the OST is compatible with the argument that once resources are recovered, they can be subject to private ownership. For example, fishermen who are fishing in the ocean could not "claim ownership to that portion of the ocean covered by the net but only to the fish that the fishermen retrieved from their nets (Weaver 1992, p. 210). This interpretation would clearly matter to a company that seeks to extract minerals from the asteroid belt but does not intend, nor does its home country intend, to establish any sort of enduring territorial/sovereign claim on the celestial body itself. Moreover, as Virgiliu Pop notes, there is no specific clause in the OST that prohibits a private individual from owning extra-terrestrial real estate (Pop 2000, p. 276). Again, just as fishermen are allowed the spoils of the fish they remove from the deep sea, any minerals mined in space are arguably the lawful property of those who mined them. This perspective, grounded in the philosophical tradition of John Locke (1632-1704), suggests that while national sovereignty is indeed prohibited, the ability of the private individual (and under U.S. law this interpretation would likely include companies) to benefit from space as the 'province of all mankind' is preserved.

Notably, Pop (2000) further argues, drawing instead from the tradition of both Jean-Jacques Rousseau (1712-1788) and Thomas Hobbes (1588-1679), that private property cannot exist outside the protective sphere of sovereignty, and as state sovereignty is outlawed by the OST any concept of private property cannot be established. From this vantage point, the OST "makes no mention of private property, but it undercuts the ability of any government to recognize or enforce a private claim" (Pop 2000, p. 279). However, though Pop might ultimately disavow private ownership via his analysis of property rights in the Enlightenment tradition, the Lockean framework does provide proponents of space exploration with a credible lever to interpret the OST as not prohibiting private resource extraction in space. In the bluntest terms, if no national entity such as a military or border patrol is controlling these space outposts and prohibiting other states and companies from conducting research or commercial activity, then asteroid mining is permissible under international law. The core legal question pivots on whether one takes a literal approach to the text or a purposive one, while the political question hinges on which country or small group of countries currently has the standing to advance this expansive view in the face of potential opposition within the international community. As John Heise argues, "asteroid mining creates tension within the OST as an activity that is prohibited by the treaty's terms but largely in line with the treaty's purpose. As such, the OST should be modified to allow for greater certainty and predictability with respect to asteroid mining" (Heise 2018, p. 205).

Finally, the question of the equitable sharing of space resources is an issue raised by the term "province of all mankind" that pertains to the question of commercialization and the overall intent of the OST. The United Nations Convention for the Law of the Sea (UNCLOS) defines the seabed of the high seas as "the common heritage of mankind," just as the OST holds all of space to be the "heritage of all mankind." Under the auspices of UNCLOS, commercial exploitation of the high seas' ocean floor required the establishment of an International Seabed Authority (ISA) to ensure that royalties and profits would be shared with developing countries (Bruhns and Haqq-Misra 2016, p. 5). With respect to outer space, the *Moon Treaty-1979* attempted to institutionalize a similar resource sharing protocol. However, as no spacefaring nation ratified the document, its provisions are not binding in the international arena (Delgado-Lopez 2015, p.7).In essence, the heritage of mankind provision of the OST may continue to ensure that developing nations are not precluded from participating in the exploitation of space but it does not seem to establish any inherent right to a share of the extracted resources. Similarly, the Moon Treaty, signed by only four countries to date, does not provide sufficient legal precedent or political heft to block the purposive interpretation of the OST.

To conclude, the OST will likely be subject to ongoing legal disputes, and resolution of its regulatory intent at the global level will likely be protracted and contentious. The slowness of international disputation and diplomacy, as compared to the urgency of establishing a commercial toehold in outer space, is one of the reasons that the United States has chosen to prioritize domestic law and bilateral agreements with collegial countries. As will be discussed fully in the subsequent section, the United States is now governed by the *U.S. Commercial Space Launch Competitiveness Act* (Public Law No. 114-90). The law explicitly states that these commercial activities do not constitute a claim of sovereignty or jurisdiction. The United States, due to its current dominance in space exploration, is thus able to have its cake and eat it in too in that the SPACE Act (2015) specifies a good-faith intent to comply with the OST while decisively moving the United States beyond the ongoing legal and diplomatic contestation around the OST at the global level. In essence, Public Law No. 114-90 acknowledges the ongoing salience of the OST but signals clearly that the United States and its public and private partners are not going to wait for further clarification of the treaty, but instead intend to move into the space economy at warp speed.

Commercializing Space Futures: Building The Extra-Terrestrial Economy

In October 2020, NASA announced a planned 2022 mission, via a SpaceX Falcon Heavy rocket, to 16 Psyche, a rare asteroid made almost entirely of metal. Astrophysicists estimate that Psyche is as old as the universe itself, hence the scientific value of the mission lies in the knowledge it might reveal about how planets form. However, what really captured the interest of the international media in this case was that the metals on Psyche are worth an estimated ten million trillion dollars (Puiu 2020). Financial estimates like this defy any normal, Earth-bound sense of even extreme wealth - it is more than two thousand times greater than the estimate by Hooke and Alati (see Chapter 1) of the gross world product of mankind from the advent of Homo habilis through 2020 - and are part of the reason that outer space, like artificial intelligence and biotechnology, is increasingly subject to the "breathless futurology" (Harrington et al. 2006, p. 3) inherent in the construction of a new economy. While the vast mineral resources of outer space are tangible, translating this space dust into a functioning economy requires an extraordinary amount of work on the ground as stakeholders vie to unlock capital investment and construct plausible narratives of success that can attract public and private stakeholders and retain their commitment over long periods of time. Headlines about asteroids worth ten million trillion dollars are hype, particularly given that reliable human settlement of space is in its technological infancy, but hype isn't frivolous in this case. Rather, hype "is constitutive, it mobilises the future into the present [and] is part of the repertoire through which a narrative path or story line is constructed for technologies" (Brown 2003, p. 6). We can put an estimated price on these minerals, and via science and technology, we have a good idea of how much mineral wealth may be out there. Ergo, just as biological samples and digital data were transformed by the late twentieth century into resources that can generate wealth and seed new industries, so now are asteroids being reconfigured into lunar capital.

Yet outer space poses technological and financial challenges to human use and financialization that even biotechnology and the computer industry did not. In June 2019, *MIT Technology Review* ran an article titled "How the asteroid-mining bubble burst" (Abrahamian 2019). The article investigated the saga of Planetary Resources, an asteroid mining start-up that began with visions of creating a trillion dollar industry and which had the backing of prominent investors such as film director James Cameron and Google executives Larry Page and Eric Schmidt. In 2018, the blockchain company ConsenSys acquired Planetary Resources, releasing few specific details about the rationale behind the acquisition or its plans for the company. Some industry observers read this story as a cautionary tale against having too much faith in the nascent space economy. Peter Marquez, former Planetary Resources policy analyst and advisor with Andart Global, for example, argues that "there was more of a focus on the religion of space than the business of space. There's the religious [segment] of space people who believe almost like manifest destiny, we're supposed to be exploring the solar system—and if we believe hard enough, it will happen" (Abrahamian 2019).

This field-of-dreams approach to the space economy will continue to churn through venture capital and entrepreneurial start-ups at a voracious rate. Nevertheless, to frame Planetary Resources as an abject failure ignores the complex political and narrative dynamics of financialization and technological disruption. As O'Riordan et al. argue, the "development and consolidation of these promissory narratives is an ongoing process. The broader imaginary remains in flux as each promissory narrative is reconfigured and moves in and out of favour with the community" (2017, p. 152). Extending from this insight, even if asteroid mining proves more difficult than anticipated, or involves a time horizon beyond 2050, the meta-imaginary of the new space economy can accommodate shifting sectoral narratives. The space economy, writ large, ranges across an extraordinary number of potential wealth-generating activities that can have exponential and synergistic effects when considered in tandem. Joe Landon and Etienne Schneider project that the five industries most likely to comprise the space economy in this century are energy, mining, transportation, construction, and hospitality and real estate. They make the essential point that "when we realize that Earth's economy is only the beginning, our concept of growth changes exponentially. For industrial firms who have the foresight to view space not as a stand-alone industry but as the next medium to conduct their business, the sky is not the limit. The only limitations are the ones we put on ourselves" (Landon and Schneider 2017). No one industry, such as asteroid mining, can expect a smooth and inexorable glide to multi-trillion-dollar capitalization. Asteroid mining, however, is as necessary as the lure of gold was in the nineteenth century for creating "communities of promise" (Martin 2008) that will underwrite the development of the broader extra-terrestrial economy.

In May 2020, ConsenSys pledged to release Planetary Resources' patents, including its plans for microsatellites and the company's proposed method for asteroid mining, to any person or company acting "in good faith" (Boyle 2020) and it continues to use its blockchain technology to support TruSat, a crowdsourced space situational awareness (SSA) project that allows individuals to report satellite positions to a network of users (Foust 2019). In disrupting the field of satellite data collection, which hitherto has been tightly controlled by government agencies such as the United States Air Force, ConsenSys, like Facebook and Google before it, is clearly seeking not to monetize data collection, but instead to aggregate and mine the data of thousands of voluntary TruSat participants in order to sell these data in bulk to private companies and public agencies. Asteroid mining, given its potential to unlock trillions of dollars of wealth, may be the sector that gets the investors inside the tent, so to speak, but even if it fails in the short to medium term, it pulls behind it other, more immediately viable ideas and ventures. As Grant Bonin, Senior Vice President for Business Development at SpaceFlight notes, "I think when DSI (Deep Space Industries) and PR got started, the headlines all said asteroid-mining [companies] were like [traditional] mining companies. But internally, we'd joke: We're not miners yet. We're the pickax and shovel or Levi's jeans of space. We're the creators of tools that were brought into existence that would support the vision, but also help a lot of other people to do a lot more" (Abrahamian 2019).

The potential financial returns of asteroid mining, in combination with the reciprocal relationship between mineral wealth and geopolitical power, has spurred the United States to clarify its position on space sovereignty and property rights. The *United States Commercial Space Launch Competitiveness Act of 2015* (Pub. L. 114-90) initially passed the House of Representatives in a bipartisan 284 – 133 vote and the Senate (with amendment) by unanimous consent. The final version of the bill was signed into law by President Obama on November 25, 2015. The international press framed the *SPACE Act 2015 (Title 1 of Pub. L. 114-90)* as the starting point of the new and commercialized space race, though we are argue that it actually represents a fulcrum between the Cold War security space imaginary and the recalibrated economic space imaginary of the present era. As early as 1985, *The Saturday Evening Post* speculated that "if we can get to the moon and set up lunar mining outposts, we can produce vast quantities of oxygen to fuel space travel, space stations, space colonies, space farms and space factories" (Dula and Ziegler 1985, p. 48). From this vantage point, the SPACE Act is not a new assertion of American intent or power but instead the logical culmination of a socio-technical imaginary for space that began in the late 1940s, as the United States turned from World War II to the new exigencies of the Cold War, and has progressed since then to encompass not only national security but also economic security and the significance of the data economy.

What is new about the SPACE Act is that it explicitly moves away both from the multilateral norms that at least rhetorically underwrote space exploration during the OST era and from aspirational notions of space exploration as intended primarily for the scientific accumulation of knowledge. The SPACE Act clearly seeks to construct a stable legislative framework within which companies and venture capitalists can, though the comparative financial risk remains extreme at this juncture, capitalize investments in the outer space economy. The SPACE Act thus accomplishes for outer space what both the Supreme Court case *Diamond v. Chakrabarty* (1980) and the *Bayh-Dole Act* (1980) did for the biotechnology industry in the last century: it clarifies the United States position on property rights and enables both companies and universities to shift decisively from strictly government-led research to private investment, public-private partnerships, and patents for intellectual property.

The SPACE Act also signals that the United States, girded by its leading position in space exploration and investment, is not willing to wait for the international community, and specifically the United Nations Committee on the Peaceful Uses of Outer Space (COPUOS), to take years and possibly decades to enact multilateral rules of the road for the outer space economy. Even within an extraordinarily partisan United States, the issue of space commercialization enjoys, as space exploration did during the Cold War, strong and consistent bipartisan support. Indeed, the Artemis Accords, the signal achievement of the Trump Administration in space policy, builds upon the framework established by the SPACE Act.

Led by NASA, whose multi-billion dollar Artemis Moon Program intends to return American astronauts to the Moon by 2024 and to build a sustained U.S. presence on the lunar South pole, the Artemis Accords actively seek collaboration with private firms and with space-faring nations on a bilateral level. The Artemis Accords also envision the establishment of "safety zones" within which U.S. companies will be protected from interference by companies or countries also operating on the Moon and further clarifies a framework for extra-terrestrial property rights under international law. Asteroid mining has finally proven to be the aspirational future that has moved the broader space economy from promissory to actionable. As Peter Marquez argues, "the horizon for asteroid mining is still a couple of decades off, but I do think we're going to do Mars missions, and we'll need resources in space. And thanks to asteroid mining, the policy framework's been established" (Abrahamian 2019).

CONCLUSION

Adams et al. argue that "one defining quality of our current moment is its characteristic state of anticipation, of thinking and living toward the future" (2009, p. 246). Outer space is the key anticipatory dimension of our era, and it is one that is proving larger and certainly more challenging to conquer than even advanced biotechnology or artificial intelligence. Perhaps only neuroscience, the study and mapping of inner space, poses scientific, technological, and ethical challenges of equal measure. Due to its leading position in space investment, as well as to the resonance of both space exploration and entrepreneurialism in both its history and national culture, we expect that the United States will continue to articulate a position that legitimizes commercial exploration of space and to work with interested parties to arrive at bilateral and public-private agreements. The United States is thinking and living towards a future of space exploration and eventual space settlement. It is unlikely to agree that the *Outer Space Treaty-1967* forestalls asteroid mining nor to reverse recent legislative and executive actions that clarify an extra-terrestrial property rights framework.

Even if the extraordinary speculative wealth of asteroid mining proves elusive, the extra-terrestrial economy is emerging via promissory discourse that includes specific legislation and Executive speeches, mass media and investment estimates of financial incentives, and the entwined American narratives of entrepreneurialism, new frontiers, and technological disruption. We conclude that as both technology and politics enable the slow shift from a promissory space economy to a tangible space economy, nation-states will have to expand the way they view and adjudicate sovereignty, just as ancient states evolved beyond the political framework of the Roman Empire. Using a rigid textual interpretation of the OST to curtail asteroid mining and other forms of space monetization is not only short sighted from the American standpoint but would also fail to secure the benefits of space for all mankind as the OST intends.

REFERENCES

Abrahamian, A. A. (2019, June 26). How the asteroid mining bubble burst. *MIT Technology Review*. https://www.technologyreview.com/2019/06/26/134510/asteroid-mining-bubble-burst-history/

Allied Market Research. (2019, June 17). Asteroid mining market to reach $3,868 mn. by 2025: New study by Allied Market Research. *GlobeNewswire.* https://www.globenewswire.com/news-release/2019/06/17/1869592/0/en/Asteroid-Mining-Market-to-Reach-3-868-9-Mn-by-2025-New-Study-by-Allied-Market-Research.html

Atkinson, N. (2015, September 12). What are asteroids made of? *Universe today: Space and astronomy news.* https://www.universetoday.com/37425/what-are-asteroids-made-of/

Boyle, A. (2020, May 1). After buying Planetary Rerouces, ConsenSys sets its space ideas free—but will sell off the hardware. *Geekwire*. https://www.geekwire.com/2020/buying-planetary-resources-consensys-gives-away-science-asteroids-will-sell-rest/

Bruhns, S., & Haqq-Misra, J. (2016). A pragmatic approach to sovereignty on Mars. *Space Policy, 38*(Supplement), C, 57–63. doi:10.1016/j.spacepol.2016.05.008

Delgado-López, L. (2015). Beyond the Moon agreement: norms of responsible behavior for private sector activities on the Moon and celestial bodies. *Space Policy, 32*(SC), 6-8.

Dula, A. M., & Ziegler, A. (1985, Mar.). American business heading into orbit. *The Saturday Evening Post,* 48, 102-104.

Harvard Law Review Association. (1961). National sovereignty of outer space. *Harvard Law Review, 74*(6), 1154–1175. doi:10.2307/1338459

Heise, J. (2018). Space, the final frontier of enterprise: Incentivizing asteroid mining under a revised international framework. *Michigan Journal of International Law*, *40*, 189–213.

Kearnes, M., & van Dooren, T. (2017). Rethinking the final frontier: Cosmo-logics and an ethic of interstellar flourishing. *GeoHumanities*, *3*(1), 178–197. doi:10.1080/2373566X.2017.1300448

Landon, J., & Schneider, E. (2017, November 24). *These 5 industries will be the first to do business in space.* World Economic Forum. https://www.weforum.org/agenda/2017/11/industries-will-make-money-in-space/

Lee, R. J. (2004). Article II of the Outer Space Treaty: Prohibition of state sovereignty, private property, or both? *Australian International Law Journal, 11*, 128–142.

Leib, K. (2015, January 2). State sovereignty in space: Current models and possible futures. *Astropolitics, 13*(1), 1–24. doi:10.1080/14777622.2015.1015112

Martin, P. (2015). Commercialising neurofutures: Promissory economies, value creation and the making of a new industry. *Biosocieties*, *10*(4), 422–443. doi:10.1057/biosoc.2014.40

Martin, P., Brown, N., & Kraft, A. (2008). From bedside to bench? Communities of promise, translational research and the making of blood stem cells. *Science as Culture*, *17*(1), 29–41. doi:10.1080/09505430701872921

McMullan, T. (2016, August 4). Moon express: First private lunar mission approved for 2017. *Alphr*. https://www.alphr.com/space/1004090/moon-express-first-private-lunar-mission-approved-for-2017/

National Aeronautics and Space Administration. (n.d.). *The Outer Space Treaty –1967*. https://history.nasa.gov/SP-4225/documentation/cooperation/treaty.htm#:~:text=Treaty%20on%20principles%20governing%20the,moon%20and%20other%20celestial%20bodies.&text=The%20moon%20and%20other%20celestial%20bodies%20shall%20be%20used%20by,Treaty%20exclusively%20for%20peaceful%20purposes

Nature Astronomy. (2019). Sustainable space mining. *Nature Astronomy, 3*(465). doi:10.103841550-019-0827-7

Pop, V. (2000). Appropriation in outer space: The relationship between land ownership and sovereignty on the celestial bodies. *Space Policy*, *16*(4), 275–282. doi:10.1016/S0265-9646(00)00037-0

Salter, A. W., & Leesen, P. T. (2014). Celestial anarchy: A threat to outer space commerce? *The Cato Journal*, *34*(3), 581–596.

Seminari, S. (2019, November 24). Op/ed: Global government space budgets continues multiyear rebound. *SpaceNews*. https://spacenews.com/op-ed-global-government-space-budgets-continues-multiyear-rebound/

Stanley, M. (2020, July 24). *Space: investing in the final frontier*. https://www.morganstanley.com/ideas/investing-in-space

Weaver, J. H. (1992). Illusion or reality, state sovereignty in outer space. *Boston University International Law Journal*, *10*(2).

ENDNOTE

1 Euroconsult's Government Space Programs Report (2019) estimates that world government expenditures for space programs totalled $70.8 billion USD. The United States accounted for approximately $41 billion USD of that amount, followed by China at $5.8 billion USD and Russia at $4.17 billion USD (Seminari 2019).

Compilation of References

2020-2022 Emerging Technologies for Enterprises. (2020). *Gartner*. https://www.gartner.com/en/information-technology/trends/emerging-technology-roadmap

AACSB. (2018). *2018 Eligibility Procedures and Accreditation Standards for Accounting Accreditation.* Available on-line on March 1st, 2021 at: https://www.aacsb.edu/-/media/aacsb/docs/accreditation/accounting/standards-and-tables/2018-accounting-standards.ashx?la=en&hash=8DCDA6CE3B0CEF6AB82D39CBF53995DA96111196

AACSB. (2020). *2020 Guiding Principles and Standards for Business School Accreditation.* Available on-line on March 1st, 2021 at: https://www.aacsb.edu/-/media/aacsb/docs/accreditation/business/standards-and-tables/2020%20business%20accreditation%20standards.ashx?la=en&hash=E4B7D8348A6860B3AA9804567F02C68960281DA2

Aaronson, S., & Christiano, P. (2012). *Quantum money from hidden subspaces*. MIT.

Abrahamian, A. A. (2019, June 26). How the asteroid mining bubble burst. *MIT Technology Review*. https://www.technologyreview.com/2019/06/26/134510/asteroid-mining-bubble-burst-history/

Accardi, L., & Boukas, A. (2007). *The quantum Black-Scholes equation*. doi:10.1145/2213977.2213983

Ageless Innovations, L. L. C. (2018). *Joy for All: Companion pets*. Retrieved from https://joyforall.com/

Age-Well Nce. (2020, August 4). *Canada, we need to talk about COVID-19, pandemics, technology and the Way forward.* YouAreUNLTD. Retrieved from https://www.youareunltd.com/2020/08/04/canada-we-need-to-talk-about-covid-19-pandemics-technology-and-moving-forward/

Agulhon, C., Petravicz, J., McMullen, A. B., Sweger, E. J., Minton, S. K., Taves, S. R., Casper, K. B., Fiacco, T. A., & McCarthy, K. D. (2008). What is the role of astrocyte calcium in neurophysiology? *Neuron*, *59*(6), 932–946. doi:10.1016/j.neuron.2008.09.004 PMID:18817732

Ahmat, H.O., Wilson, W., Sorelli, L., Tolnai, B., & Hamou, A. T. (2019). *Nanocellulose for improved concrete performance*. researchgate.net/publication/331134845_Nanocellulose_for_improved_concrete_performance_A_macro-to-micro_investigation_for_disclosing_the_effects_of_cellulose_filaments_on_strength_of_cement_systems

Ahmed, N. (2020, March 18). *Portable dialysis machines, artificial kidneys to revolutionize treatment.* HealthUnits. Retrieved from https://healthunits.com/news/portable-dialysis-machines-artificial-kidneys-to-revolutioniz e-treatment/

Albrecht, W. S., & Sack, R. J. (2001). The Perilous Future of Accounting Education. *The CPA Journal*, *71*(3), 16–23.

Albuquerque, D., Nóbrega, C., Manco, L., & Padez, C. (2017). The contribution of genetics and environment to obesity. *British Medical Bulletin*, *123*(1), 159–173. doi:10.1093/bmb/ldx022 PMID:28910990

Albus, J. S., & Meystel, A. M. (2001). *Engineering of Mind*. John Wiley & Sons, Inc.

Alibaba. (2020). *Speeding Up CT Scans to Detect COVID-19.* https://www.alibabacloud.com/blog/speeding-up-ct-scans-to-detect-covid-19_596124

Alivisatos, A. P., Andrews, A. M., Boyden, E. S., Chun, M., Church, G. M., Deisseroth, K., Donoghue, J. P., Fraser, S. E., Lippincott-Schwartz, J., Looger, L. L., Masmanidis, S., McEuen, P. L., Nurmikko, A. V., Park, H., Peterka, D. S., Reid, C., Roukes, M. L., Scherer, A., Schnitzer, M., . . . Zhuang, X. (2013). Nanotools for neuroscience and brain activity mapping. *American Chemical Society Nano, 7*(3), 1850-1866. https://dx.doi.org/10.1021/nn401288k

Allen, P. G. (2020, April 2). Paul Allen: The Singularity Isn't Near. *MIT Technology Review.* https://www.technologyreview.com/2011/10/12/190773/paul-allen-the-singularity-isnt-near/

Allied Market Research. (2019, June 17). Asteroid mining market to reach $3,868 mn. by 2025: New study by Allied Market Research. *GlobeNewswire.* https://www.globenewswire.com/news-release/2019/06/17/1869592/0/en/Asteroid-Mining-Market-to-Reach-3-868-9-Mn-by-2025-New-Study-by-Allied-Market-Research.html

Ambrose, S. H. (2001). Paleolithic Technology and Human Evolution. *Science, 291*(5509), 1748–1753. doi:10.1126cience.1059487 PMID:11249821

American Association of Retired Persons. (2016). *Caregivers & technology: What they want and need.* https://www.aarp.org/content/dam/aarp/research/surveys_statistics/ltc/2018/caregivers-t echnology-needs.doi.10.26419-2Fres.00191.002.pdf

American Association of Retired Persons. (2019). *The Longevity Economy® outlook: How people age 50 and older are fueling economic growth, stimulating jobs, and creating opportunities for all.* Retrieved from https://www.aarp.org/content/dam/aarp/research/surveys_statistics/econ/2019/longevity-economy-outlook.doi.10.26419-2Fint.00042.001.pdf

American Association of Retired Persons. (2020). *The economic impact of age discrimination: How discriminating against older workers could cost the U.S. economy $850 billion.* Retrieved from https://www.aarp.org/content/dam/aarp/research/surveys_statistics/econ/2020/impac t-of-age-discrimination.doi.10.26419-2Fint.00042.003.pdf

American Association of Retired Persons. (n.d.). *The Longevity Economy® outlook.* AARP. Retrieved from https://longevityeconomy.aarp.org/?CMP=RDRCT-PRI-OTHER-LONGECON-111319

American Institute of Certified Public Accountants (AICPA). (2019). *Accounting Graduates Supply and Demand Report.* Available on-line on December 15, 2020 at: https://www.aicpa.org/content/dam/aicpa/interestareas/accountingeducation/newsandpublications/downloadabledocuments/2019-trends-report.pdf

American Institute of Certified Public Accountants (AICPA). (2020). *CPA Evolution.* Available on-line on December 12, 2020. https://www.evolutionofcpa.org/

Anderson, J., DiCicco, D., Ginder, J., Kramer, U., Leone, T., Raney-Pablo, H., & Wallington, T. (2012). High octane number ethanol–gasoline blends: Quantifying the potential benefits in the United States. *Fuel, 97,* 585–594. doi:10.1016/j.fuel.2012.03.017

Andrew Ng: Artificial Intelligence is the New Electricity. (2017, February 2). [Video]. YouTube. https://www.youtube.com/watch?v=21EiKfQYZXc

Andrews, J. G., Buzzi, S., Choi, W., Hanly, S. V., Lozano, A., Soong, A. C., & Zhang, J. C. (2014). What will 5G be? *IEEE Journal on Selected Areas in Communications, 32*(6), 1065–1082.

Anzola, M. (2004). Hepatocellular carcinoma: Role of hepatitis B and hepatitis C viruses proteins in hepatocarcinogenesis. *Journal of Viral Hepatitis, 11*(5), 383–393. doi:10.1111/j.1365-2893.2004.00521.x PMID:15357643

APCollege. (n.d.). Full-sized Drone Ambulances – the future vision. *Australian Paramedical College.* https://apcollege.edu.au/full-sized-drone-ambulances-future-vision/

Araque, A., & Navarrete, M. (2010). Glial cells in neuronal network function. *Philosophical Transactions of the Royal Society of London. Series B, Biological Sciences*, 2375–2381.

Arranz-Otaegui, A., Carretero, L. G., Ramsey, M. N., Fuller, D. Q., & Richter, T. (2018). Archaeobotanical evidence reveals the origins of bread 14,400 years ago in northeastern Jordan. *Proceedings of the National Academy of Sciences of the United States of America*, *115*(31), 7925–7930. doi:10.1073/pnas.1801071115 PMID:30012614

Asher, J. (2012). *Brain's code for visual working memory deciphered in monkeys NIH-funded study*. NIH Press Release. https://www.nimh.nih.gov/news/science-news/2012/in-sync-brain-waves-hold-memory-of-objects-just-seen.shtml

Ashton, K. (2009). That 'Internet of Things' Thing. *RFID Journal*. https://www.rfidjournal.com/that-internet-of-things-thing?4986

Ashton, T. S. (1998). *The Industrial Revolution, 1760-1830*. Oxford University Press.

Asimov, I. (1950). *I Robot*. Doubleday.

Asimov, I. (1994). *I, Robot*. Bantam Books. (Original work published 1950)

Atkinson, N. (2015, September 12). What are asteroids made of? *Universe today: Space and astronomy news*. https://www.universetoday.com/37425/what-are-asteroids-made-of/

Austin, J. H. (2014). *Zen-Brain Horizons: Toward a Living Zen*. The MIT Press. doi:10.7551/mitpress/10060.001.0001

Australian Network on Disability. (2020). *Disability statistics*. Australian Network on Disability. https://www.and.org.au/pages/disability-statistics.html

Australian Seaweed Institute. (n.d.). *Seaweed Products R&D*. https://www.australianseaweedinstitute.com.au /about-us

Ava. (2019, April 4). *How to use Ava? The accessibility mobile app for deaf and hard-of-hearing people* [Video]. YouTube. https://www.youtube.com/watch?v=PNoSWNUz1Z0

Ava. (2020). *Pricing*. Ava. https://www.ava.me/pricing

Baaquie, B. E., Coriano, C., & Srikant, M. (2002). *Quantum mechanics, path integrals and option pricing: Reducing the complexity of finance*. National U Singapore.

Bahri, Y., Kadmon, J., Pennington, J., Schoenholz, S. S., Sohl-Dickstein, J., & Ganguli, S. (2020). Statistical mechanics of deep learning. *Annual Review of Condensed Matter Physics*.

Bajorek, J. P. (2018, October 22). *Voice first verses the multimodal user interfaces of the future*. UXMatters. Retrieved from https://www.uxmatters.com/mt/archives/2018/10/voice-first-versus-the-multimodal-us er-interfaces-of-the-future.php

Ballow, M. (2002). Primary immunodeficiency disorders: Antibody deficiency. *The Journal of Allergy and Clinical Immunology*, *109*(4), 581–591. doi:10.1067/mai.2002.122466 PMID:11941303

Banaclocha, M. A. M., Bo'okkon, I., & Banaclocha, H. M. (2010). Long-term memory in brain magnetite. *Medical Hypotheses*, *74*(2), 254–257.

Bandivadekar, D., & Berquand, A. (2021). Five ways artificial intelligence can help space exploration. *The Conversation*. https://theconversation.com/five-ways-artificial-intelligence-can-help-space-exploration-153664

Bankart, P. C. (2007). *Talking Cures: A History of Western and Eastern Psychotherapies*. Cengage Learning.

Barkho, G. (2019, August 12). *The wearable tech industry is expected to hit $54 billion by 2023*. Observer. https://observer.com/2019/08/wearable-tech-industry-hit-54-billion-by-2 023/

Barr, A., & Feigenbaum, E. A. (Eds.). (2014). *The Handbook of Artificial Intelligence* (Vol. 2). Butterworth-Heinemann.

Bartkiewicz, K., Cernoch, A., Chimczak, G., Lemr, K., Miranowicz, A., & Nori, F. (2016). *Experimental quantum forgery of quantum optical money*. Adam Mickiewicz University.

Bas, E. (2015). The Future Internet: Alternative Visions (Public Administration and Information Technology (17). In R. Ono & J. Winter (Eds.), *The Liquid Self: Exploring the Ubiquitous Nature of the Future Internet and Its Pervasive Consequences on Social Life* (1st ed., pp. 193–215). Springer.

Bas, E., & Guillo, M. (2015). Participatory foresight for social innovation. FLUX-3D method (Forward Looking User Experience), a tool for evaluating innovations. *Technological Forecasting and Social Change*, *101*, 275–290. doi:10.1016/j.techfore.2015.06.016

Bassett, R. K. (2002). *To the digital age: Research labs, start-up companies, and the rise of MOS technology*. Johns Hopkins University Press.

Bauman, Z. (2006). *Liquid Times: Living in an Age of Uncertainty* (1st ed.). Polity.

Beck, J., Stern, M., & Haugsjaa, E. (1996). *Applications of AI in Education*. XRDS.

Becquemont, L. (2009). Pharmacogenomics of adverse drug reactions: Practical applications and perspectives. *Pharmacogenomics*, *10*(6), 961–969. doi:10.2217/pgs.09.37 PMID:19530963

Behring, E. v. (1890). Ueber das zustandekommen der diphtherie-immunität und der tetanus-immunität bei thieren. *Deutsche medizinsche Wochenschrift*, *16*, 1113–1114.

Bejan, A. (2020). Human evolution is biological & technological evolution. *Bio Systems*, *195*, 104156. doi:10.1016/j.biosystems.2020.104156 PMID:32492467

Bellinger, S. (2005). Modeling calcium wave oscillations in astrocytes. *Neurocomputing*, *65*(66), 843–850.

Belmaker, R. (2004). Bipolar disorder. *The New England Journal of Medicine*, *351*(5), 476–486. doi:10.1056/NEJMra035354 PMID:15282355

Belton, P. (2016, January 28). *The tech giving people power to deal with disability.* BBC. https://www.bbc.com/news/business-35427933

Ben-Achour, S. (2020, January 9). *Bringing down the cost of bionic arms*. Marketplace. https://www.marketplace.org/2020/01/09/bringing-down-the-cost-of-bionic-arms/

Bengtson, V. L., & Whittington, F. J. (2014). From ageism to the longevity revolution: Robert Butler, pioneer. *The Gerontologist*, *54*(6), 1064–1069. doi:10.1093/geront/gnu100

Benioff, P. (1980). The computer as a physical system: A microscopic quantum mechanical Hamiltonian model of computers as represented by Turing machines. *Journal of Statistical Physics*, *22*(5), 563–591.

Ben-Sasson, E., Chiesa, A., Genkin, G., Tromer, E., & Vizra, M. (2013). *SNARKs for C: Verifying Program Executions Succinctly and in Zero Knowledge*. https://eprint.iacr.org/2013/507.pdf

Berghs, M., Atkin, K., Graham, H., Hatton, C., & Thomas, C. (2016). Implications for public health research of models and theories of disability: A scoping study and evidence synthesis. *Public Health Research*, *4*(8), 1–166. doi:10.3310/phr04080 PMID:27512753

Bergveld, P. (1985). The impact of MOSFET-based sensors. *Sensors and Actuators*, *8*(2), 109–127. doi:10.1016/0250-6874(85)87009-8

Berlin, L. R. (2001). Robert Noyce and Fairchild Semiconductor, 1957-1968. *Business History Review*, 63–101.

Berners-Lee, T., Hendler, J., & Lassila, O. (2001). The Semantic Web. *Scientific American, 284*(5), 34–43. doi:10.1038cientificamerican0501-34 PMID:11323639

Bienvenu, O. J., & Ginsburg, G. S. (2007). Prevention of anxiety disorders. *International Review of Psychiatry (Abingdon, England), 19*(6), 647–654. doi:10.1080/09540260701797837 PMID:18092242

Binstock, R. H. (2003). The war on "anti-aging medicine". *The Gerontologist, 43*(1), 4–14. doi:10.1093/geront/43.1.4 PMID:12604740

Blau, N., Van Spronsen, F. J., & Levy, H. L. (2010). Phenylketonuria. *Lancet, 376*(9750), 1417–1427. doi:10.1016/S0140-6736(10)60961-0 PMID:20971365

Bloomer, W. M. (Ed.). (2015). *A companion to ancient education* (Vol. 120). John Wiley & Sons.

Blum, M., & Blum, L. (2020). *A Theoretical Computer Science Perspective on Consciousness*. https://arxiv.org/ftp/arxiv/papers/2011/2011.09850.pdf

Bolton, R. (2020). Uni students prefer online learning: study. *Financial Review*. https://www.afr.com/work-and-careers/education/uni-students-prefer-online-learning-study-20201011-p563yr

Bolton, D. (2008). *What is mental disorder? An essay in philosophy, science, and values*. Oxford University Press. doi:10.1093/med/9780198565925.001.0001

Bong, K.-W., Utreras-Alarcon, A., Ghafari, F., Liang, Y.-C., Tischler, N., Cavalcanti, E. G., Pryde, G. J., & Wiseman, H. M. (2020). A strong no-go theorem on the Wigner's friend paradox. *Nature Physics, 17*(Aug), 1745–2481.

Bordenave, G. (2003). Louis Pasteur (1822–1895). *Microbes and Infection, 5*(6), 553–560. doi:10.1016/S1286-4579(03)00075-3 PMID:12758285

Bosselut, A., Rashkin, H., Sap, M., Malaviya, C., Celikyilmaz, A., & Choi, Y. (2019). *COMET: Commonsense Transformers for Automatic Knowledge Graph Construction*. https://arxiv.org/pdf/1906.05317.pdf

Bostrom, N. (2005). A History of Transhumanist Thought. *Journal of Evolution and Technology / WTA, 14*(1).

Bostrom, N. (2005). *Superintelligence: Paths, Dangers, Strategies*. Oxford University Press.

Botsman, R. (2018). *Who Can You Trust: How Technology Brought Us Together and Why It Might Drive Us Apart*. Penguin. https://www.penguin.com.au/books/who-can-you-trust-9780241296189

Boudreau, G., Panicker, J., Guo, N., Chang, R., Wang, N., & Vrzic, S. (2009). Interference coordination and cancellation for 4G networks. *IEEE Communications Magazine, 47*(4), 74–81.

Bower, J. L., & Christensen, C. M. (1995, January–February). Disruptive Technologies: Catching the Wave. *Harvard Business Review*, 43–53.

Bowers, J. Z. (1951). Acute radiation syndrome: A problem for practitioners. *Journal of the American Medical Association, 145*(2), 63–65. doi:10.1001/jama.1951.02920200003002 PMID:14794399

Boyle, A. (2020, May 1). After buying Planetary Rerouces, ConsenSys sets its space ideas free—but will sell off the hardware. *Geekwire*. https://www.geekwire.com/2020/buying-planetary-resources-consensys-gives-away-science-asteroids-will-sell-rest/

Breithaupt, H. (2006). The engineer's approach to biology. *EMBO Reports, 7*(1), 21–23. doi:10.1038j.embor.7400607 PMID:16391532

Brennen, J. S., & Kreiss, D. (2016). Digitalization. *The international encyclopedia of communication theory and philosophy*, 1-11.

Brent, G. A. (2008). Graves' disease. *The New England Journal of Medicine*, *358*(24), 2594–2605. doi:10.1056/NEJMcp0801880 PMID:18550875

Brewer, E., Brueggemann, B. J., Hetrick, N., & Yergeau, M. (2012). Introduction, background, and history. In B. Brueggemann & G. L. Albrecht (Eds.), *Disability key issues and future directions: Arts and humanities* (pp. 1–62). SAGE Publications.

BrightSign. (2020). *Why BrightSign?* BrightSign. https://www.brightsignglove.com/

British Telecom. (2005). *Technology Timeline*. London, UK: British Telecom. www.bt.com/technologytimeline

Brody, D. C., & Meier, D. M. (2018). How to model fake news, *Cornell University*, https://arxiv.org/abs/1809.00964

Bruhns, S., & Haqq-Misra, J. (2016). A pragmatic approach to sovereignty on Mars. *Space Policy*, *38*(Supplement), C, 57–63. doi:10.1016/j.spacepol.2016.05.008

Buchanan, R., & Layton, N. (2019). Innovation in assistive technology: Voice of the user. *Societies (Basel, Switzerland)*, *9*(48), 1–10. doi:10.3390oc9020048

Budd, J. W. (2011). *The thought of work*. ILR Press. doi:10.7591/9780801462658

Buehler, M., Iagnemma, K., & Singh, S. (Eds.). (2009). *The DARPA Urban Challenge: Autonomous Vehicles in City Traffic*. Springer-Verlag. doi:10.1007/978-3-642-03991-1

Burdick, L., Wang, M., Ignat, O., Wilson, S., Zhang, Y., Wei, Y., Mihalcea, R., & Deng, J. (2019). *Building a Flexible Knowledge Graph to Capture Real-World Events*. https://lauraburdick. github.io/papers/AIDA_UMich_Princeton_2019.pdf

Burns, L. D., & Shulgan, C. (2018). *Autonomy: The Quest to Build the Driverless Car—And How It Will Reshape Our World*. Ecco Publ.

Bushnell, D. M. (2020). The Coming Digital Reality. In *Aerospace America*. AIAA.

Bushnell, D. M., & Moses, R. W. (2018). *Commercial Space in The Age of New Space, Reusable Rockets and The Ongoing Tech Revolutions*. NASA/TM.

Bushnell, D. M., & Moses, R. W. (2019). *Reliability, Safety and Performance for Two Aerospace Revolutions – UAS/ODM And Commercial Deep Space*. NASA/TM.

Bushnell, D. M., & Moses, R. W. (2020). *Prospectives in Deep Space Infrastructures, Development and Colonization*. NASA/TM.

Butalia, S., Kaplan, G. G., Khokhar, B., & Rabi, D. M. (2016). Environmental risk factors and type 1 diabetes: Past, present, and future. *Canadian Journal of Diabetes*, *40*(6), 586–593. doi:10.1016/j.jcjd.2016.05.002 PMID:27545597

Butler, R. N. (1963). The façade of chronological age: An interpretative summary. *The American Journal of Psychiatry*, *199*(8), 19–28. doi:10.1176/ajp.119.8.721 PMID:14017385

Butler, R. N., Fossel, M., Harman, S. M., Heward, C. B., Olshansky, S. J., Perls, T. T., ... West, M. D. (2002). Is there an antiaging medicine? *The Journals of Gerontology. Series A, Biological Sciences and Medical Sciences*, *57*(9), B333–B338. doi:10.1093/gerona/57.9.B333 PMID:12196485

Buzzetti, R., Zampetti, S., & Maddaloni, E. (2017). Adult-onset autoimmune diabetes: Current knowledge and implications for management. *Nature Reviews. Endocrinology*, *13*(11), 674–686. doi:10.1038/nrendo.2017.99 PMID:28885622

Byrne, E. (1979). Technology and Human Existence. *The Southwestern Journal of Philosophy*, *10*(1), 55–69. doi:10.5840wjphil19791015

Calvin, W. (2004). *A Brief History of the Mind: From Apes to Intellect and Beyond*. Oxford University Press.

Carrington, D. (2013). Seaweed biofuels: a green alternative that might just save the planet. *The Guardian*. http://marineagronomy.org/sites/default/files/Seaweed%20biofuels_%20a%20green%20alternative%20that%20might%20just%20save%20the%20planet%20_%20Environment%20_%20The%20Guardian.pdf

Carter, R. J., Sorenson, G., Heffernan, R., Kiehlbauch, J. A., Kornblum, J. S., Leggiadro, R. J., Nixon, L. J., Wertheim, W. A., Whitney, C. G., & Layton, M. (2005). Failure to Control an Outbreak of Multidrug-Resistant Streptococcus pneumoniae in a Long-Term–Care Facility Emergence and Ongoing Transmission of a Fluoroquinolone-Resistant Strain. *Infection Control and Hospital Epidemiology*, *26*(3), 248–255. doi:10.1086/502534 PMID:15796275

Castells, M. (2010). *The Rise of the Network Society*. Wiley-Blackwell.

Cattiau, J. (2019, May 7). *How AI can improve products for people with impaired speech*. Google: The Keyword. https://www.blog.google/outreach-initiatives/accessibility/impaired-speech-recognition/

Cavalcanti, A., Shirinzadeh, B., Freitas, R. A. Jr, & Hogg, T. (2007). Nanorobot architecture for medical target identification. *Nanotechnology*, *19*(1), 015103. doi:10.1088/0957-4484/19/01/015103

Centers for Medicare & Medicaid Services. (2019, April 5). *CMS finalizes policies to bring innovative telehealth benefit to Medicare Advantage*. CMS. Retrieved from https://www.cms.gov/newsroom/press-releases/cms-finalizes-policies-bring-innovative-te

CGI. (n.d.). *Empowered Learning by CGI*. https://www.cgi.com/uk/en-gb/brochure/education /empowered-learning

Cha, A. E. (2015, April 3). *Peter Thiel's quest to find the key to eternal life*. https://www.washingtonpost.com/business/on-leadership/peter-thiels-life-goal-to-extend-our-time-on-this-earth/2015/04/03/b7a1779c-4814-11e4-891d-713f052086a0_story.html

Chaisson, E. (2005). *Epic of Evolution: Seven Ages of the Cosmos*. Columbia University Press. doi:10.7312/chai13560

Chalmers, D. J. (2010). The singularity: A philosophical analysis. *Journal of Consciousness Studies*, *17*(9–10), 7–65.

Chamie, J. (2020, February 11). World Population: 2020 overview. *YaleGlobal Online*. Retrieved from https://yaleglobal.yale.edu/content/world-population-2020-overview

Chamola, V., Hassija, V., Gupta, V., & Guizani, M. (2020). A comprehensive review of the COVID-19 pandemic and the role of IoT, drones, AI, blockchain, and 5G in managing its impact. *IEEE Access: Practical Innovations, Open Solutions*, *8*, 90225–90265.

Chander, D., & Ming, V. (2019). Exponential Ethics. *SingularityU Portugal Summit Cascais*. https://www.youtube.com/watch?v=KchPtF09PZ0

Charlton, J. (1998). *Nothing about us without us*. University of California Press. doi:10.1525/9780520925441

Chomsky, N. (2013). *Noam Chomsky on AI: The Singularity is Science Fiction!* https://www.youtube.com/watch?v=0kICLG4Zg8s

Chomsky, N. (2015). *Noam Chomsky on Adam Smith's Invisible Hands*. https://www.youtube.com/watch?v=_qfX0ASvg3Y

Chomsky, N. (2016). *Noam Chomsky – Mind, Consciousness & AI*. https://www. youtube.com/watch?v=0ORHGa-vQp0&t=3s

Chomsky, N. (2017). *The Five Filters of the Mass Media*. https://www.openculture. com/2017/03/an-animated-introduction-to-noam-chomskys-manufacturing-consent.html

Chomsky, N. (2010). Some simple evo devo theses: how true might they be for language? In *The Evolution of Human Language* (pp. 45–62). Cambridge University Press. doi:10.1017/CBO9780511817755.003

Christensen, C. M. (1997). *The Innovator's Dilemma*. Harvard Business Review Press.

Chui, M., & Chris, W. (2019). AI Ethics in today's world. *McKinsey Podcast*. https://www.mckinsey.com/featured-insights/artificial-intelligence/ai-ethics-podcast

Chu, S., & Man, H. (2012). History of Accounting Education. *Journal of Higher Education Theory and Practice*, *12*(1), 119–128.

Cisco. (2014). *Cisco Live!* [Video]. IBM Watson Media. http://www.ustream.tv/ciscolive

Cisco. (2015). *Cisco Live!* [Video]. IBM Watson Media. https://www.cisco.com/c/r/en/us/internet-of-everythingioe/index.html

Clark, A. (2004). Natural-Born Cyborgs: Minds, Technologies, and the Future of Human Intelligence (Illustrated ed.). Oxford University Press.

Clark, A., & Chalmers, D. (1998). The Extended Mind. *Analysis*, *58*(1), 7–19. doi:10.1093/analys/58.1.7

Clarke, A. C. (1984). *Profiles of the Future: An Inquiry into the Limits of the Possible*. Henry Holt and Company.

Clark, L. C. Jr, & Lyons, C. (1962). Electrode systems for continuous monitoring in cardiovascular surgery. *Annals of the New York Academy of Sciences*, *102*(1), 29–45. doi:10.1111/j.1749-6632.1962.tb13623.x PMID:14021529

CNBC. (2020, February 23). *The ultra-rich are investing in companies trying to reverse aging. Is it going to work?* https://www.cnbc.com/2020/02/19/the-ultra-rich-are-investing-in-companies-trying-to-reverse-aging.html

Cohen, A. M., & Hersh, W. R. (2005). A survey of current work in biomedical text mining. *Briefings in Bioinformatics*, *6*(1), 57–71. doi:10.1093/bib/6.1.57 PMID:15826357

Cohen, S. N., Chang, A. C., Boyer, H. W., & Helling, R. B. (1973). Construction of biologically functional bacterial plasmids in vitro. *Proceedings of the National Academy of Sciences of the United States of America*, *70*(11), 3240–3244. doi:10.1073/pnas.70.11.3240 PMID:4594039

Coleman, J. S. (1988). *Supplement: Organizations and Institutions: Sociological and Economic Approaches to the Analysis of Social Structure*. American Journal of Sociology, 94. https://www.jstor.org/stable/2780243

Columbia University's Zuckerman Institute. (2018). *A new party trick: A hearing aid that reads minds* [Video]. YouTube. https://www.youtube.com/watch?v=8P3QalFi73U

Condorcet, Marie-Jean-Antoine-Nicolas de Caritat. (1979). *Sketch for a historical picture of the progress of the human mind*. Westport, CT: Greenwood Press.

Congressional Research Service. (2019), Defense Primer: U.S. Policy on Lethal Autonomous Weapon Systems Updated, Washington, DC: Author.

Conway, J., & Kochen, S. (2006). *The free will theorem*. Princeton U. J.

Coolidge, F., & Wynn, T. (2011). *The Rise of Homo sapiens: The Evolution of Modern Thinking*. Wiley-Blackwell.

Cooper, D. K., Gollackner, B., & Sachs, D. H. (2002). Will the pig solve the transplantation backlog? *Annual Review of Medicine*, *53*(1), 133–147. doi:10.1146/annurev.med.53.082901.103900 PMID:11818467

Cordeiro, J. L. (2006). *Energy 2020: A Vision of the Future in Creating Global Strategies for Humanity's Future*. World Future Society.

Cordeiro, J. L. (2010). *Telephones and Economic Development: A Worldwide Long-Term Comparison*. Lambert Academic Publishing.

Cordeiro, J. L. (2010). *The Future of Energy and the Energy of the Future*. Singularity Institute for Artificial Intelligence.

Cordeiro, J. L. (2018). *La Muerte de la Muerte*. Editorial Planeta, Deusto.

Coughlin, J. (2018, December 7). *The longevity economy: Why seniors are a fast-growing emerging market.* Barron's. Retrieved from https://www.barrons.com/articles/seniors-have-spending-power-especially-as-life-spans-lengthen-1544226021

Coughlin, J. F., & Yoquinto, L. (2018, October 15). Technology for older people doesn't have to be ugly. *Wall Street Journal.*

Crippen, D. (2015, October 19). The science surrounding cryonics. *MIT Technology Review*. https://www.technologyreview.com/2015/10/19/109714/the-science-surrounding-cryonics/

Crofting Commission. (2020). *Facts & Figures*. https://www.crofting.scotland.gov.uk/facts-and-figures

Cuello, J. L. (2019). *One Giant Leap: Space Programs Spin Off Vertical Farming*. https:// urbanagnews.com/blog/news/one-giant-leap-space-programs-spin-off-vertical-farming/

Daemmrich, A. (2017). Invention, Innovation Systems, and the Fourth Industrial Revolution. *Technology and Innovation*, *18*(4), 257–265. doi:10.21300/18.4.2017.257

Daniell, H., Streatfield, S. J., & Wycoff, K. (2001). Medical molecular farming: Production of antibodies, biopharmaceuticals and edible vaccines in plants. *Trends in Plant Science*, *6*(5), 219–226. doi:10.1016/S1360-1385(01)01922-7 PMID:11335175

Danowitz, A., Kelley, K., Mao, J., Stevenson, J. P., & Horowitz, M. (2012). CPU DB: Recording microprocessor history. *Communications of the ACM*, *55*(4), 55–63.

Darwiche, A. (2020). Three Modern Roles for Logic in AI, *Cornell University*. https://arxiv.org/ abs/2004.08599

Darwin, C. (1968). *On the origin of species by means of natural selection*. John Murray.

Darwin, C. (2003). *The Origin of the Species*. Fine Creative Media. (Original work published 1859)

Dator, J. (1996). Futures studies as applied knowledge. In R. A. Slaughter (Ed.), *New Thinking for a New Millennium* (pp. 105–115). Routledge.

Daugherty, G. (2020, April 21). *Talking to the dead: How the 1918 pandemic spurred a spiritualism craze.* History. https://www.history.com/news/flu-pandemic-wwi-ouija-boards-spiritualism

Davenport, C. (2015). Military Pushes for Emergency Robots as Skeptics Worry About Lethal Uses. *Washington Post.*

David, D., Zoizner, G., & Werner, P. (2018). Self-stigma and age-related hearing loss: A qualitative study of stigma formation and dimensions. *American Journal of Audiology*, *27*(1), 126–136.

Davidson, A., & Diamond, B. (2001). Autoimmune diseases. *The New England Journal of Medicine*, *345*(5), 340–350. doi:10.1056/NEJM200108023450506 PMID:11484692

Davisson, C., & Germer, L. H. (1927). Diffraction of electrons by a crystal of nickel. *Physical Review*, *30*, 705–741.

Dawson, W. K., Maciejczyk, M., Jankowska, E. J., & Bujnicki, J. M. (2016). Coarse-grained modeling of RNA 3D structure. *Methods (San Diego, Calif.)*, *103*, 138–156. doi:10.1016/j.ymeth.2016.04.026 PMID:27125734

de Chardin, T. P. (1964). The Future of Man. New York, NY: Harper & Row.

De Chardin, P. (1955). *The Phenomenon of Man*. Harper.

De Clercq, E. (2009). Anti-HIV drugs: 25 compounds approved within 25 years after the discovery of HIV. *International Journal of Antimicrobial Agents*, *33*(4), 307–320. doi:10.1016/j.ijantimicag.2008.10.010 PMID:19108994

De Pitta, M., Volman, V., Berry, H., Parpura, V., Volterra, A., & Ben-Jacob, E. (2012). Computational quest for understanding the role of astrocyte signaling in synaptic transmission and plasticity. *Frontiers in Computational Neuroscience*, *6*(98), 1–25.

Deal, J. A., Reed, N. S., Kravetz, A. D., Weinreich, H., Yeh, C., Lin, F. R., & Altan, A. (2019). Incident hearing loss and comorbidity: A longitudinal administrative claims study. *JAMA Otolaryngology-Head & Neck Surgery*, *145*(1), 36–43.

Degener, T. (2016). Disability in a human rights context. *Laws*, *5*(3), 1–24. doi:10.3390/laws5030035

DeLeo, F. R., & Hinnebusch, B. J. (2005). A plague upon the phagocytes. *Nature Medicine*, *11*(9), 927–928. doi:10.1038/nm0905-927 PMID:16145573

Delgado-López, L. (2015). Beyond the Moon agreement: norms of responsible behavior for private sector activities on the Moon and celestial bodies. *Space Policy, 32*(SC), 6-8.

Dennis, B., & Mooney, C. (2018, Dec. 6). Global Carbon Reaches Record. *Washington Post*.

Dhar, P. (2020). Understanding Causality Is the Next Challenge for Machine Learning. *IEEE Spectrum*. https://spectrum.ieee.org/tech-talk/artificial-intelligence/machine-learning/understanding-causality-is-the-next-challenge-for-machine-learning

Diamandis, P., & Kotler, S. (2012). Abundance: The Future is Better than You Think. Free Press (Simon & Schuster).

Diamandis, P., & Kotler, S. (2020). *The Future is Faster Than You Think*. Simon & Schuster.

Dietrich, O., Heun, M., Notroff, J., Schmidt, K., & Zarnkow, M. (2012). The role of cult and feasting in the emergence of Neolithic communities. New evidence from Göbekli Tepe, south-eastern Turkey. *Antiquity*, *86*(333), 674–695. doi:10.1017/S0003598X00047840

Dillet, R. (2017, July 12). *The OrCam MyEye helps visually impaired people read and identify things*. TechCrunch. https://techcrunch.com/2017/07/12/the-orcam-myeye-helps-visually-impaired-people-rea d-and-identify-things/

Dingyu, Z. (2018). Cloud computing technology and its application in robot control. *Iop Conference*, *392*(6).

Disabled World. (2020, March 30). *Disability statistics: Information, charts, graphs and tables*. Disabled World. https://www.disabled-world.com/disability/statistics/

Domingo, J. L., & Bordonaba, J. G. (2011). A literature review on the safety assessment of genetically modified plants. *Environment International*, *37*(4), 734–742. doi:10.1016/j.envint.2011.01.003 PMID:21296423

Dominguez-Rodrigo, M. (2002). Hunting and Scavenging by Early Humans: The State of the Debate. *Journal of World Prehistory*, *16*(1), 1–54. doi:10.1023/A:1014507129795

Dondorp, A. M., Nosten, F., Yi, P., Das, D., Phyo, A. P., Tarning, J., ... Lee, S. J. (2009). Artemisinin resistance in Plasmodium falciparum malaria. *The New England Journal of Medicine*, *361*(5), 455–467. doi:10.1056/NEJMoa0808859 PMID:19641202

Donlan, A. (2020, March 17). *Trump administration announces more relaxed restrictions on telehealth services during COVID-19 emergency.* Home Health Care News. Retrieved from https://homehealthcarenews.com/2020/03/trump-administration-announces-more-relaxed-restrictions-on-telehealth-services-during-covid-19-emergency/

Donnelly, M. (2017, January 11). Cryogenics' chilling prospect of return of the frozen dead. *The Daily Telegraph.* https://www.dailytelegraph.com.au/news/cryogenics-chilling-prospect-of-return-of-the-frozen-dead/news-story/afd172f56e6d10decbda80e01afac75c

Doraiswamy, P. M. (2020). Therapeutic and diagnostic apps and bots are almost here. *Scientific American.* https://www.scientificamerican.com/article/digital-medicine-can-diagnose-and-treat-what-ails-you/

Dos Santos, A. D. P., Ferrari, A. L. M., Medola, F. O., & Sandnes, F. E. (2020). Aesthetics and the perceived stigma of assistive technology for visual impairment. *Disability and Rehabilitation. Assistive Technology*, 1–7. doi:10.1080/17483107.2020.1768308 PMID:32501732

Dot Incorporation. (2020). *Dot Watch.* Dot. https://www.dotincorp.com/

Dow, K., Jacknis, N., & Watson, M. (2021). A Framework and Resources to Create a Data Analytics-Infused Accounting Curriculum. *Issues in Accounting Education*, ISSUES-19-071. Advance online publication. doi:10.2308/ISSUES-19-071

Drexler, K. E. (1987). *Engines of Creation.* Anchor Books. www.e-drexler.com/d/06/00/EOC/EOC_Cover.html

Drexler, K. E. (2013). *Radical Abundance: How a Revolution in Nanotechnology Will Change Civilization.* Public Affairs.

Dudley, M. E., Wunderlich, J. R., Shelton, T. E., Even, J., & Rosenberg, S. A. (2003). Generation of tumor-infiltrating lymphocyte cultures for use in adoptive transfer therapy for melanoma patients. *Journal of Immunotherapy, 26*(4), 332-342.

Dula, A. M., & Ziegler, A. (1985, Mar.). American business heading into orbit. *The Saturday Evening Post,* 48, 102-104.

Durodié, B. (2020). Handling uncertainty and ambiguity in the COVID-19 pandemic. *Psychological Trauma: Theory, Research, Practice, and Policy*, *12*(S1), S61.

Dyson, L., & Alden, G. A. (2017). *Evaluating Machine Learning Algorithms for Fake News Detection, Final Project for CAPP 30255 Advanced Machine Learning for Public Policy.* U of Chicago.

Edelman, G. M., & Tononi, G. (2000). *A Universe of Consciousness.* Basic Books.

Edmond, M. B., Ober, J. F., Weinbaum, D. L., Pfaller, M. A., Hwang, T., Sanford, M. D., & Wenzel, R. P. (1995). Vancomycin-resistant Enterococcus faecium bacteremia: Risk factors for infection. *Clinical Infectious Diseases*, *20*(5), 1126–1133. doi:10.1093/clinids/20.5.1126 PMID:7619987

Edwards, N. L., Magilavy, D. B., Cassidy, J. T., & Fox, I. H. (1978). Lymphocyte ecto-5′-nucleotidase deficiency in agammaglobulinemia. *Science*, *201*(4356), 628–630. doi:10.1126cience.27864 PMID:27864

Embodied Labs. (2020). *Content library.* Embodied Labs. Retrieved from https://embodiedlabs.com/content-library/

Emerging trends in social media. (2014, July 3). *MIT News.* https://news.mit.edu/2014/emerging-trends-social-media

e-NABLE. (2020). *FAQs.* Enabling the Future. https://enablingthefuture.org/faqs/

Enright, M. C., Robinson, D. A., Randle, G., Feil, E. J., Grundmann, H., & Spratt, B. G. (2002). The evolutionary history of methicillin-resistant Staphylococcus aureus (MRSA). *Proceedings of the National Academy of Sciences of the United States of America*, *99*(11), 7687–7692. doi:10.1073/pnas.122108599 PMID:12032344

Essl Foundation. (2019, June 10). *Artificial Intelligence and the potential to increase inclusion, participation and independence for people with disabilities*. Panel presented at the United Nations Conference of States Parties to the Convention on the Rights of Persons with Disabilities, New York, NY. https://www.un.org/development/desa/dspd/wp-content/uploads/sites/22/2019/05/10-June-4.45_Essl-Foundation_CN_AI-and-the-Potential.pdf

European Commission. (2018). *Pantheon: robots giving a helping hand to hazelnut farmers*. https://ec.europa.eu/digital-single-market/en/news/pantheon-robots-giving-helping-hand-hazelnut-farmers

Evans, D. (2011). *The Internet of Things: How the Next Evolution of the Internet Is Changing Everything*. Academic Press.

Evans, A. S. (1976). Causation and disease: The Henle-Koch postulates revisited. *The Yale Journal of Biology and Medicine*, *49*(2), 175–195. PMID:782050

Evans, W. E., & Relling, M. V. (1999). Pharmacogenomics: Translating functional genomics into rational therapeutics. *Science*, *286*(5439), 487–491. doi:10.1126cience.286.5439.487 PMID:10521338

Ezezika, O. C., & Singer, P. A. (2010). Genetically engineered oil-eating microbes for bioremediation: Prospects and regulatory challenges. *Technology in Society*, *32*(4), 331–335. doi:10.1016/j.techsoc.2010.10.010

Facchini, F. (2003). *A Day with Homo habilis: Life 2,000,000 Years Ago*. Amazon.

Facchi, P., Lidar, D. A., & Pascazio, S. (2004). Unification of dynamical decoupling and the quantum Zeno effect. *Physical Review A.*, *69*(032314), 1–6.

Facchi, P., & Pascazio, S. (2008). Quantum Zeno dynamics: Mathematical and physical aspects. *Journal of Physics. A. Mathematical Nuclear and General*, *41*(493001), 1–45.

Factbook.World (2021). CIA.

Fang, X.-T., Zeng, P., Liu, H., Zou, M., Wu, W., Tang, Y.-L., Sheng, Y.-J., Xiang, Y., Zhang, W., Li, H., Wang, Z., You, L., Li, M.-J., Chen, H., Chen, Y.-A., Zhang, Q., Peng, C.-Z., Ma, X., Chen, T.-Y., & Pan, J.-W. (2020). Implementation of quantum key distribution surpassing the linear rate-transmittance bound. *Nature Photonics*, *14*, 422–425.

FAO. (n.d.). *Introduction to Commercial Seaweeds*. http://www.fao.org/3/y4765e/y4765e04.htm

FDA. (2018). *FDA Approves First Drug Comprised of an Active Ingredient Derived from Marijuana to Treat Rare, Severe Forms of Epilepsy*. https://www.fda.gov/NewsEvents/ Newsroom/PressAnnouncements/ucm611046.htm

Feng, X. (2018). AI will spell the end of capitalism. *The Washington Post*. https://www. washingtonpost.com/news/theworldpost/wp/2018/05/03/end-of-capitalism/

Feng, P., Li, Q., Zhang, P., & Chen, Z. (2018). Logical acquisition method based on data migration for Android mobile devices. *Digital Investigation*, *26*, 55–62.

Ferrante, R. J., Andreassen, O. A., Jenkins, B. G., Dedeoglu, A., Kuemmerle, S., Kubilus, J. K., Kaddurah-Daouk, R., Hersch, S. M., & Beal, M. F. (2000). Neuroprotective effects of creatine in a transgenic mouse model of Huntington's disease. *The Journal of Neuroscience: The Official Journal of the Society for Neuroscience*, *20*(12), 4389–4397. doi:10.1523/JNEUROSCI.20-12-04389.2000 PMID:10844007

Festing, M. F. (2010). Inbred strains should replace outbred stocks in toxicology, safety testing, and drug development. *Toxicologic Pathology*, *38*(5), 681–690. doi:10.1177/0192623310373776 PMID:20562325

Feynman, R. (1982). Simulating physics with computers. *International Journal of Theoretical Physics*, *21*(6/7), 467–488. https://dx.doi.org/10.1007/BF02650179

Feynman, R. P. (1948). Space-time approach to non-relativistic quantum mechanics. *Reviews of Modern Physics*, *20*(2).

Feynman, R. P. (1949). Space-time approach to quantum electrodynamics. *Physical Review*, *76*(6).

Financial Times. (2018a). *How Adam Smith predicted the power of Big Tech.* https://www.ft.com/content/3a1b0632-a61e-11e8-a1b6-f368d365bf0e

Financial Times. (2018b). *Artificial Intelligence must avoid a backlash.* https://www.ft.com/content/d94d07e4-b278-11e8-8d14-6f049d06439c

Finkelstein, R. (2016). *Autonomous Cars: Opportunities and Threats.* Presented at the University Seminar On Reflexive Systems, The George Washington University, Washington, DC.

Finkelstein, R. (2016). *Autonomous Robots: The Next Big Thing That Will Change The World.* Presented at the Czech Technical University, Czech Institute of Informatics, Robotics, and Cybernetics, Prague, Czech Republic.

Finkelstein, R. (2019). *Autonomous Intelligent Robots for Hospitals, Nursing Home, and Assisted Living Facilities.* White Paper, Potomac, MD.

Finkelstein, R. (2006), Military Robotics: Malignant Machines or the Path to Peace? Presented at the *Military Robotics Conference: Advancing Autonomy to Reduce Human Casualties, Institute for Defense and Government Advancement*, Georgetown University Conference Center, Washington, DC.

Finkelstein, R. (2010), Military Robotics: Malignant Machines or the Path to Peace? Presented at the *First AUVSI-IEEE Joint Conference on Unmanned Vehicles, Ben-Gurion University*, Beer-Sheva, Israel.

Finkelstein, R. (2020). Evolutionary Psychology and Robotics. In T. K. Shackelford (Ed.), *SAGE Handbook of Evolutionary Psychology*. Sage Publications.

Finkelstein, R., & Shaker, S. (1994). *Unmanned Vehicle Systems: Military and Civil Robots for the 21st Century & Beyond.* Pasha Publications, Inc.

Fisher, R., & Ury, W. (1981). *Getting to Yes: Negotiating Agreement Without Giving In.* Penguin Group.

Flaherty, R. E. (Ed.). (1998). Accounting Education Series: Vol. 14. *Accounting Education Change Commission Grant Experience: A Summary*. AAA.

Folegatti, P. M., Ewer, K. J., Aley, P. K., Angus, B., Becker, S., Belij-Rammerstorfer, S., ... Clutterbuck, E. A. (2020). Safety and immunogenicity of the ChAdOx1 nCoV-19 vaccine against SARS-CoV-2: A preliminary report of a phase 1/2, single-blind, randomised controlled trial. *Lancet*, *396*(10249), 467–478. doi:10.1016/S0140-6736(20)31604-4 PMID:32702298

Foley, A., & Ferri, B. A. (2012). Technology for people, not disabilities: Ensuring access and inclusion. *Journal of Research in Special Educational Needs*, *12*(4), 192–200.

Food and Drug Administration. (2020). *Approved Cellular and Gene Therapy Products.* Retrieved from https://www.fda.gov/vaccines-blood-biologics/cellular-gene-therapy-products/approved-cellular-and-gene-therapy-products

Forterre, P. (2003). The great virus comeback—From an evolutionary perspective. *Research in Microbiology*, *154*(4), 223–225. doi:10.1016/S0923-2508(03)00111-6 PMID:12901366

Foundation for the Future. (2002). *The Next Thousand Years*. Bellevue, WA: Foundation for the Future. www.future-foundation.org/documents/nty_projdesc.pdf

Fowler, L. (1985). *From Outer Space to Inner Odyssey*. New Age Journal.

Francisco, J. (2017, September 27). *Satya and Anu Nadella open up about their family life*. Good Housekeeping. https://www.goodhousekeeping.com/life/inspirational-stories/a46221/satya-anu-nadella-microsoft/

Franz. (2018). *Franz and Semantic Web Co. Partner to Create a Noam Chomsky Knowledge Graph*. https://allegrograph.com/franz-and-semantic-web-company-partner-to-create-a-noam-chomsky-knowledge-graph/

Frazer, A. (1997). Antidepressants. *The Journal of Clinical Psychiatry*, *58*(SUPPL. 6), 9–25. PMID:9227669

Freedman, A. H., & Wayne, R. K. (2017). Deciphering the origin of dogs: From fossils to genomes. *Annual Review of Animal Biosciences*, *5*(1), 281–307. doi:10.1146/annurev-animal-022114-110937 PMID:27912242

Freeman, D., Reeve, S., Robinson, A., Ehlers, A., Clark, D., Spanlang, B., & Slater, M. (2017). Virtual reality in the assessment, understanding, and treatment of mental health disorders. *Psychological Medicine*, *47*(14), 2393–2400. doi:10.1017/S003329171700040X PMID:28325167

Frie, F. (2018). How to build trust. *TED*. https://www.ted.com/talks/frances_frei_how_to_build_and_rebuild_trust

Friedlander, H. (1995). *The origins of Nazi genocide: From euthanasia to the final solution*. University of North Carolina Press.

Front Porch Communities and Services. (2018, July 23). *Front Porch Center for Innovation and Wellbeing releases results of Amazon Alexa pilot project in San Diego retirement community*. Front Porch Center for Innovation and Wellbeing. Retrieved from https://fpciw.org/news/front-porch-center-for-innovation-and-wellbeing-releases-results-of-amazon-alexa-pilot-project-in-san-diego-retirement-community/

Fukuyama, F. (1989). The End of History? *The National Interest*, (16), 3-18.

Fukuyama, F. (1995). *Trust: The Social Virtues and the Creation of Prosperity*. Free Press. https://books.google.com.au/books/about/Trust.html?id=BaS2AAAAIAAJ

Fumento, M. (2003). *BioEvolution: How Biotechnology is Changing the World*. Encounter Books.

Futia, G., & Vetro, A. (2020). Information, On the Integration of Knowledge Graphs into Deep Learning Models for a More Comprehensible AI—Three Challenges for Future Research. *Information (Basel)*, *11*(2), 122. doi:10.3390/info11020122

Gabbin, A. L. (2002, April). The Crisis in Accounting Education. *Journal of Accountancy*, *193*(4), 81–86.

Gabow, P. A. (1993). Autosomal dominant polycystic kidney disease. *The New England Journal of Medicine*, *329*(5), 332–342. doi:10.1056/NEJM199307293290508 PMID:8321262

Gabrielli, A., Avvedimento, E. V., & Krieg, T. (2009). Scleroderma. *The New England Journal of Medicine*, *360*(19), 1989–2003. doi:10.1056/NEJMra0806188 PMID:19420368

Gale, J., Wandel, A., & Hill, H. (2019). Will recent advances in AI result in a paradigm shift in Astrobiology and SETI? *International Journal of Astrobiology*, *19*(3), 295–298. doi:10.1017/S1473550419000260

Galer, D. (2012). Disabled capitalists: Exploring the intersections of disability and identity formation in the world of work. *Disability Studies Quarterly*, *32*(3). https://dsq-sds.org/article/view/3277/3122

Gandhi, N. R., Nunn, P., Dheda, K., Schaaf, H. S., Zignol, M., Van Soolingen, D., Jensen, P., & Bayona, J. (2010). Multidrug-resistant and extensively drug-resistant tuberculosis: A threat to global control of tuberculosis. *Lancet, 375*(9728), 1830–1843. doi:10.1016/S0140-6736(10)60410-2 PMID:20488523

Garreau, J. (2005). *Radical Evolution: The Promise and Peril of Enhancing Our Minds, Our Bodies—and What It Means to Be Human*. Doubleday.

Gaskell, G., Bauer, M. W., Durant, J., & Allum, N. C. (1999). Worlds apart? The reception of genetically modified foods in Europe and the US. *Science, 285*(5426), 384–387. doi:10.1126cience.285.5426.384 PMID:10411496

Gazzaniga, M. S. (2011). *Who's in Charge: Free Will and the Science of the Brain*. Harper Collins.

Gershon, D., & Lehrman, S. (1994). Genetically altered tomato set to get the green light. *Nature, 368*(6472), 574–574. doi:10.1038/368574a0 PMID:8145830

Giacosa, G., & Pagliara, G. (2014). Quantum Zeno effect by general measurements. *Physical Review A., 052107*, 1–5.

Gill, K. S. (2016). *Artificial super intelligence: beyond rhetoric*. Academic Press.

Gillbard, E. (2020). *Farmer's Weekly*. https://www.fwi.co.uk/arable/land-preparation/ soils/fermenting-organic-matter-better-for-soil-health-than-composting

Glenn, J. (2019). *State of the Future 19.1*. The Millennium Project. www.StateOfTheFuture.org

Goertzel, B. (2020). *Say Hello to the Singularity*. https://www.coindesk.com/say-hello-to-the-singularity

Goggin, G., & Newell, C. (2003). *Digital disability: The social construction of disability in new media*. Rowman & Littlefield.

Goldberg, M., De Pitta', M., Volman, V., Berry, H., & Ben-Jacob, E. (2010). Nonlinear Gap Junctions Enable Long-Distance Propagation of Pulsating Calcium Waves in Astrocyte Networks. *Public Library of Science Computational Biology., 6*(8), 1–14.

Goman, A. M., Reed, N. S., & Lin, F. R. (2017). Addressing estimated hearing loss in adults in 2060. *JAMA Otolaryngology-Head & Neck Surgery, 143*(7), 733–734.

Goodman, L. (2018). *The universe is a hologram and other mind-blowing theories in theoretical physics*. https://phys.org/news/2018-03-universe-hologram-mind-blowing theories-theoretical.html

Gould, G. M. (1891). Immorality. *The Monist, 1*(3), 372–392. doi:10.5840/monist18911339

Government of Western Australia. (n.d.). *What is disability?* Government of Western Australia Department of Communities Disability Services. http://www.disability.wa.gov.au/understanding-disability1/understanding-disability/what-is-disability/

Government Office for Science. (2016). *Distributed Ledger Technology: beyond block chain*. https://assets.publishing.service.gov.uk/government/uploads/system/uploads/attachment_data/file/492972/gs-16-1-distributed-ledger-technology.pdf

Gozes, O., Frid-Adar, M., Greenspan, H., Browning, P. D., Zhang, H., Ji, W., . . . Siegel, E. (2020). *Rapid ai development cycle for the coronavirus (covid-19) pandemic: Initial results for automated detection & patient monitoring using deep learning ct image analysis*. arXiv preprint arXiv:2003.05037.

Graham, J. (2020, July 23). *Seniors who struggle with technology face telehealth challenges and social isolation*. CNN. Retrieved from https://edition.cnn.com/2020/07/23/health/seniors-technology-telehealth-wellness-partner/index.html

Graham, S. (2020). *Dendra Systems*. https://www.dendra.io/post/if-we-want-a-future-on-earth-were-going-to-have-to-restore-2-billion-hectares-of-land-lets-get-on-with-it

GrandPad. (2020). *GrandPad.* https://www.grandpad.net/

Gray, J. (2011). *The immortalization commission: Science and the strange quest to cheat death.* Farrar, Strauss, & Giroux.

Graziano, M. S. (2013). *Consciousness and the Social Brain.* Oxford University Press.

Greenfield Robotics. (2020). https://www.greenfieldrobotics.com/

Grignolio, A., Mishto, M., Faria, A. M., Garagnani, P., Franceschi, C., & Tieri, P. (2014). Towards a liquid self: How time, geography, and life experiences reshape the biological identity. *Frontiers in Immunology*, *5*, 153. doi:10.3389/fimmu.2014.00153 PMID:24782860

Grimm, E. A., Mazumder, A., Zhang, H., & Rosenberg, S. (1982). Lymphokine-activated killer cell phenomenon. Lysis of natural killer-resistant fresh solid tumor cells by interleukin 2-activated autologous human peripheral blood lymphocytes. *The Journal of Experimental Medicine*, *155*(6), 1823–1841. doi:10.1084/jem.155.6.1823 PMID:6176669

Grine, F., Fleagle, J., & Leakey, R. (2006). *The First Humans – Origin and Early Evolution of the Genus Homo.* Springer Science & Business Media.

Grover, L. K. (1997). *Quantum computers can search rapidly by using almost any transformation.* Bell Labs.

Grünfeld, J.-P. (1985). The clinical spectrum of hereditary nephritis. *Kidney International*, *27*(1), 83–92. doi:10.1038/ki.1985.14 PMID:3981875

Guarente, L., & Picard, F. (2005). Calorie restriction—The SIR2 connection. *Cell*, *120*(4), 473–482. doi:10.1016/j.cell.2005.01.029 PMID:15734680

Gür, T. M. (2018). Review of electrical energy storage technologies, materials and systems: Challenges and prospects for large-scale grid storage. *Energy & Environmental Science*, *11*(10), 2696–2767.

Guy, C. T., Cardiff, R., & Muller, W. J. (1992). Induction of mammary tumors by expression of polyomavirus middle T oncogene: A transgenic mouse model for metastatic disease. *Molecular and Cellular Biology*, *12*(3), 954–961. doi:10.1128/MCB.12.3.954 PMID:1312220

Gyongyosi, L., & Imre, S. (2019). A survey on quantum computing technology. *Computer Science Review*, *31*, 51–71.

Haenlein, M., & Kaplan, A. (2019). A brief history of artificial intelligence: On the past, present, and future of artificial intelligence. *California Management Review*, *61*(4), 5–14.

Haensch, S., Bianucci, R., Signoli, M., Rajerison, M., Schultz, M., Kacki, S., ... Achtman, M. (2010). Distinct clones of Yersinia pestis caused the black death. *PLoS Pathogens*, *6*(10), e1001134. doi:10.1371/journal.ppat.1001134 PMID:20949072

Halal, W. (2004). The Life Cycle of Evolution: A Meta-Technological Analysis of Civilization's Progress. *Journal of Future Studies, 9*(1).

Halal, W. (2008). *Technology's Promise: Expert Knowledge on the Transformation of Business and Society.* Palgrave Macmillan. doi:10.1057/9780230582538

Haldane, J. B. S. (1924). *Daedalus; or, Science and the Future.* K. Paul, Trench, Trubner & Co.

Hall, J. (2019, November 18). *Wearable tech is failing people with disabilities.* OneZero. https://onezero.medium.com/wearable-tech-is-failing-people-with-disabilities-673f67d65724

Hanifan, L. J. (1916). The Rural School Community Center. *The American Academy of Political and Social Science.* https://journals.sagepub.com/doi/10.1177/000271621606700118

Hanis, C., Boerwinkle, E., Chakraborty, R., Ellsworth, D., Concannon, P., Stirling, B., ... Gogolin-Ewens, K. (1996). A genome–wide search for human non–insulin–dependent (type 2) diabetes genes reveals a major susceptibility locus on chromosome 2. *Nature Genetics, 13*(2), 161–166. doi:10.1038/ng0696-161 PMID:8640221

Hanson, R. (2000). *Long Term Growth as a Sequence of Exponential Models*. Carow Hall.

Hao, K. (2020). This could lead to the next big breakthrough in common sense AI. *MIT Technology Review*. https://www.technologyreview.com/2020/11/06/ 1011726/ai-natural-language-processing-computer-vision/

Harari, Y. N. (2018). *Twenty-One Lessons for the Twenty-first Century*. Spiegel & Grau.

Haraway, D. (1991). A Cyborg Manifesto. In *Simians, Cyborgs and Women: The Reinvention of Nature*. Routledge.

Harris, S. (2012). *Free Will*. Free Press.

Harris, S. (2014). *Waking Up: A Guide to Spirituality Without Religion*. Simon & Schuster.

Harvard Law Review Association. (1961). National sovereignty of outer space. *Harvard Law Review, 74*(6), 1154–1175. doi:10.2307/1338459

Haslberger, A. G. (2003). Codex guidelines for GM foods include the analysis of unintended effects. *Nature Biotechnology, 21*(7), 739–741. doi:10.1038/nbt0703-739 PMID:12833088

Hawking, S. (2016, Dec. 1). This Is the Most Dangerous Time for Our Planet. *The Guardian*.

Hawking, S. (2002). *The Theory of Everything: The Origin and Fate of the Universe*. New Millennium Press.

Heise, J. (2018). Space, the final frontier of enterprise: Incentivizing asteroid mining under a revised international framework. *Michigan Journal of International Law, 40*, 189–213.

He, J., Baxter, S. L., Xu, J., Xu, J., Zhou, X., & Zhang, K. (2019). The practical implementation of artificial intelligence technologies in medicine. *Nature Medicine, 25*(1), 30–36.

Heller, M. J. (2002). DNA microarray technology: Devices, systems, and applications. *Annual Review of Biomedical Engineering, 4*(1), 129–153. doi:10.1146/annurev.bioeng.4.020702.153438 PMID:12117754

Hendrix, M. (2015, September 15). The false science of cryonics. *MIT Technology Review*. https://www.technologyreview.com/2015/09/15/109906/the-false-science-of-cryonics/

Herman, E. S., & Chomsky, N. (1988). *Manufacturing Consent: The Political Economy of the Mass Media*. Pantheon Books. https://books.google.com.au/books/about/ Manufacturing_consent.html?id=Up5sAAAAIAAJ&redir_esc=y

Hiatt, A., Caffferkey, R., & Bowdish, K. (1989). Production of antibodies in transgenic plants. *Nature, 342*(6245), 76–78. doi:10.1038/342076a0 PMID:2509938

Hicks, S. (Speaker)., & Lenzi, M. (Host). (2019, August 8). *HealthRedesigned: Oxsight: Hacking reality for the visually impaired* [Audio podcast]. Hanno. https://hanno.co/healthredesigned/oxsight-smart-glasses/

Hiday, V. A. (1995). The social context of mental illness and violence. *Journal of Health and Social Behavior, 36*(2), 122–137. doi:10.2307/2137220 PMID:9113138

Hobbs, H. H., Brown, M. S., & Goldstein, J. L. (1992). Molecular genetics of the LDL receptor gene in familial hypercholesterolemia. *Human Mutation, 1*(6), 445–466. doi:10.1002/humu.1380010602 PMID:1301956

Hoekman, S. K. (2009). Biofuels in the US–challenges and opportunities. *Renewable Energy, 34*(1), 14–22. doi:10.1016/j.renene.2008.04.030

Holley, P. (2019). George Mason Students Have a New Dining Option: Food Delivered by Robots. *Washington Post.*

Holt-Lunstad, J., Smith, T. B., Baker, M., Harris, T., & Stephenson, D. (2015). Loneliness and social isolation as risk factors for mortality: A meta-analytic review. *Perspectives on Psychological Science*, *10*(2), 227–237.

Hooke, A. (2019). *Global Economic Development: Past, Present and Future*. Lakeland House.

Horsley, P. (2019, September 13). *The 3 big trends influencing assistive technology.* Remarkable. https://remarkable.org.au/trends-influencing-assistive-technology/

Hosny, A., Parmar, C., Quackenbush, J., Schwartz, L. H., & Aerts, H. J. W. L. (2018). Artificial intelligence in radiology. *Nature Reviews. Cancer*, *18*, 500–510.

Hoyer, L. W. (1994). Hemophilia a. *The New England Journal of Medicine*, *330*(1), 38–47. doi:10.1056/NEJM199401063300108 PMID:8259143

Huang, H.-M. (Ed.). (2008). *Autonomy Levels for Unmanned Systems (ALFUS) Framework, Volume I: Terminology, Version 2.0.* NIST Special Publication 1011-I-2.0, National Institute of Standards and Technology.

Huerta, E., & Jensen, S. (2017). An Accounting Information Systems Perspective on Data Analytics and Big Data. *Journal of Information Systems*, *31*(3), 101–114. doi:10.2308/isys-51799

Huff, C. (2020, March 11). How artificial kidneys and miniaturized dialysis could save millions of lives. *Nature*, *579*, 186–188. https://www.nature.com/articles/d41586-020-00671-8

Hughes, J. (2004). *Citizen Cyborg: Why Democratic Societies Must Respond to the Redesigned Human of the Future*. Westview Press.

Hu, H., & Xiong, L. (2014). Genetic engineering and breeding of drought-resistant crops. *Annual Review of Plant Biology*, *65*(1), 715–741. doi:10.1146/annurev-arplant-050213-040000 PMID:24313844

Humanity+ (2009). *Transhumanist declaration.* https://humanityplus.org/philosophy/transhumanist-declaration/

Hurley, L. H. (2002). DNA and its associated processes as targets for cancer therapy. *Nature Reviews. Cancer*, *2*(3), 188–200. doi:10.1038/nrc749 PMID:11990855

Huxley, J. (1957). Transhumanism. In *New Bottles for New Wine*. Chatto & Windus. (Original work published 1927)

IDC. (2020). Worldwide Spending on AI. *IDC*. https://www.idc.com/getdoc.jsp? containerId=prUS46794720

IEA. (2020). *Direct Air Capture*. https://www.iea.org/reports/direct-air-capture

IMFBlog. (2018). *Chart of the Week: The Rise of Corporate Giants*. https://blogs.imf.org/ 2018/06/06/chart-of-the-week-the-rise-of-corporate-giants/

iN2L. (2020). *iN2L Tablet*. Retrieved from https://in2l.com/our-solution/in2l-tablet/

Ingber, L. (1993). *Adaptive Simulated Annealing (ASA). Global optimization C-code*. Caltech Alumni Association.

Ingber, L. (2011). Computational algorithms derived from multiple scales of neocortical processing, In *Pointing at Boundaries: Integrating Computation and Cognition on Biological Grounds*. Springer. doi:10.100712559-011-9105-4

Ingber, L. (2012a). Adaptive Simulated Annealing, In *Stochastic global optimization and its applications with fuzzy adaptive simulated annealing*. Springer. https://www.ingber.com/asa11_options.pdf

Ingber, L. (2012b). Columnar EEG magnetic influences on molecular development of short-term memory, In *Short-Term Memory: New Research.* Nova. https://www.ingber.com/smni11_stm_scales.pdf

Ingber, L. (2012c). Influence of macrocolumnar EEG on Ca waves. *Current Progress Journal, 1*(1), 4-8. https://www.ingber.com/smni12_vectpot.pdf

Ingber, L. (2015). Calculating consciousness correlates at multiple scales of neocortical interactions. In *Horizons in Neuroscience Research.* https://www.ingber.com/smni15_calc_conscious.pdf

Ingber, L. (2018a). Quantum calcium-ion interactions with EEG.Sci. 1(7), 1-21.

Ingber, L. (2018b). Quantum Variables in Finance and Neuroscience. In *Lecture Plates 2018: QVFN.* Physical Studies Institute. https://www.ingber.com/path18_qpathint_lecture.pdf

Ingber, L. (1982). Statistical mechanics of neocortical interactions. I. Basic formulation. *Physica D. Nonlinear Phenomena, 5*, 83–107.

Ingber, L. (1983). Statistical mechanics of neocortical interactions. Dynamics of synaptic modification. *Physical Review A., 28*, 395–416.

Ingber, L. (1984). Statistical mechanics of neocortical interactions. Derivation of short-term-memory capacity. *Physical Review A., 29*, 3346–3358.

Ingber, L. (1994). Statistical mechanics of neocortical interactions: Path-integral evolution of short-term memory. *Physical Review. E, 49*(5B), 4652–4664.

Ingber, L. (1997). Statistical mechanics of neocortical interactions: Applications of canonical momenta indicators to electroencephalography. *Physical Review. E, 55*(4), 4578–4593. https://www.ingber.com/smni97_cmi.pdf

Ingber, L. (1998). Statistical mechanics of neocortical interactions: Training and testing canonical momenta indicators of EEG. *Mathematical and Computer Modelling, 27*(3), 33–64.

Ingber, L. (2000). High-resolution path-integral development of financial options. *Physica A, 283*(3-4), 529–558.

Ingber, L. (2016a). Path-integral quantum PATHTREE and PATHINT algorithms. *International Journal of Innovative Research in Information Security., 3*(5), 1–15.

Ingber, L. (2016b). Statistical mechanics of neocortical interactions: Large-scale EEG influences on Molecular processes. *Journal of Theoretical Biology, 395*, 144–152.

Ingber, L. (2017a). Evolution of regenerative Ca-ion wave-packet in neuronal-firing fields: Quantum path- integral with serial shocks. *International Journal of Innovative Research in Information Security., 4*(2), 14–22.

Ingber, L. (2017b). Options on quantum money: Quantum path-integral with serial shocks. *International Journal of Innovative Research in Information Security, 4*(2), 7–13.

Ingber, L., Fujio, H., & Wehner, M. F. (1991). Mathematical comparison of combat computer models to Exercise data. *Mathematical and Computer Modelling, 15*(1), 65–90.

Ingber, L., & Nunez, P. L. (2010). Neocortical Dynamics at Multiple Scales: EEG Standing Waves, Statistical Mechanics, and Physical Analogs. *Mathematical Biosciences, 229*, 160–173.

Ingber, L., Pappalepore, M., & Stesiak, R. R. (2014). Electroencephalographic field influence on calcium Momentum waves. *Journal of Theoretical Biology, 343*, 138–153.

Ingber, L., Wehner, M. F., Jabbour, G. M., & Barnhill, T. M. (1991). Application of statistical mechanics methodology to term-structure bond-pricing models. *Mathematical and Computer Modelling*, *15*(11), 77–98.

Ingber, L., & Wilson, J. K. (2000). Statistical mechanics of financial markets: Exponential modifications to Black-Scholes. *Mathematical and Computer Modelling*, *31*(8/9), 167–192.

Innocenti, B., Parpura, V., & Haydon, P. G. (2000). Imaging extracellular waves of glutamate during calcium signaling in cultured astrocytes. *The Journal of Neuroscience: The Official Journal of the Society for Neuroscience*, *20*(5), 1800–1808.

Insel, T. R. (2009). Disruptive insights in psychiatry: Transforming a clinical discipline. *The Journal of Clinical Investigation*, *119*(4), 700–705. doi:10.1172/JCI38832 PMID:19339761

Institute of Management Accountants (IMA). (2018). *Management Accounting Competencies: Fit for Purpose in the Digital Age?* Available on-line on January 21, 2021 at: https://www.imanet.org/insights-and-trends/the-future-of-management-accounting/management-accounting-competencies---fit-for-purpose-in-a-digital-age?ssopc=1

Intel. (2020). *Intel, MIT and Georgia Tech Deliver Improved Machine-Programming Code Similarity System.* https://newsroom.intel.com/news/intel-mit-georgia-tech-machine-programming-code-similarity-system/#gs.r77d21

Irving, P., Beamish, R., & Burstein, A. (2014). *Silver to Gold: The business of aging.* Milken Institute. Retrieved from https://milkeninstitute.org/sites/default/files/reports-pdf/FINAL-Silver-to-Gold-0226.pdf

Islam, K. T., Mujtaba, G., Raj, R. G., & Nweke, H. F. (2017). Handwritten digits recognition with artificial neural network. In IEEE international conference on engineering technology and technopreneurship (pp. 1–4). Kuala Lumpur, Malaysia: IEEE.

Istvan, Z. (2019, May 14). Death could soon become a curable disease. *Metro.* https://metro.co.uk/2019/05/14/death-could-soon-become-a-curable-disease-9191393/

Jackson, D. A., Symons, R. H., & Berg, P. (1972). Biochemical method for inserting new genetic information into DNA of Simian Virus 40: Circular SV40 DNA molecules containing lambda phage genes and the galactose operon of Escherichia coli. *Proceedings of the National Academy of Sciences of the United States of America*, *69*(10), 2904–2909. doi:10.1073/pnas.69.10.2904 PMID:4342968

Jackson, J. D. (1962). *Classical Electrodynamics*. Wiley & Sons.

Jaenisch, R., & Mintz, B. (1974). Simian virus 40 DNA sequences in DNA of healthy adult mice derived from preimplantation blastocysts injected with viral DNA. *Proceedings of the National Academy of Sciences of the United States of America*, *71*(4), 1250–1254. doi:10.1073/pnas.71.4.1250 PMID:4364530

Jaffe, E. S., Blattner, W. A., Blayney, D. W., Bunn, P. A. Jr, Cossman, J., Robert-Guroff, M., & Gallo, R. C. (1984). The pathologic spectrum of adult T-cell leukemia/lymphoma in the United States. Human T-cell leukemia/lymphoma virus-associated lymphoid malignancies. *The American Journal of Surgical Pathology*, *8*(4), 263–275. doi:10.1097/00000478-198404000-00003 PMID:6324600

Jasanoff, S., & Kim, S.-h. (2015). *Dreamscapes of modernity: Sociotechnical imaginaries and the fabrication of power.* The University of Chicago Press.

Javaid, A., Steinberg, D. H., Buch, A. N., Corso, P. J., Boyce, S. W., Pinto Slottow, T. L., ... Torguson, R. (2007). Outcomes of coronary artery bypass grafting versus percutaneous coronary intervention with drug-eluting stents for patients with multivessel coronary artery disease. *Circulation*, *116*(11, supplement), I-200–I-206. doi:10.1161/CIRCULATIONAHA.106.681148 PMID:17846304

Jeffrey, C. (2019, April 4). *Open Bionics' 3D-printed 'Hero Arm' is now available in the US.* TechSpot. https://www.techspot.com/news/79515-open-bionics-3d-printed-hero-arm-now-available.html

Jepsen, M. L. (2018). *How we can use light to see deep inside our bodies and brains.* https://www.ted.com/talks/mary_lou_jepsen_how_we_can_use_light_to_see_deep_inside_our_bodies_and_brains#t-7676

Jesimani, V., Du, H., Aslam, M., Pei, P., & Huang, N. (2019). *Potential Use of Seaweed Bioactive Compounds in Skincare—A Review.* https://www.ncbi.nlm.nih.gov/pmc/ articles/PMC6950024/

Jewell, C. (2019, October). *BrightSign: a smart glove that gives a voice to those who cannot speak.* WIPO Magazine. https://www.wipo.int/wipo_magazine/en/2019/05/article_0005.html#:~:text=BrightSign%2C%20an%20AI%2Dbased%20smart%20glove%2C%20allows%20sign%20language,photo%3A%20Courtesy%20of%20BrightSign)

Jing, M. (2018). Is Xi Jinping's iron grip better than Adam Smith's invisible hand for technological innovation? *South China Morning Post.* https://www.scmp.com/tech/article /2173128/xi-jinpings-iron-grip-better-adam-smiths-invisible-hand-technology-innovation

Jogenfors, J. (2016). *Quantum bitcoin: An anonymous and distributed currency secured by the no-cloning theorem of quantum mechanics.* Linkoping U.

Johnson, I. S. (1983). Human insulin from recombinant DNA technology. *Science, 219*(4585), 632–637. doi:10.1126cience.6337396 PMID:6337396

Johnson, N. P., & Mueller, J. (2002). Updating the accounts: Global mortality of the 1918-1920" Spanish" influenza pandemic. *Bulletin of the History of Medicine, 76*(1), 105–115. doi:10.1353/bhm.2002.0022 PMID:11875246

Johnstone, D. (2012). *An introduction to disability studies* (2nd ed.). Taylor and Francis.

Joint Center for Housing Studies of Harvard University. (2018). *Housing America's older adults, 2018.* Retrieved from https://www.jchs.harvard.edu/sites/default/files/Harvard_JCHS_Housing_Americas_Older_Adults_2018_1.pdf

Joy, B. (2018, October 2). Why the Future Doesn't Need Us. *Wired.* https://www.wired.com/2000/04/joy-2/

Kabat-Zinn, J. (2018). *Meditation Is Not What You Think: Mindfulness and Why It Is So Important.* Hachette Book Group.

Kafer, A. (2013). *Feminist, queer, crip.* Indiana University Press.

Kahneman, D., Malhotra, D., Meyer, E., & Bazerman, M. H. (2019). HBR's 10 Must Reads on Negotiation. *Harvard Business Review.* https://www.booktopia.com.au/hbr-s-10-must-reads-on-negotiation-harvard-business-review/book/9781633697751.html

Kahn, S. E., Hull, R. L., & Utzschneider, K. M. (2006). Mechanisms linking obesity to insulin resistance and type 2 diabetes. *Nature, 444*(7121), 840–846. doi:10.1038/nature05482 PMID:17167471

Kaku, M. (2005). *Parallel Worlds: The Science of Alternative Universes and Our Future in the Cosmos.* Doubleday.

Kaku, M. (2012). *Physics of the Future: How Science Will Shape Human Destiny and Our Daily Lives by the Year 2100.* Anchor.

Kaku, M. (2014). *The Future of the Mind: The Scientific Quest to Understand, Enhance, and Empower the Mind.* Doubleday.

Kaku, M. (2018). *The Future of Humanity: Terraforming Mars, Interstellar Travel, Immortality, and Our Destiny Beyond Earth.* Doubleday.

Kalis, B., Collier, M., & Fu, R. (2018). 10 promising AI applications in health care. *Harvard Business Review.*

Kardashev, N. (1984). On the Inevitability and the Possible Structures of Supercivilizations. *The Search for Extraterrestrial Life: Recent Developments; Proceedings of the Symposium.*

Kato, M., Takahashi, M., Akhand, A. A., Liu, W., Dai, Y., Shimizu, S., Iwamoto, T., Suzuki, H., & Nakashima, I. (1998). Transgenic mouse model for skin malignant melanoma. *Oncogene*, *17*(14), 1885–1888. doi:10.1038j.onc.1202077 PMID:9778055

Katsanis, S. H., & Katsanis, N. (2013). Molecular genetic testing and the future of clinical genomics. *Nature Reviews. Genetics*, *14*(6), 415–426. doi:10.1038/nrg3493 PMID:23681062

Kay, A. (2000). Overview of 'allergy and allergic diseases: With a view to the future'. *British Medical Bulletin*, *56*(4), 843–864. doi:10.1258/0007142001903481 PMID:11359624

Kearnes, M., & van Dooren, T. (2017). Rethinking the final frontier: Cosmo-logics and an ethic of interstellar flourishing. *GeoHumanities*, *3*(1), 178–197. doi:10.1080/2373566X.2017.1300448

Keech, C., Albert, G., Cho, I., Robertson, A., Reed, P., Neal, S., ... Zhou, H. (2020). Phase 1–2 trial of a SARS-CoV-2 recombinant spike protein nanoparticle vaccine. *The New England Journal of Medicine*, *383*(24), 2320–2332. doi:10.1056/NEJMoa2026920 PMID:32877576

Keeney, J. (2015). *To beat death and become immortals, we first must defeat entropy.* Inverse. https://www.inverse.com/article/8867-to-beat-death-and-become-immortals-we-first-must-defeat-entropy

Kelly, K. (2016). *The Inevitable: Understanding the 12 Technological Forces That Will Shape Our Future*. Viking.

Ker, J., Wang, L., Rao, J., & Lim, T. (2017). Deep learning applications in medical image analysis. *IEEE Access: Practical Innovations, Open Solutions*, *6*, 9375–9389.

Kesteven, S. (2016). *Feeding cows seaweed could slash global greenhouse gas emissions*. https://www.abc.net.au/news/2016-10-19/environmental-concerns-cows-eating-seaweed/7946630

Kidd, E. (2020, May 12). My staff don't want to return to work – coming back after COVID-19. *Human Resources Director*.

Kiefer, M. C., Schmid, C., Waldvogel, M., Schläpfer, I., Futo, E., Masiarz, F., Green, K., Barr, P. J., & Zapf, J. (1992). Characterization of recombinant human insulin-like growth factor binding proteins 4, 5, and 6 produced in yeast. *The Journal of Biological Chemistry*, *267*(18), 12692–12699. doi:10.1016/S0021-9258(18)42332-0 PMID:1377672

Kite Power Systems. (2020). http://www.kps.energy/

Kmiecik, S., Gront, D., Kolinski, M., Wieteska, L., Dawid, A. E., & Kolinski, A. (2016). Coarse-grained protein models and their applications. *Chemical Reviews*, *116*(14), 7898–7936. doi:10.1021/acs.chemrev.6b00163 PMID:27333362

Knowles, K. (2018). *BrightSign: Incredible sign-to-speech smart glove startup is raising $1.4 million.* Forbes. https://www.forbes.com/sites/kittyknowles/2018/04/30/brightsign-incredible-sign-to-spee ch-smart-glove-startup-is-raising-1-million/#341ca8b773be

Koch, C. (2011) Christof Koch at Singularity Summit 2011: 'The neurobiology and mathematics of consciousness.' *Exponential Times*. http://www.exponentialtimes.net/videos/christof-koch-singularitysummit-2011-neurobiology-and-mathematics-consciousness

Koeleman, J., Parlevliet, G., Dijkshoorn, L., Savelkoul, P., & Vandenbroucke-Grauls, C. (1997). Nosocomial outbreak of multi-resistant Acinetobacter baumannii on a surgical ward: Epidemiology and risk factors for acquisition. *The Journal of Hospital Infection*, *37*(2), 113–123. doi:10.1016/S0195-6701(97)90181-X PMID:9364260

Koller, D. (2011). Death knell for the lecture: Technology as a passport to personalized education. *New York Times*, 5.

Kort, H. S. M., Woolrych, R., & van Bronswijk, J. E. M. H. (2014). Applying the gerontechnology matrix for research involving ageing adults. In L. Pecchia, L.L. Chen, C. Nugent, & J. Bravo (Eds.), *Lecture Notes in Computer Science: Vol. 8868. Ambient assisted living and daily activities: 6th international work-conference, IWAAL 2014, Belfast, UK, December 2-5, 2014, proceedings.* Springer.

Kostić, V. S., Dobričić, V., Stanković, I., Ralić, V., & Stefanova, E. (2014). C9orf72 expansion as a possible genetic cause of Huntington disease phenocopy syndrome. *Journal of Neurology, 261*(10), 1917–1921. doi:10.100700415-014-7430-8 PMID:25034271

Kozin, S. H. (2001). Syndactyly. *Journal of the American Society for Surgery of the Hand, 1*(1), 1–13. doi:10.1053/jssh.2001.21778

Kozlowski, W., Caballero-Benitez, S. F., & Mekhov, I. B. (2015). *Non-Hermitian dynamics in the quantum Zeno limit.* U Oxford.

KPMG Master of Accounting with Data and Analytics Program. (n.d.). Available on-line on January 21, 2021 at:https://audit.kpmg.us/kpmg-master-accounting-data-analytics-program.html

Krabbendam, L., & Van Os, J. (2005). Schizophrenia and urbanicity: A major environmental influence—conditional on genetic risk. *Schizophrenia Bulletin, 31*(4), 795–799. doi:10.1093chbulbi060 PMID:16150958

Kreisberg, J. C. (2018, January 17). A Globe, Clothing Itself with a Brain. *Wired.* https://www.wired.com/1995/06/teilhard/

Kumar, N. (2018). *Global Public Health: Blockchain for Contagious Disease Relief.* https:// medium.com/swlh/global-public-health-blockchain-for-contagious-disease-relief-453df74f38da

Kurian, G. T., & Molitor, G. T. T. (1996). *Encyclopedia of the Future.* Macmillan.

Kurita, K. (2006). Chitin and chitosan: Functional biopolymers from marine crustaceans. *Marine Biotechnology (New York, N.Y.), 8*(3), 203–226. doi:10.100710126-005-0097-5 PMID:16532368

Kurzweil, R. (2001, March 7). *The Law of Accelerating Returns.* Kurzweil Accelerating Intelligence. https://www.kurzweilai.net/the-law-of-accelerating-returns

Kurzweil, R. (2005). *TED talk.* https://www.ted.com/talks/ray_Kurzweil_the_accelerating_power_of_technology?language=eu

Kurzweil, R. (1999). Spiritual Machines, the merging of man and machine. *The Futurist, 33*(9), 16–21.

Kurzweil, R. (1999). *The Age of Spiritual Machines.* Penguin Books., www.kurzweilai.net

Kurzweil, R. (2005). *The Singularity is Near.* Penguin Putnam Inc.

Kurzweil, R. (2005). *The Singularity is Near: When Humans Transcend Biology.* Viking., www.singularity.com

Kurzweil, R. (2006). *The Singularity Is Near: When Humans Transcend Biology.* Penguin Books.

Kurzweil, R. (2012). *How to Create a Mind: The Secret of Human Thought Revealed.* Penguin Books.

Lagasse, J. (2020, June 3). *Telehealth claim lines increased more than 4,000% in the past year.* Healthcare Finance. Retrieved from https://www.healthcarefinancenews.com/news/telehealth-claim-lines-increased-more-4000-past-year

Lahoz-Beltra, R. (2014, May 14). *The "crisis of noosphere" as a limiting factor to achieve the point of technological singularity.* https://arxiv.org/abs/1405.3378

Laitner, J. (2004). *Energy Impact of Emerging Technologies.* Environmental Protection Agency.

Landon, J., & Schneider, E. (2017, November 24). *These 5 industries will be the first to do business in space.* World Economic Forum. https://www.weforum.org/agenda/2017/11/industries-will-make-money-in-space/

Lane, N. (2015). The unseen world: reflections on Leeuwenhoek (1677) 'Concerning little animals'. *Philosophical Transactions of the Royal Society B: Biological Sciences, 370*(1666), 20140344.

Langouche, F., Roekaerts, D., & Tirapegui, E. (1979). Functional integral methods for stochastic fields. *Physica A, 95*, 252–274.

Langston, J. (2020). *To the moon and beyond: How HoloLens 2 is helping build NASA's Orion spacecraft.* https://news.microsoft.com/innovation-stories/hololens-2-nasa-orion-artemis/

Last, C. (2020). Atechnogenesis and Technocultural Evolution. *World-Systems Evolution and Global Futures*, 165–188. doi:10.1007/978-3-030-46966-5_9

Last, C. (2015). Big Historical Foundations for Deep Future Speculations: Cosmic Evolution, Atechnogenesis, and Technocultural Civilization. *Foundations of Science*, *22*(1), 39–124. doi:10.100710699-015-9434-y

Leavitt, H. J., & Whisler, T. L. (1958). Management in the 1980's. *Harvard Business Review*, *14*, 41–48.

Lee, R. J. (2004). Article II of the Outer Space Treaty: Prohibition of state sovereignty, private property, or both? *Australian International Law Journal*, *11*, 128–142.

Leib, K. (2015, January 2). State sovereignty in space: Current models and possible futures. *Astropolitics*, *13*(1), 1–24. doi:10.1080/14777622.2015.1015112

Leifeld, J., & Menichetti, L. (2018). The underappreciated potential of peatlands in global climate change mitigation strategies. *Nature Communications*, *9*(1), 1071. doi:10.103841467-018-03406-6 PMID:29540695

Lepore, J. (2015, January 25). The iceman. *The New Yorker*. https://www.newyorker.com/magazine/2010/01/25/the-iceman

Lieberman, J. A., & First, M. B. (2018). Psychotic disorders. *The New England Journal of Medicine*, *379*(3), 270–280. doi:10.1056/NEJMra1801490 PMID:30021088

Liftware. (2020). *Liftware by Verify.* Verily. https://www.liftware.com/

Ligon, B. L. (2004). Penicillin: Its discovery and early development. *Seminars in Pediatric Infectious Diseases*, *15*(1), 52–57. doi:10.1053/j.spid.2004.02.001 PMID:15175995

Li, M., Ona, V. O., Guégan, C., Chen, M., Jackson-Lewis, V., Andrews, L. J., ... Przedborski, S. (2000). Functional role of caspase-1 and caspase-3 in an ALS transgenic mouse model. *Science*, *288*(5464), 335–339. doi:10.1126cience.288.5464.335 PMID:10764647

Limbitless Solutions. (2019). *What we do.* Limbitless Solutions. https://limbitless-solutions.org/ourWork

Li, N., Franceschi, S., Howell-Jones, R., Snijders, P. J., & Clifford, G. M. (2011). Human papillomavirus type distribution in 30,848 invasive cervical cancers worldwide: Variation by geographical region, histological type and year of publication. *International Journal of Cancer*, *128*(4), 927–935. doi:10.1002/ijc.25396 PMID:20473886

Lin, P., Abney, K., & Bekey, G. A. (2012). *Robot Ethics: The Ethical and Social Implications of Robotics.* MIT Press.

Linstone, H., & Turoff, M. (1975). *The Delphi Method.* Addison-Wesley.

Lipson, H., & Kurman, M. (2016). *Driverless: Intelligent Cars on the Road Ahead.* MIT Press.

Listek, V. (2020). *3D Printed Vegan Salmon to be Commercialised with Help of FELIXprinters*. https://3dprint.com/270667/3d-printed-vegan-salmon-to-be-commercialized-with-help-of-felixprinters/

Loeb, J. (1913). *Artificial parthenogenesis and fertilization*. University of Chicago Press.

Logan, R. K. (2013). McLuhan extended and the Extended mind thesis (EMT). Avant: Journal of Philosophical-Interdisciplinary Vanguard, 4(2-3), 45-58. 12849/40202013.0709.0003

Lovelock, J., & Margulis, L. (1974). Atmospheric homeostasis by and for the biosphere: The gaia hypothesis. *Tellus*, *26*(1–2), 2–10. doi:10.1111/j.2153-3490.1974.tb01946.x

Lufkin, B. (2020, February 7). What the world can learn from Japan's robots. *BBC*. Retrieved from https://www.bbc.com/worklife/article/20200205-what-the-world-can-learn-from-japans-robots

Lyon, D. (2009, September 16). Ray Kurzweil Wants to Be a Robot. *Newsweek*. https://web.archive.org/web/20100413081045/http://www.newsweek.com/id/197812

MacDonald, R., & MacRae, D. (2019). Maximising Scotland's Well-Being by Bravely Innovating. *HIAlba-IDEA*. http://box5817.temp.domains/~hialbaor/wpcontent/uploads/2019/04/HIAlba-IDEA-Overview-16a.pdf

MacDonald, R., & MacRae, D. (2019). *Maximising Scotland's Well-Being by Bravely Innovating*. http://box5817.temp.domains/~hialbaor/wp-content/uploads/2019/04/HIAlba-IDEA-Overview-16a.pdf

Madakam, S., Lake, V., Lake, V., & Lake, V. (2015). Internet of Things (IoT): A literature review. *Journal of Computer and Communications*, *3*(05), 164.

Maeda, Y., Yoshino, T., Matsunaga, T., Matsumoto, M., & Tanaka, T. (2018). Marine microalgae for production of biofuels and chemicals. *Current Opinion in Biotechnology*, *50*, 111–120. doi:10.1016/j.copbio.2017.11.018 PMID:29232601

Mageit, S. (2020, July 23). *Technology to the rescue - COVID-19 as an effective accelerator of digital health adoption*. Healthcare IT News. Retrieved from https://www.healthcareitnews.com/news/europe/technology-rescue-covid-19-effective-accelerator-digital-health-adoption

Maharani, A., Dawes, P., Nazroo, J., Tampubolon, G., & Pendleton, N. (2018). Longitudinal relationship between hearing aid use and cognitive function in older Americans. *Journal of the American Geriatrics Society*, *66*(6), 1130–1136.

Manyika, J., Jake, S., & Presten, B. (2019). What Do We Do About the Biases in AI? *Harvard Business Review*. https://hbr.org/2019/10/what-do-we-do-about-the-biases-in-ai

Ma, Q.-P. (2012). Government and individual strategy for dealing with the increasing post-retirement life expectancy. In A. Tavidze (Ed.), *Progress in Economics Research* (Vol. 26, pp. 1–29). Nova Science Publisher.

Ma, Q.-P. (2017). Pension provision in the robotic age. In A. Webb (Ed.), *Pensions: Global issues, perspectives and challenges* (pp. 1–38). Nova Science Publisher.

Marcus, G., & Davis, E. (2019). *Rebooting AI: Building Artificial Intelligence We Can Trust*. Penguin Random House. https://www.penguinrandomhouse.com/books/603982/rebooting-ai-by-gary-marcus-and-ernest-davis/

Marshall, K. P. (1999). Has technology introduced new ethical problems? *Journal of Business Ethics*, *19*(1), 81–90. doi:10.1023/A:1006154023743

Martin, P. (2015). Commercialising neurofutures: Promissory economies, value creation and the making of a new industry. *Biosocieties*, *10*(4), 422–443.

Martin, P., Brown, N., & Kraft, A. (2008). From bedside to bench? Communities of promise, translational research and the making of blood stem cells. *Science as Culture*, *17*(1), 29–41. doi:10.1080/09505430701872921

Masala, C., & Petretto, D. R. (2008). From disablement to enablement: Conceptual models of disability in the 20th century. *Disability and Rehabilitation*, *30*(17), 1233–1244.

Mathers, C. D., Boerma, T., & Ma Fat, D. (2009). Global and regional causes of death. *British Medical Bulletin*, *92*(1), 7–32. doi:10.1093/bmb/ldp028 PMID:19776034

Matignon, L. d G. (2019). *Space Tourism Legal Aspects*. Space Legal Issues. https://www.spacelegalissues.com/space-law-space-tourism-legal-aspects/

Matney, L. (2016, November 22). *Ava gives the deaf and hard-of-hearing a more present voice in group conversations.* TechCrunch. https://techcrunch.com/2016/11/21/ava-gives-the-hearing-impaired-a-more-present-voice-in-group-conversations/

Matsuyama, S., Akagiri, K., & Suginuma, K. (2005). *U.S. Patent Application No. 10/985,729*. US Patent Office.

Mattingly, M., & Finkelstein, R. (1992), Combat Robotics: From Kaiser to the New World Order. In Defence Yearbook. Royal United Services Institute and Brassey's Publisher.

Mayer, R. E. (2019). Thirty years of research on online learning. *Applied Cognitive Psychology*, *33*(2), 152–159.

Mazoyer, M., & Roudart, L. (2005). *A History of World Agriculture*. New York Press.

Mazur, M. (2016). *Six Ways Drones Are Revolutionizing Agriculture*. https://www. technologyreview. com/2016/07/20/158748/six-ways-drones-are-revolutionizing-agriculture/

McCusker, C., & Gardiner, D. M. (2011). The axolotl model for regeneration and aging research: A mini-review. *Gerontology*, *57*(6), 565–571. doi:10.1159/000323761 PMID:21372551

McEvedy, C., & Jones, R. (1978). *Atlas of World Population History*. Puffin.

McGowan, J. E. Jr. (2001). Economic impact of antimicrobial resistance. *Emerging Infectious Diseases*, *7*(2), 286–292. doi:10.3201/eid0702.010228 PMID:11294725

McInnes, I. B., & Schett, G. (2011). The pathogenesis of rheumatoid arthritis. *The New England Journal of Medicine*, *365*(23), 2205–2219. doi:10.1056/NEJMra1004965 PMID:22150039

McLuhan, M. (1994). *Understanding Media: The Extensions of Man*. The MIT Press.

McMullan, T. (2016, August 4). Moon express: First private lunar mission approved for 2017. *Alphr*. https://www.alphr.com/space/1004090/moon-express-first-private-lunar-mission-approved-for-2017/

MedRec. (n.d.). *What is Medrec?* https://medrec.media.mit.edu/

Melnick, M., & Shields, E. D. (1975). Orofaciodigital syndrome, type I: A phenotypic and genetic analysis. *Oral Surgery, Oral Medicine, and Oral Pathology*, *40*(5), 599–610. doi:10.1016/0030-4220(75)90370-9 PMID:1059061

Mercado, N. B., Zahn, R., Wegmann, F., Loos, C., Chandrashekar, A., Yu, J., ... Tostanoski, L. H. (2020). Single-shot Ad26 vaccine protects against SARS-CoV-2 in rhesus macaques. *Nature*, *586*(7830), 583–588. doi:10.103841586-020-2607-z PMID:32731257

Meyer, K. (2009). *Extending and simulating the quantum binomial options pricing model* (Thesis). U Manitoba. http://hdl.handle.net/1993/3154

Meystel, A. M., & Albus, J. S. (2002). *Intelligent Systems: Architecture, Design, and Control*. John Wiley & Sons, Inc. doi:10.6028/NIST.IR.6576

Microsoft. (2016, March 30). *Seeing AI 2016 Prototype - A Microsoft research project* [video]. YouTube. https://www.youtube.com/watch?v=R2mC-NUAmMk

Microsoft. (2018, October 31). *How Eye Control empowers people with disabilities (audio description)* [Video]. YouTube. https://www.youtube.com/watch?v=NXOzDf26FnU

Microsoft. (2020a). *AI for accessibility*. Microsoft. https://www.microsoft.com/en-us/ai/ai-for-accessibility

Microsoft. (2020b). *Inclusive hiring for people with disabilities*. Microsoft. https://www.microsoft.com/en-us/diversity/inside-microsoft/cross-disability/hiring.aspx

Microsoft. (2020c). *Seeing AI*. Microsoft. https://www.microsoft.com/en-us/ai/seeing-ai

Ming, V. (2015). The Workforce of the Future. *TedMegBlog*. https://blog.tedmed. com /tag/vivienne-ming/

Ming, V. (2016). The Hidden Tax on Being Different. *HR Magazine*. https://www.hrmagazine.co.uk/article-details/the-hidden-tax-on-being-different

Mingus, M. (2011). Access intimacy: The missing link. *Leaving Evidence*. https://leavingevidence.wordpress.com/2011/05/05/access-intimacy-the-missing-link/

Minsky, M. (1994). Will robots inherit the Earth? *Scientific American*. www.ai.mit.edu/people/minsky/papers/sciam.inherit.txt

Minsky, M. (2016). *Chomsky's theories of language were irrelevant*. https://www.youtube.com/watch?v=wH98yW1SMAo

Minsky, M. (2016). *Kurzweil Interviews Minsky: Is Singularity Near?* https://www.kurzweilai.net/kurzweil-interviews-minsky-is-singularity-near

Minsky, M. (1987). *The Society of Mind*. Simon and Schuster.

Miraz, M. H., Ali, M., Excell, P. S., & Picking, R. (2015, September). A review on Internet of Things (IoT), Internet of Everything (IoE) and Internet of Nano Things (IoNT). *2015 Internet Technologies and Applications (ITA)*. doi:10.1109/ITechA.2015.7317398

Miraz, M. H., Ali, M., Excell, P. S., & Picking, R. (2015). *A review on Internet of Things (IoT), Internet of Everything (IoE) and Internet of Nano Things (IoNT). 2015 Internet Technologies and Applications*. ITA. doi:10.1109/itecha.2015.7317398

Mittler, R., & Blumwald, E. (2010). Genetic engineering for modern agriculture: Challenges and perspectives. *Annual Review of Plant Biology*, *61*(1), 443–462. doi:10.1146/annurev-arplant-042809-112116 PMID:20192746

Mixon, T. (2020). *USAF to Use Four-Legged Robots for Security at Florida Base*. The News Herald.

Mok, C., & Lau, C. (2003). Pathogenesis of systemic lupus erythematosus. *Journal of Clinical Pathology*, *56*(7), 481–490. doi:10.1136/jcp.56.7.481 PMID:12835292

Monash. (2019, October 27). *3D-printed prosthetic limbs: how a social entrepreneur is changing lives*. Monash Life Magazine. https://lens.monash.edu/@monash-life/2019/10/27/1377036/3d-printed-prosthetic-limbs-changing-lives

Moon, N. W., Baker, P. M. A., & Goughnour, K. (2019). Designing wearable technologies for users with disabilities: Accessibility, usability, and connectivity factors. *Journal of Rehabilitation and Assistive Technologies Engineering*, *6*, 1–12.

Moravec, H. (1988). *Mind Children*. Harvard University Press.

Moravec, H. (1998). *Robot: Mere Machine to Transcendent Mind*. Oxford University Press. www.frc.ri.cmu.edu/~hpm/book97

More, M. (2003). *The Principles of Extropy*. The Extropy Institute. www.extropy.org/principles.htm

Moritz, C. (2018). A giant step for spinal cord injury research. *Nature Neuroscience, 21*, 1647–1648.

Morris, A., Kellner, J. D., & Low, D. E. (1998). The superbugs: Evolution, dissemination and fitness. *Current Opinion in Microbiology, 1*(5), 524–529. doi:10.1016/S1369-5274(98)80084-2 PMID:10066531

Moses, R. W. (2018). *Maintaining Human Health for Humans on Mars*. AIAA.

Mulhall, D. (2002). *Our Molecular Future*. Prometheus Books.

Muller, M. M., Gherardini, S., & Caruso, F. (2016). *Quantum Zeno Dynamics through stochastic protocols*. U Florence.

Müller-Oerlinghausen, B., Berghöfer, A., & Bauer, M. (2002). Bipolar disorder. *Lancet, 359*(9302), 241–247. doi:10.1016/S0140-6736(02)07450-0 PMID:11812578

Murillo, S. (2020). *Review of the Dot, the world's first braille smartwatch*. The Chicago Lighthouse. https://chicagolighthouse.org/sandys-view/review-of-the-dot-the-worlds-first-braille-smartwatch/

Mykytyn, C. E. (2006). Anti-aging medicine: A patient/practitioner movement to redefine aging. *Social Science & Medicine, 62*(3), 643–653. doi:10.1016/j.socscimed.2005.06.021 PMID:16040177

Nagi, S. Z. (1991). Disability concepts revised: Implications for prevention. In A. M. Pope & A. R. Tarlov (Eds.), *Disability in America: Toward a national agenda for prevention*. National Academy Press.

Nanoracks. (2020). *Space Growth Crops for Earth Farms*. https://nanoracks.com/space-grown-crops-earth-farms/

Nanowerk. (2017). *Seaweed: From superfood to superconductor*. https://www.nanowerk.com/nanotechnology-news/newsid=46330.php

NASA. (2015). *NASA, Microsoft Collaborate to Bring Science Fiction to Science Fact*. https://www.nasa.gov/press-release/nasa-microsoft-collaborate-to-bring-science-fiction-to-science-fact

NASA. (2019). *NASA Opens International Space Station To New Commercial Opportunities, Private Astronauts*. https://www.nasa.gov/press-release/nasa-opens-international-space-station-to-new-commercial-opportunities-private

National Academies of Sciences Engineering Medicine. (2017). Addressing Continuous Threats: HIV/AIDS, Tuberculosis, and Malaria. In *Global Health and the Future Role of the United States*. National Academies Press (US).

National Aeronautics and Space Administration. (n.d.). *The Outer Space Treaty –1967*. https://history.nasa.gov/SP-4225/documentation/cooperation/treaty.htm#:~:text=Treaty%20on%20principles%20governing%20the,moon%20and%20other%20celestial%20bodies.&text=The%20moon%20and%20other%20celestial%20bodies%20shall%20be%20used%20by,Treaty%20exclusively%20for%20peaceful%20purposes

National Council On Aging. (n.d.). *Fall management facts*. National Council on Aging. Retrieved from https://www.ncoa.org/news/resources-for-reporters/get-the-facts/falls-prevention-facts/

Nature Astronomy. (2019). Sustainable space mining. *Nature Astronomy, 3*(465). doi:10.103841550-019-0827-7

Nauman, A., Qadri, Y. A., Amjad, M., Zikria, Y. B., Afzal, M. K., & Kim, S. W. (2020). Multimedia Internet of Things: A comprehensive survey. *IEEE Access: Practical Innovations, Open Solutions, 8*, 8202–8250.

Nazaretyan, A. P. (2014). Singularity Hypotheses: A Scientific and Philosophical Assessment. Minds and Machines, 24(2), 245–248. doi:10.100711023-014-9340-z

Neilsen, A., & Wilson, R. (2019). Combining e-mental health intervention development with human computer interaction (HCI) design to enhance technology-facilitated recovery for people with depression and/or anxiety conditions: An integrative literature review. *International Journal of Mental Health Nursing*, *28*(1), 22–39. doi:10.1111/inm.12527 PMID:30133096

Nemeroff, C. B. (2003). Anxiolytics: Past, present, and future agents. *The Journal of Clinical Psychiatry*. PMID:12662127

Nerlich, S. (2020). 2020 by the numbers. *International Education Association of Australia*. https://www.ieaa.org.au/blog/2020-by-the-numbers

Nicolelis, M. (2020). *The True Creator of Everything: How the Human Brain Shaped the Universe as We Know It*. Yale University Press. doi:10.2307/j.ctvt1sggf

Nicolia, A., Manzo, A., Veronesi, F., & Rosellini, D. (2014). An overview of the last 10 years of genetically engineered crop safety research. *Critical Reviews in Biotechnology*, *34*(1), 77–88. doi:10.3109/07388551.2013.823595 PMID:24041244

Nilsson, N. J. (2014). *Principles of artificial intelligence*. Morgan Kaufmann.

Nott, W. (2020). UK: Face-to-face teaching vital for international students. *Professionals in International Education*. https://thepienews.com/news/uk-face-to-face-teaching-vital-for-intl-students

Nutten, S. (2015). Atopic dermatitis: Global epidemiology and risk factors. *Annals of Nutrition & Metabolism*, *66*(Suppl. 1), 8–16. doi:10.1159/000370220 PMID:25925336

NVIDIA. (2020). *NVIDIA Announces Cloud-AI Video-Streaming Platform to Better Connect Millions Working and Studying Remotely*. https://nvidianews.nvidia.com/news/nvidia-announces-cloud-ai-video-streaming-platform-to-better-connect-millions-working-and-studying-remotely

Odell, G. (2013). *Stone Tools: Insights into Pre-History*. Springer-Verlag.

Oeppen, J., & Vaupel, J. W. (2002). *Broken limits to life expectancy*. American Association for the Advancement of Science.

Office of Science and Technology Policy. (2019, March). *Emerging technologies to support an aging population*. https://www.whitehouse.gov/wp-content/uploads/2019/03/Emerging-Tech-to-Support-Aging-2019.pdf

Olshansky, S. J., Carnes, B. A., & Désesquelles, A. (2001). Prospects for human longevity. *Science*, *291*(5508), 1491–1492. doi:10.1126cience.291.5508.1491 PMID:11234076

Olson, E. T. (2011). The Extended Self. *Minds and Machines*, *21*(4), 481–495. doi:10.100711023-011-9258-7

Open Bionics. (2018). *FAQ*. Open Bionics. https://openbionics.com/faq/

OrCam. (2020). *OrCam MyEye: For people who are blind or visually impaired*. OrCam. https://www.orcam.com/en/myeye2/

Orimaye, S. O., Wong, J. S., Golden, K. J., Wong, C. P., & Soyiri, I. N. (2017). Predicting probable Alzheimer's disease using linguistic deficits and biomarkers. *BMC Bioinformatics*, *18*(34), 1–13.

Orlov, L. (2020, March). *Technology for aging: 2020 market overview*. Aging and Health Technology Watch. Retrieved from https://www.ageinplacetech.com/files/aip/Market%20Overview%202020%20Final-March-2020.pdf

Oussous, A., Benjelloun, F. Z., Lahcen, A. A., & Belfkih, S. (2018). Big Data technologies: A survey. *Journal of King Saud University-Computer and Information Sciences*, *30*(4), 431–448.

OxSight. (2019, December 20). *OXSIGHT: Best of 2019*. OxSight. https://www.oxsight.co.uk/news/oxsight-best-of-2019/

OxSight. (2020). *OXSIGHT aids for the visually impaired*. OxSight. https://www.oxsight.co.uk/products/

Pai, S., Lin, Y., Macaes, B., Meneshian, A., Hung, C.-F., & Wu, T. C. (2006). Prospects of RNA interference therapy for cancer. *Gene Therapy*, *13*(6), 464–477. doi:10.1038j.gt.3302694 PMID:16341059

Palermo, A. E., Maher, J. L., Baunsgaard, C. B., & Nash, M. S. (2017). Clinician-focused overview of bionic exoskeleton use after spinal cord injury. *Topics in Spinal Cord Injury Rehabilitation*, *23*(3), 234–244.

Pas, J. F. (1995). *Visions of Sukhavati: Shan-tao's Commentary on the Kuan wu-liang-shou-Fo ching*. SUNY Press.

Pasteur, L. (1857). Mémoire sur la fermentation alcoolique. *Comptes Rendus. Chimie*, *45*(6), 1032–1036.

Pasteur, L. (1858). Nouveaux faits concernant l'histoire de la fermentation alcoolique. *Comptes Rendus. Chimie*, *47*, 1011–1013.

Patel, A. B., Rothman, D. L., Cline, G. W., & Behard, K. L. (2001). Glutamine is the major precursor for GABA synthesis in rat neocortex in vivo following acute GABA-transaminase inhibition. *Brain Research*, *919*(2), 207–220.

Patel, S., & Saunders, K. (2018). Apps and wearables in the monitoring of mental health disorders. *British Journal of Hospital Medicine (London, England)*, *79*(12), 672–675. doi:10.12968/hmed.2018.79.12.672 PMID:30526097

Pathak, A., Redmond, J. A., Allen, M., & Chou, K. L. (2013). A noninvasive handheld assistive device to accommodate essential tremor: A pilot study. *Movement Disorders*, *29*(6), 838–842.

Patil, Y. S., Chakram, S., & Vengalattore, M. (2015). Measurement-induced localization of an ultracold lattice gas. *Physical Review Letters*, *115*(140402), 1–5.

Patterson, K. B., & Runge, T. (2002). Smallpox and the native American. *The American Journal of the Medical Sciences*, *323*(4), 216–222. doi:10.1097/00000441-200204000-00009 PMID:12003378

Paul, G. S., & Cox, E. (1996). *Beyond Humanity: Cyberevolution and Future Minds*. Charles River Media.

Pauling, L., Itano, H. A., Singer, S. J., & Wells, I. C. (1949). Sickle cell anemia, a molecular disease. *Science*, *110*(2865), 543–548. doi:10.1126cience.110.2865.543 PMID:15395398

Pauly, P. J. (1987). *Controlling life: Jacques Loeb and the engineering ideal in biology*. Oxford University Press.

Penn, J. (2018). AI thinks like a corporation - and that's worrying. *The Economist*. https://www.economist.com/open-future/2018/11/26/ai-thinks-like-a-corporation-and-thats-worrying

Pereira, A. Jr, & Furlan, F. A. (2009). On the role of synchrony for neuron-astrocyte interactions and perceptual conscious processing. *Journal of Biological Physics*, *35*(4), 465–480.

Petch, M. (2020). 3D Printing Responds to Covid-19 and Coronavirus Resources. *3D Printing Industry*. https://3dprintingindustry.com/news/3d-printing-community-responds-to-covid-19-and-coronavirus-resources-169143/

Peters, T. (2018). Radical life extension, cybernetic immortality, and techno-salvation. Really? *Dialog*, *57*(4), 250–256. https://doi.org/10.1111/dial.12432

Phelan, D. (2020, August 11). Apple Watch Series 6: Patents reveal game-changing hardware upgrades. *Forbes*. Retrieved from https://www.forbes.com/sites/davidphelan/2020/08/11/apple-watch-series-6-patents-reveals-game-changing-hardware-upgrades/#5a364f5354e6

Philippidis, A. (2020). Moderna's COVID-19 Vaccine Speeds to Phase III Trial After More Positive Data. *Clinical OMICs*, *7*(4), 10–11. doi:10.1089/clinomi.07.04.17

Phillips, S.D. (2009). "There are no invalids in the USSR!": A missing Soviet chapter in the new disability history. *Disability Studies Quarterly, 29*(3).

Phillips, J. C., Hardy, D. J., Maia, J. D., Stone, J. E., Ribeiro, J. V., Bernardi, R. C., ... Tajkhorshid, E. (2020). Scalable molecular dynamics on CPU and GPU architectures with NAMD. *The Journal of Chemical Physics*, *153*(4), 044130.

PHYS.ORG. (2018). *New type of opal formed by common seaweed discovered.* https://phys.org/news/2018-04-opal-common-seaweed.html#jCp

Piotrowski, E. W., Schroeder, M., & Zambrzycka, A. (2005). Quantum extension of European option pricing based on the Ornstein-Uhlenbeck process. *Physica A*, *368*(1), 176–182.

Pious, N., & Ingole, S. (2020). Race for COVID-19 Vaccine. *Trends in Biomaterials & Artificial Organs*, *34*(S3), 62–65.

Piperno, D. R., Ranere, A. J., Holst, I., Iriarte, J., & Dickau, R. (2009). Starch grain and phytolith evidence for early ninth millennium BP maize from the Central Balsas River Valley, Mexico. *Proceedings of the National Academy of Sciences of the United States of America*, *106*(13), 5019–5024. doi:10.1073/pnas.0812525106 PMID:19307570

Polivy, J., & Herman, C. P. (2002). Causes of eating disorders. *Annual Review of Psychology*, *53*(1), 187–213. doi:10.1146/annurev.psych.53.100901.135103 PMID:11752484

Pop, V. (2000). Appropriation in outer space: The relationship between land ownership and sovereignty on the celestial bodies. *Space Policy*, *16*(4), 275–282. doi:10.1016/S0265-9646(00)00037-0

Preskill, J. (2015). *Quantum Mechanics. Lecture Notes*. Caltech. http://www.theory.caltech.edu/people/preskill/ph219/

Preston, S. (2017). *Homo Erectus as a Man*. CreateSpace Independent Publishing Platform.

Pullin, G. (2009). *Design meets disability*. MIT Press.

Pultarova, T. (2021). *Could space greenhouses solve Earth's food crisis?* https://www.space.com/space-greenhouses-nanoracks-food-crisis

Purdy, C. (2018). *Welcome to the first luxury space hotel*. Quartz. https://qz.com/1247091/orion-span-claims-its-luxury-space-hotel-will-open-to-its-first-guests-in-2022/

Purugganan, M. D., & Fuller, D. Q. (2009). The nature of selection during plant domestication. *Nature*, *457*(7231), 843–848. doi:10.1038/nature07895 PMID:19212403

Putnam, R. (2000). *Bowling Alone*. New York: Simon & Schuster. http://bowlingalone.com/

QuoteB. (2020a). https://www.brainyquote.com/quotes/alfred_north_whitehead_108058

QuoteB. (2020b). https://www.brainyquote.com/quotes/thomas_bailey_aldrich_141405

QuoteB. (2020c). https://www.brainyquote.com/quotes/isaac_bashevis_singer_141018

Radhakrishna, U., Wild, A., Grzeschik, K.-H., & Antonarakis, S. E. (1997). Mutation in GLI3 in postaxial polydactyly type A. *Nature Genetics*, *17*(3), 269–271. doi:10.1038/ng1197-269 PMID:9354785

Rajeshwari, M., & Krishna Prasad, K. (2020). IBM Watson Industry Cognitive Education Methods. *International Journal of Case Studies in Business, IT, and Education*, *4*(1), 38–50.

Ran, F. A., Hsu, P. D., Wright, J., Agarwala, V., Scott, D. A., & Zhang, F. (2013). Genome engineering using the CRISPR-Cas9 system. *Nature Protocols*, *8*(11), 2281–2308. doi:10.1038/nprot.2013.143 PMID:24157548

Rao, K. R. (2017). *Foundations of Yoga Psychology*. Springer. doi:10.1007/978-981-10-5409-9

Rao, K. R., Paranjpe, A. C., & Dalal, A. K. (2008). *Handbook of Indian Psychology*. Cambridge University Press India Pvt. Ltd. doi:10.1017/UPO9788175968448

Rao, V. R., Cohen, G. B., & Oprian, D. D. (1994). Rhodopsin mutation G90D and a molecular mechanism for congenital night blindness. *Nature*, *367*(6464), 639–642. doi:10.1038/367639a0 PMID:8107847

Rapoport, S. I., Basselin, M., Kim, H.-W., & Rao, J. S. (2009). Bipolar disorder and mechanisms of action of mood stabilizers. *Brain Research. Brain Research Reviews*, *61*(2), 185–209. doi:10.1016/j.brainresrev.2009.06.003 PMID:19555719

Reaven, G. (2002). Metabolic syndrome: Pathophysiology and implications for management of cardiovascular disease. *Circulation*, *106*(3), 286–288. doi:10.1161/01.CIR.0000019884.36724.D9 PMID:12119239

Reeves, G., & Ainsworth, E. (1995). Description of the chronic radiation syndrome in humans irradiated in the former Soviet Union. *Radiation Research*, *142*(2), 242–243. doi:10.2307/3579035 PMID:7724741

Refolo, L. M., Pappolla, M. A., Malester, B., LaFrancois, J., Bryant-Thomas, T., Wang, R., Tint, G. S., Sambamurti, K., & Duff, K. (2000). Hypercholesterolemia accelerates the Alzheimer's amyloid pathology in a transgenic mouse model. *Neurobiology of Disease*, *7*(4), 321–331. doi:10.1006/nbdi.2000.0304 PMID:10964604

Regalado, A. (2018, May 9). A stealthy Harvard startup wants to reverse aging in dogs, and humans could be next. *MIT Technology Review*. https://www.technologyreview.com/2018/05/09/142971/a-stealthy-harvard-startup-wants-to-reverse-aging-in-dogs-and-humans-could-be-next/

Regenerative International. (2016). https://regenerationinternational.org/regenerative-not-climate-smart-agriculture-needed-to-feed-the-world-and-cool-the-planet

Rehman, A. (2019, July 26). *How virtual reality can transform dementia care*. The Academy of International Extended Reality. Retrieved from https://aixr.org/insights/how-virtual-reality-can-transform-dementia-care/

Reilly, C. (2020, July 9). *Cryonics, brain preservation, and the weird science of cheating death*. Cnet. https://www.cnet.com/features/cryonics-brain-preservation-and-the-weird-science-of-cheating-death-alcor/

Remes, J., Linzer, K., Singhal, S., Dewhurst, M., Dash, P., Woetzel, J., Smit, S., Evers, M., Wilson, M., Rutter, K., & Ramdorai, A. (2020). *Report - Prioritizing health: A prescription for prosperity*. McKinsey Global Institute. https://www.mckinsey.com/industries/healthcare-systems-and-services/our-insights/prioritizing-health-a-prescription-for-prosperity

Rendever, Inc. (2020). *Rendever*. https://rendever.com/

Reyes, R. C., & Parpura, V. (2009). The trinity of Ca2+ sources for the exocytotic glutamate release from astrocytes. *Neurochemistry International*, *55*(3), 1–14.

Reyzelman, A., Koelewyn, K., Murphy, M., Shen, X., Pillai, R., Fu, J., Scholten, H. J., & Ma, R. (2018). Continuous temperature-monitoring socks for home use in patients with diabetes: Observational study. *Journal of Medical Internet Research*, *20*(12), 1–14.

Riedel, S. (2005). Edward Jenner and the history of smallpox and vaccination. *Baylor University Medical Center Proceedings*, *18*(1), 21–25. doi:10.1080/08998280.2005.11928028 PMID:16200144

Rightmire, G. (1993). *The Evolution of Homo erectus*. Cambridge University Press.

Risérus, U., Willett, W. C., & Hu, F. B. (2009). Dietary fats and prevention of type 2 diabetes. *Progress in Lipid Research, 48*(1), 44–51. doi:10.1016/j.plipres.2008.10.002 PMID:19032965

Robinson, D., & Medlock, N. (2005). Diamond v. Chakrabarty: a retrospective on 25 years of biotech patents. *Intellectual Property & Technology Law Journal, 17*(10), 12-15.

Rocklin, R. E., Sheffer, A. L., Greineder, D. K., & Melmon, K. L. (1980). Generation of antigen-specific suppressor cells during allergy desensitization. *The New England Journal of Medicine, 302*(22), 1213–1219. doi:10.1056/NEJM198005293022201 PMID:6154241

Roco, M. C., & Montemagno, C. D. (Eds.). (2004). The Coevolution of Human Potential and Converging Technologies. New York, NY: New York Academy of Sciences.

Roco, M. C., & Bainbridge, W. S. (Eds.). (2003). *Converging Technologies for Improving Human Performance*. Kluwer. doi:10.1007/978-94-017-0359-8

Rogers, C., & Cook, C. (2020). Where to now? Blended futures. *Deloitte*. https://www2.deloitte.com/au/en/pages/public-sector/articles/where-now-blended-futures.html

Rolls-Royce. (2020). Rolls-Royce Has Achieved A Breakthrough in Ethics & Trustworthiness. *Rolls-Royce*. https://www.carandbike.com/news/rolls-royce-has-achieved-a-breakthrough-in-ai-ethics-trustworthiness-2291937

Rony, M. R. A. H., & Alam, M. M. (2019). Hand gesture recognition using image segmentation and deep neural network. *Eleventh International Conference on Machine Vision, 2018. Proceedings of the Society for Photo-Instrumentation Engineers, 11041*, 1–6.

Rose, N. (2007). Molecular biopolitics, somatic ethics, and the spirit of biocapital. *Social Theory & Health, 5*(1), 3–29.

Ross, W. N. (2012). Understanding calcium waves and sparks in central neurons. *Nature Reviews. Neuroscience, 13*, 157–168.

Rothenberg, D. (1993). *Hand's End. Technology and the limits of nature*. University California Press. doi:10.1525/9780520913202

Russell, M., & Malhotra, R. (2002). Capitalism and disability. *Socialist Register, 38*, 211–228.

Russey, C. (2019, July 30). *How smart bandages are transforming medicine and healthcare.* Wearable Technologies. Retrieved from https://www.wearable-technologies.com/2019/07/how-smart-bandages-are-transforming-medicine-and-healthcare/

Rutschman, A. S. (2018). *U.S. Health Care Companies Begin Exploring Blockchain Technologies.* https://www.scientificamerican.com/article/u-s-health-care-companies-begin-exploring-blockchain-technologies/?

RxFunction, Inc. (2020). *RxFunction.* https://rxfunction.com/

Rydning, D. R. J. G. J. (2018). *The digitization of the world from edge to core*. International Data Corporation.

Sagan, C. (1973). *The Cosmic Connection*. Doubleday.

Sagan, C. (1977). *The Dragons of Eden: Speculations on the Evolution of Human Intelligence*. Random House.

Salazar, R. F., Dotson, N. M., Bressler, S. L., & Gray, C. M. (2012). Content-specific fronto-parietal synchronization during visual working memory. *Science, 338*(6110), 1097–1100.

Salque, M., Bogucki, P. I., Pyzel, J., Sobkowiak-Tabaka, I., Grygiel, R., Szmyt, M., & Evershed, R. P. (2013). Earliest evidence for cheese making in the sixth millennium BC in northern Europe. *Nature*, *493*(7433), 522–525. doi:10.1038/nature11698 PMID:23235824

Salter, A. W., & Leesen, P. T. (2014). Celestial anarchy: A threat to outer space commerce? *The Cato Journal*, *34*(3), 581–596.

Sanger-Katz, M., Kliff, S., & Parlapiano, A. (2020, March 17). These places could run out of hospital beds as coronavirus spreads. *The New York Times.* Retrieved from https://www.nytimes.com/interactive/2020/03/17/upshot/hospital-bed-shortages-coronavirus.html

Sapci, A. H., & Sapci, H. A. (2019). Innovative assisted living tools, remote monitoring technologies, artificial intelligence-driven solutions, and robotic systems for aging societies: Systematic review. *JMIR Aging*, *2*(2), 1–16.

Saravanan, R., & Sujatha, P. (2018). A state of art techniques on machine learning algorithms: A perspective of supervised learning approaches in data classification. In *2018 Second International Conference on Intelligent Computing and Control Systems (ICICCS)* (pp. 949–5). Madurai, India: Academic Press.

Sateia, M. J. (2014). International classification of sleep disorders. *Chest*, *146*(5), 1387–1394. doi:10.1378/chest.14-0970 PMID:25367475

Saxena, D., Flores, S., & Stotzky, G. (1999). Insecticidal toxin in root exudates from Bt corn. *Nature*, *402*(6761), 480–480. doi:10.1038/44997 PMID:10591205

Scemes, E., & Giaume, C. (2006). Astrocyte calcium waves: What they are and what they do. *Glia*, *54*(7), 716–725.

Schalk, S. (2017). Critical disability studies as methodology. *Lateral*, *6*(1).

Scharping, N. (2016, October 6). *Humans may not be able to live past 125*. Discover. https://www.discovermagazine.com/health/humans-may-not-be-able-to-live-past-125

Schneider, N. (2018). An Internet of ownership: Democratic design for the online economy. *The Sociological Review*, *66*(2), 320–340.

Schneider, S. L., & Council, M. L. (2020). Distance learning in the era of COVID-19. *Archives of Dermatological Research*, 1–2.

Schöning, M. J., & Poghossian, A. (2002). Recent advances in biologically sensitive field-effect transistors (BioFETs). *Analyst (London)*, *127*(9), 1137–1151. doi:10.1039/B204444G PMID:12375833

Schranz, M., Umlauft, M., Sende, M., & Elmenreich, W. (2020). Swarm Robotic Behaviors and Current Applications. *Frontiers in Robotics and AI*, *7*, 36. Advance online publication. doi:10.3389/frobt.2020.00036 PMID:33501204

Schulman, L. S. (1981). *Techniques and Applications of Path Integration*. J. Wiley & Sons.

Schwab, K. (2017). The Fourth Industrial Revolution (Illustrated ed.). Currency.

ScienceDaily. (2019). *New holographic technique opens the way for quantum computation*. https://www.sciencedaily.com/releases/2019/05/190504130303.htm

Seal, B., Otero, T., & Panitch, A. (2001). Polymeric biomaterials for tissue and organ regeneration. *Materials Science and Engineering R Reports*, *34*(4-5), 147–230. doi:10.1016/S0927-796X(01)00035-3

Selgelid, M. J. (2005). Ethics and infectious disease 1. *Bioethics*, *19*(3), 272–289. doi:10.1111/j.1467-8519.2005.00441.x PMID:16167406

Seminari, S. (2019, November 24). Op/ed: Global government space budgets continues multiyear rebound. *SpaceNews*. https://spacenews.com/op-ed-global-government-space-budgets-continues-multiyear-rebound/

Serviceplan Group. (2018, April 6). *DOT: First braille smartwatch* [Video]. YouTube. https://www.youtube.com/watch?v=2ax-NG-YVGw

Shah, R., & Chircu, A. (2018). IoT and AI in healthcare: A systematic literature review. *Issues in Information Systems, 19*(3).

Shaker, S. M., & Wise, A. R. (1988). *War without Men: Robots on the Future Battlefield*. Pergamon-Brassey's International Defense Publishers, Inc.

Shannon, C. E. (1948). A mathematical theory of communication. *The Bell System Technical Journal, 27*(3), 379–423.

Sheldon, A. (2004). Changing technology. In J. Swain, S. French, C. Barnes, & C. Thomas (Eds.), *Disabling barriers: Enabling environments* (pp. 155–160). SAGE Publications.

Sherrington, R., Rogaev, E., Liang, Y., Rogaeva, E., Levesque, G., Ikeda, M., ... Holman, K. (1995). Cloning of a gene bearing missense mutations in early-onset familial Alzheimer's disease. *Nature, 375*(6534), 754–760. doi:10.1038/375754a0 PMID:7596406

Shimizu Corp of Japan. (2020). https://www.shimz.co.jp/en/topics/dream/content01/

Shishehgar, M., Kerr, D., & Blake, J. (2018). A systematic review of research into how robotic technology can help older people. *Smart Health, 7-8*, 1–18.

Shor, J., & Emanuel, D. (2019, August 13). *Project Euphonia's personalized speech recognition for non-standard speech.* Google AI Blog. https://ai.googleblog.com/2019/08/project-euphonias-personalized-speech.html

Shor, P. W. (1994) Algorithms for quantum computation: discrete logarithms and factoring. *Proceedings 35th Annual Symposium on Foundations of Computer Science, IEEE Xplore*, 124-134. https://ieeexplore.ieee.org/document/365700 and https://dx.doi.org/10.1109/SFCS.1994.365700

Shrestha, P., Cooper, B. S., Coast, J., Oppong, R., Thuy, N. D. T., Phodha, T., ... Lubell, Y. (2018). Enumerating the economic cost of antimicrobial resistance per antibiotic consumed to inform the evaluation of interventions affecting their use. *Antimicrobial Resistance and Infection Control, 7*(1), 98. doi:10.118613756-018-0384-3 PMID:30116525

Shuttleworth, J. (2019). Automated-Driving Graphic Update, *SAE. Standardization News, J3016*. https://www.sae.org/news/2019/01/sae-updates-j3016-automated-driving-graphic

Sicherer, S. H., & Sampson, H. A. (2014). Food allergy: epidemiology, pathogenesis, diagnosis, and treatment. *Journal of Allergy and Clinical Immunology, 133*(2), 291-307.

Sidhu, M. (2020). *WVTOX, This Is Why Kelp Leather Is the Best Vegan Leather Alternative To Animal Skin.* https://wtvox.com/fashion/kelp-leather/

Sim, A. Y., Minary, P., & Levitt, M. (2012). Modeling nucleic acids. *Current Opinion in Structural Biology, 22*(3), 273–278. doi:10.1016/j.sbi.2012.03.012 PMID:22538125

Singh, S. (2019). The role of speech technology in biometrics, forensics and man-machine interface. *Iranian Journal of Electrical and Computer Engineering, 9*(1), 281–288.

SISSA. (2020). *Black holes? They are like a hologram.* https://www.sissa.it/news/black-holes-they-are-hologram

Skaugrud, Ø., Hagen, A., Borgersen, B., & Dornish, M. (1999). Biomedical and pharmaceutical applications of alginate and chitosan. *Biotechnology & Genetic Engineering Reviews*, *16*(1), 23–40. doi:10.1080/02648725.1999.10647970 PMID:10819076

Slamon, D. J. (1987). Proto-oncogenes and human cancers. Mass Medical Soc. doi:10.1056/NEJM198710083171509

Small Robot Company. (2020). https://www.smallrobotcompany.com/

Smalley, E. (2017). *AI-powered drug discovery captures pharma interest*. Academic Press.

Snooks, G. D. (1989). *The Dynamic Society*. Routledge.

Snyder, T., & Byrd, G. (2017). The internet of everything. *IEEE Computer Architecture Letters*, *50*(06), 8–9.

Spitzer, R. L., Kroenke, K., Williams, J. B., & Löwe, B. (2006). A brief measure for assessing generalized anxiety disorder: The GAD-7. *Archives of Internal Medicine*, *166*(10), 1092–1097. doi:10.1001/archinte.166.10.1092 PMID:16717171

Squassina, A., Manchia, M., Manolopoulos, V. G., Artac, M., Lappa-Manakou, C., Karkabouna, S., Mitropoulos, K., Zompo, M. D., & Patrinos, G. P. (2010). Realities and expectations of pharmacogenomics and personalized medicine: Impact of translating genetic knowledge into clinical practice. *Pharmacogenomics*, *11*(8), 1149–1167. doi:10.2217/pgs.10.97 PMID:20712531

Srinivasan, R., Winter, W. R., Ding, J., & Nunez, P. L. (2007). EEG and MEG coherence: Measures of functional connectivity at distinct spatial scales of neocortical dynamics. *Journal of Neuroscience Methods*, *166*(1), 41–52.

Staff, W. (2017, June 5). The Year 2020, Explained. *Wired*. https://www.wired.com/2000/07/the-year-2020-explained/

Stambler, I. (2014). *A history of life-extensionism in the twentieth century*. CreateSpace Independent Publishing Platform.

Stanford Medicine. (2020). *'Smart toilet' monitors for signs of disease*. Stanford Medicine News Center. Retrieved from http://med.stanford.edu/news/all-news/2020/04/smart-toilet-monitors-for-signs-of-disease.html

Stanley, M. (2020, July 24). *Space: investing in the final frontier*. https://www.morganstanley.com/ideas/investing-in-space

Stern Law, P. L. L. C. (2020). *Eye tracking communication for individuals that are non-verbal*. My Child at CerebalPalsy.org. https://www.cerebralpalsy.org/inspiration/technology/eye-tracking

Stevens Institute of Technology. (2020). *A.I. Tool Promises Faster, More Accurate Alzheimer's Diagnosis*. https://www.stevens.edu/news/ai-tool-promises-faster-more-accurate-alzheimers-diagnosis

Stock, G. (2002). *Redesigning Humans: Our Inevitable Genetic Future*. Houghton Mifflin Company.

Stoeger T., Gerlach M., Morimoto R.I., & Amaral L.A.N (2018). *Large-scale investigation of the reasons why potentially important genes are ignored*. doi:10.1371/journal.pbio.2006643

Stone, D. (2018, February 21). These people believe death is only temporary. *National Geographic*. https://www.nationalgeographic.com/photography/proof/2018/february/nucci-transhumanists-technology/

Stout, D., Passingham, R., Frith, C., Apel, J., & Chaminade, T. (2011). Technology, expertise and social cognition in human evolution. *The European Journal of Neuroscience*, *33*(7), 1328–1338. doi:10.1111/j.1460-9568.2011.07619.x PMID:21375598

Stout, D., Toth, N., Schick, K., & Chaminade, T. (2008). Neural correlates of Early Stone Age toolmaking: Technology, language and cognition in human evolution. *Philosophical Transactions of the Royal Society of London. Series B, Biological Sciences*, *363*(1499), 1939–1949. doi:10.1098/rstb.2008.0001 PMID:18292067

Stracuzzi, M. (2020). *4 Revolutionary Use Cases of 5G in Healthcare*. https://www.telit.com/blog/4-revolutionary-use-cases-5g-healthcare/

Strauer, B. E., & Kornowski, R. (2003). Stem cell therapy in perspective. *Circulation, 107*(7), 929–934. doi:10.1161/01.CIR.0000057525.13182.24 PMID:12600901

Talks, T. E. D. (2006, November 14). *The accelerating power of technology* [Video]. Kurzweil. https://www.ted.com/talks/ray_kurzweil_the_accelerating_power_of_technology?language=eu doi:10.1111/j.1365-2966.2007.11895.x

Tan, J. (2018). In Machines We Trust? – Verifying Authenticity by AI and Blockchain. *10 Plus Brand*. https://10plusbrand.com/2018/10/21/in-machines-we-trust-verifying-authenticity-by-ai-and-blockchain-part-1-of-2/

Tango Technologies. (2019). *Tango: Life in Motion*. https://www.tangobelt.com/

Tao, M. (2019). *Top 25 vertical farming companies*. https://roboticsandautomationnews.com/2019/05/03/top-25-vertical-farming-companies/22181/

Taylor, T. (2010). *The Artificial Ape: How Technology Changed the Course of Human Evolution (MacSci)* (1st ed.). St. Martin's Press.

TEDx Talks. (2017). *Don't silence your anger, find your purpose with it | Thibault Duchemin | TEDxSoMa* [video]. YouTube. https://www.youtube.com/watch?v=YPPUXDMTbuk

Tegmark, M. (2017). *Life 3.0: Being Human in the Age of Artificial Intelligence*. Knopf.

Tekin, C., Braun, J., & van der Schaar, M. (2015, April). eTutor: Online learning for personalized education. In *2015 IEEE International Conference on Acoustics, Speech and Signal Processing (ICASSP)* (pp. 5545-5549). IEEE.

Terri, P. (2020). *Advancing artificial intelligence research*. MIT Schwarzman College of Computing. https://news.mit.edu/2020/advancing-artificial-intelligence-research-1118

Thagard, P. (2020). Cognitive Science. In *The Stanford Encyclopedia of Philosophy*. https://plato.stanford.edu/archives/win2020/entries/cognitive-science/

Thavasi, R., Jayalakshmi, S., & Banat, I. M. (2011). Effect of biosurfactant and fertilizer on biodegradation of crude oil by marine isolates of Bacillus megaterium, Corynebacterium kutscheri and Pseudomonas aeruginosa. *Bioresource Technology, 102*(2), 772–778. doi:10.1016/j.biortech.2010.08.099 PMID:20863694

The Economist. (2016). Adding ages. *The Economist, 420*(9002), 14-16.

The Economist. (2020). *Economists are turning to culture to explain wealth and poverty*. https://www.economist.com/schools-brief/2020/09/03/economists-are-turning-to-culture-to-explain-wealth-and-poverty

The Scotsman. (2018). *Leader comment: Learning the lessons of Scottish suicides*. https://www.scotsman.com/news/politics/leader-comment-learning-lesson-scottish-suicides-267281

The Union of the Physically Impaired Against Segregation. (1976). *Fundamental principles of disability*. Retrieved from https://disability-studies.leeds.ac.uk/wp-content/uploads/sites/40/library/UPIAS-fundamental-principles.pdf

The World Bank. (2020, May 15). *Disability inclusion*. The World Bank. https://www.worldbank.org/en/topic/disability

Thein, M., Goh, L., & Phua, K. (1988). The smallpox story: From variolation to victory. *Asia-Pacific Journal of Public Health, 2*(3), 203–210. doi:10.1177/101053958800200313 PMID:3052542

Thomsen, B., Narayanan, D., & Rowstron, A. (2020). *In search for future of cloud storage, researchers look to holographic storage solutions.* https://www.microsoft.com/en-us/research/blog/in-search-for-future-of-cloud-storage-researchers-look-to-holographic-storage-solutions/

Tian, Y., Ma, J., Gong, Q., Sengupta, S., Chen, Z., Pinkerton, J., & Zitnick, L. (2019, May). Elf opengo: An analysis and open reimplementation of alphazero. In *International Conference on Machine Learning* (pp. 6244-6253). PMLR.

TimTek. (n.d.). *Redefining the Engineered Lumber Industry.* http://timtekllc.com/about.html

Tipler, F. J. (2003). Intelligent life in cosmology. *International Journal of Astrobiology, 2*(2), 141–148. doi:10.1017/S1473550403001526

Tobii Dynavox. (2020a). *The new I-Series.* Tobii Dynavox. https://www.tobiidynavox.com/products/i-series/

Tobii Dynavox. (2020b). *Tobii Dynavox in brief.* Tobii Dynavox. https://www.tobiidynavox.com/about/about-us/tobii-dynavox-in-brief/

Tobii, A. B. (2020). *This is eye tracking.* Tobii. https://www.tobii.com/group/about/this-is-eye-tracking/

Topol, E. J. (2019). High-performance medicine: The convergence of human and artificial intelligence. *Nature Medicine, 25*(1), 44–56.

Torgersen, S., Lygren, S., Øien, P. A., Skre, I., Onstad, S., Edvardsen, J., Tambs, K., & Kringlen, E. (2000). A twin study of personality disorders. *Comprehensive Psychiatry, 41*(6), 416–425. doi:10.1053/comp.2000.16560 PMID:11086146

Toumazou, C., & Georgiou, P. (2011). Piet Bergveld-40 years of ISFET technology: From neuronal sensing to DNA sequencing. *Electronics Letters, 47*(26), S7–S12. doi:10.1049/el.2011.3231

Tubb, C., & Seba, T. (2019). *Rethinking Food & Agriculture.* https://www.rethinkx.com/food-and-agriculture

Turing, A. M. (1937). On computable numbers, with an application to the entscheidungsproblem. *Proceedings of the London Mathematical Society, 42*(1), 230–265.

Tweed, A. (2019, December 12). *Ethics, machine learning and disabilities.* AbilityNet. https://abilitynet.org.uk/news-blogs/ethics-machine-learning-and-disabilities

U.S. Food and Drug Administration. (2020, June). *Enforcement policy for non-invasive remote monitoring devices used to support patient monitoring during the coronavirus disease 2019 (COVID-19) public health emergency (revised): Guidance for Food and Drug Administration staff.* Retrieved from https://www.fda.gov/media/136290/download

U.S. General Accounting Office. (2003). *Issues Facing the Army's Future Combat Systems Program.* Author.

Unblocked. (2018). *Can Blockchain save the NHS?* https://un-blocked.co.uk/events/ can-blockchain-save-nhs/

United Nations Office for Outer Space Affairs. (n.d.). *Benefits of Space for Humankind.* https://www.unoosa.org/oosa/en/benefits-of-space/sustainable-development.html

United Nations Treaties & Principles on Outer Space. (1967). https://www. unoosa.org/pdf/publications/STSPACE11E.pdf

United Nations. (2019). *World population ageing 2019: Highlights.* (Report No. ST/ESA/SER.A/430). New York, NY: United Nations. Retrieved from https://www.un.org/en/development/desa/population/publications/pdf/ageing/WorldPopulationAgeing2019-Highlights.pdf

University of California. Berkeley. (2020). *Transcence app makes group conversations between deaf and hearing possible.* Berkeley Innovators. https://innovators.berkeley.edu/spotlights/transcence-app-makes-group-conversations-between-deaf-and-hearing-possible

UNOOSA. (n.d.). *Space Law*. https://www.unoosa.org/oosa/en/ourwork/spacelaw/index.html

Utroske, D. (2016). *Researchers at work on new kelp source for natural cosmetics*. https://www.cosmeticsdesign.com/Article/2016/02/10/Researchers-at-work-on-new-kelp-source-for-natural-cosmetics

Vaishya, R., Javaid, M., Khan, I. H., & Haleem, A. (2020). Artificial Intelligence (AI) applications for COVID-19 pandemic. *Diabetes & Metabolic Syndrome*, *14*(4), 337–339.

Van Leeuwen, J., & Wiedermann, J. (2001). The Turing machine paradigm in contemporary computing. In *Mathematics unlimited—2001 and beyond* (pp. 1139–1155). Springer.

Van Tulder, M., Malmivaara, A., & Koes, B. (2007). Repetitive strain injury. *Lancet*, *369*(9575), 1815–1822. doi:10.1016/S0140-6736(07)60820-4 PMID:17531890

Vance, A. (2010, June 14). In the Singularity Movement, Humans Are So Yesterday. *The New York Times*. https://www.nytimes.com/2010/06/13/business/13sing.html

Varo, R., Crowley, V. M., Sitoe, A., Madrid, L., Serghides, L., Kain, K. C., & Bassat, Q. (2018). Adjunctive therapy for severe malaria: A review and critical appraisal. *Malaria Journal*, *17*(1), 1–18. doi:10.118612936-018-2195-7 PMID:29361945

Vegan Organic Network. (n.d.). https://veganorganic.net/plant-based-permaculture/

Venter, J. C. (2008). *A Life Decoded: My Genome: My Life*. Penguin.

Venter, J. C. (2013). *Life at the speed of light: From the double helix to the dawn of digital life*. Penguin Books.

Venter, J. C. (2013). *Life at the Speed of Light: From the Double Helix to the Dawn of Digital Life*. Viking Adult.

Verbeke, R., Lentacker, I., Smedt, S. C. D., & Dewitte, H. (2019). Three decades of messenger RNA vaccine development. *Nano Today*, *28*, 100766. doi:10.1016/j.nantod.2019.100766

Verma, I. M., Naldini, L., Kafri, T., Miyoshi, H., Takahashi, M., Blömer, U., & Gage, F. (2000). Gene therapy: promises, problems and prospects. In *Genes and Resistance to Disease* (pp. 147–157). Springer. doi:10.1007/978-3-642-56947-0_13

Vernadsky, V. I. (1998). *The Biosphere – Complete Annotated Edition*. Copernicus / Springer Verlag.

Viehmann A., Kersting C., Thielmann A., & Weltermann B. (2017). *Prevalence of chronic stress in general practitioners and practice assistants: Personal, practice and regional characteristics*. . doi:10.1371/journal.pone.0176658

Vinge, V. (1993). The Coming Technological Singularity: How to Survive in the Post-Human Era. In NASA (ed.), Vision-21: Interdisciplinary Science and Engineering in the Era of Cyberspace (pp. 11-22), Cleveland, OH: NASA Lewis Research Center.

Vinge, V. (1993, Winter). The Coming Technological Singularity. *Whole Earth Review*.

Volterra, A., Liaudet, N., & Savtchouk, I. (2014). Astrocyte Ca2+ signalling: An unexpected complexity. *Nature Reviews. Neuroscience*, *15*, 327–335.

Waldmann, T. A. (1991). Monoclonal antibodies in diagnosis and therapy. *Science*, *252*(5013), 1657–1662. doi:10.1126cience.2047874 PMID:2047874

Walker, L., LeVine, H., & Jucker, M. (2006). Koch's postulates and infectious proteins. *Acta Neuropathologica*, *112*(1), 1–4. doi:10.100700401-006-0072-x PMID:16703338

Walls, A. B., Waagepetersen, H. S., Bak, L. K., Schousboe, A., & Sonnewald, U. (2015). The glutamine- glutamate/GABA cycle: Function, regional differences in glutamate and GABA production and effects of interference with GABA metabolism. *Neurochemical Research*, *40*(2), 402–409.

Waltz, E. (2016). GM salmon declared fit for dinner plates. *Nature Biotechnology*, *34*(1), 7–8. doi:10.1038/nbt0116-7a PMID:26744960

Wang, F. Y., Zhang, J. J., Zheng, X., Wang, X., Yuan, Y., Dai, X., . . . Yang, L. (2016). Where does AlphaGo go: From church-turing thesis to AlphaGo thesis and beyond. *IEEE/CAA Journal of Automatica Sinica, 3*(2), 113-120.

Wang, Z. L. (2012). *Piezoelectric Nanogenerators for Self-Powered Nanosensors and Nanosystems*. John Wiley Sons. https://dx.doi.org/10.1002/047134608X.W8136

Warwick, K. (2003). *I, Cyborg*. Garnder's.

Wattles, J. (2020). *SpaceX Teams Up With Space Tourism Agency To Sell Rides Aboard Its' Spacecraft.* CNN Business. https://www.cnn.com/2020/02/18/tech/spacex-tourism-space-adventures/index.html

Wave Networking Group. (2020). *6G*. Retrieved from http://mmwave.dei.unipd.it/research/6g/

Weaver, J. H. (1992). Illusion or reality, state sovereignty in outer space. *Boston University International Law Journal*, *10*(2).

Wehner, M. F., & Wolfer, W. G. (1983a). Numerical evaluation of path-integral solutions to Fokker-Planck equations. I. *Physical Review A.*, *27*, 2663–2670.

Wehner, M. F., & Wolfer, W. G. (1983b). Numerical evaluation of path-integral solutions to Fokker-Planck equations. II. Restricted stochastic processes. *Physical Review A.*, *28*, 3003–3011.

Wehner, M. F., & Wolfer, W. G. (1987). Numerical evaluation of path integral solutions to Fokker-Planck equations. III. Time and functionally dependent coefficients. *Physical Review A.*, *35*, 1795–1801.

Weiling, F. (1991). Historical study: Johann Gregor Mendel 1822–1884. *American Journal of Medical Genetics*, *40*(1), 1–25. doi:10.1002/ajmg.1320400103 PMID:1887835

Weinberg, R. A. (1991). Tumor suppressor genes. *Science*, *254*(5035), 1138–1146. doi:10.1126cience.1659741 PMID:1659741

Weiner, L. M., Surana, R., & Wang, S. (2010). Monoclonal antibodies: Versatile platforms for cancer immunotherapy. *Nature Reviews. Immunology*, *10*(5), 317–327. doi:10.1038/nri2744 PMID:20414205

Weintraub, K. (2016, December 15). Aging is reversible—at least in human cells and live mice. *Scientific American*. https://www.scientificamerican.com/article/aging-is-reversible-at-least-in-human-cells-and-live-mice/

Weissberger, A. (2014). TiECon 2014 Summary-Part 1: Qualcomm Keynote & IoT Track Overview. *IEEE ComSoc*. Available online: https://community.comsoc.org/blogs/alanweissberger/tiecon-2014-summarypart-1-qualcomm-keynote-iot-track-overview

Welker, B. (2017). Will the Big Four change hiring practices in the coming years? *CPA Practice Advisor*. Available on-line on January 26, 2021 at: https://www.cpapracticeadvisor.com/article/12296978/will-the-big-four-change-hiring-practices-in-the-coming-years

Wells, H. G. (1902). The Discovery of the Future. *Nature*, *65*. www.geocities.com/yokelcraig/hgwells1.html

Wheatley, L. M., & Togias, A. (2015). Allergic rhinitis. *The New England Journal of Medicine*, *372*(5), 456–463. doi:10.1056/NEJMcp1412282 PMID:25629743

White, J. (2000). Bites and stings from venomous animals: A global overview. *Therapeutic Drug Monitoring*, *22*(1), 65–68. doi:10.1097/00007691-200002000-00014 PMID:10688262

Wiener, N. (1965). Cybernetics, Second Edition: or the Control and Communication in the Animal and the Machine (2nd ed.). The MIT Press.

Wigner, E. P. (1961). Remarks on the mind-body problem. In I. J. Good (Ed.), *The Scientist Speculates* (pp. 171–184). Heineman.

Wikipedia. (2021). *List of close asteroid approaches to Earth in 2020*. https://en. wikipedia. org/wiki/List_of_asteroid_close_approaches_to_Earth_in_2020

Wilcke, X., Bloem, P., & de Boer, V. (2017). The knowledge graph as the default data model for learning on heterogeneous knowledge. *Data Science*, *1*(1-2), 39-57. https://content.iospress.com/articles/data-science/ds007

Willis, N. (1997). Edward Jenner and the eradication of smallpox. *Scottish Medical Journal*, *42*(4), 118–121. doi:10.1177/003693309704200407 PMID:9507590

Winzer, P. J., Neilson, D. T., & Chraplyvy, A. R. (2018). Fiber-optic transmission and networking: The previous 20 and the next 20 years. *Optics Express*, *26*(18), 24190–24239.

Wolfers, J., & Zitezwitz, E. (2004). Prediction Markets. *The Journal of Economic Perspectives*, *18*(2), 107–126. doi:10.1257/0895330041371321

Wong, B. (2011). Color blindness. *Nature Methods, 8*(6), 441-442.

Wong, S. (2017). *Defibrillator drones could save lives before ambulance arrives*. https://www.newscientist.com/article/2134473-defibrillator-drones-could-save-lives-before-ambulance-arrives/

Wood, C. (2019). *Hologram Within a Hologram Hints at Fate of Black Holes*. https://www.quantamagazine.org/hologram-within-a-hologram-hints-at-solution-to-black-hole-information-paradox-20191119/

Woodcock, A., Moody, L., McDonagh, D., Jain, A., & Jain, L. C. (Eds.). (2020). *Design of assistive technology for ageing populations*. Springer.

Wood, R. D., Mitchell, M., Sgouros, J., & Lindahl, T. (2001). Human DNA repair genes. *Science*, *291*(5507), 1284–1289. doi:10.1126cience.1056154 PMID:11181991

Woods, S. W. (2003). Chlorpromazine equivalent doses for the newer atypical antipsychotics. *The Journal of Clinical Psychiatry*, *64*(6), 663–667. doi:10.4088/JCP.v64n0607 PMID:12823080

World Bank. (2018). *Blockchain & Distributed Ledger Technology*. The World Bank. https://www.worldbank.org/en/topic/financialsector/brief/blockchain-dlt

World Health Organization. (2011). *World report on disability*. Author.

World Health Organization. (2018, May 18). *Assistive technology*. World Health Organization. https://www.who.int/news-room/fact-sheets/detail/assistive-technology

World Health Organization. (2019, October 8). *Blindness and vision impairment*. World Health Organization. https://www.who.int/news-room/fact-sheets/detail/blindness-and-visual-impairment

World Health Organization. (2019, September 19). *Dementia.* World Health Organization. https://www.who.int/en/news-room/fact-sheets/detail/dementia

World Health Organization. (2020a). *Assistive devices and technologies.* World Health Organization. https://www.who.int/disabilities/technology/en/

World Health Organization. (2020b). *Deafness and hearing loss.* World Health Organization. https://www.who.int/news-room/fact-sheets/detail/deafness-and-hearing-loss

World Transhumanist Association (Humanity+). (2002). *The Transhumanist Declaration.* World Transhumanist Association. www.transhumanism.org/declaration.htm

Wright, E. M. (1998). I. Glucose galactose malabsorption. *American Journal of Physiology. Gastrointestinal and Liver Physiology, 275*(5), G879–G882. doi:10.1152/ajpgi.1998.275.5.G879 PMID:9815014

Wu, S. L., Wang, L. C., & Yi, X. X. (2012). Time-dependent decoherence-free subspace. *Journal of Physics. A. Mathematical Nuclear and General, 405305*, 1–11.

XLSemanal. (2020, November 3). Roger Penrose, Nobel de Física: "El reto de la física es explicar cómo funciona la conciencia." *XLSemanal.* https://www.xlsemanal.com/personajes/20201006/roger-penrose-nobel-fisica-agujeros-negros.html

Yalçın, Ö. N., & DiPaola, S. (2020). Modeling empathy: Building a link between affective and cognitive processes. *Artificial Intelligence Review, 53*(4), 2983–3006. doi:10.100710462-019-09753-0

Yalom, I. D. (2017). *Becoming Myself: A Psychiatrist's Memoir.* Basic Books.

You, E. (2016). *Anatomy of next: Utopia/Bio.* Founders Fund. https://foundersfund.com/2016/06/anatomy-next-utopia-episode-2-bio/

Zeder, R. (2020). *The Sectors of the Economy.* https://quickonomics.com/category/microeconomics/

Zhang, B., Xia, H., & Lim, H. O. (2019, December). Development of an AI based teaching assisting system. In *Proceedings of the International Conference on Artificial Intelligence, Information Processing and Cloud Computing* (pp. 1-5). Academic Press.

Zhang, P., Ai, Q., Li, Y., Xu, D., & Sun, C. (2014). Dynamics of quantum Zeno and anti-Zeno effects in an open system. *Science China. Physics, Mechanics & Astronomy, 57*(2), 194–207.

Zhang, X. L., Begleiter, H., & Porjesz, B. (1997). Do chronic alcoholics have intact implicit memory? An ERP study. *Electroencephalography and Clinical Neurophysiology, 103*, 457–473.

Zhang, X. L., Begleiter, H., Porjesz, B., & Litke, A. (1997). Electrophysiological evidence of memory impairment in alcoholic patients. *Biological Psychiatry, 42*, 1157–1171.

Zhang, X. L., Begleiter, H., Porjesz, B., Wang, W., & Litke, A. (1995). Event related potentials during object recognition tasks. *Brain Research Bulletin, 38*(6), 531–538.

Zhu, N., Zhang, D., Wang, W., Li, X., Yang, B., Song, J., . . . Lu, R. (2020). A novel coronavirus from patients with pneumonia in China, 2019. *New England Journal of Medicine.*

Zindell, D. (1994). *The Broken God.* Acacia Press.

Zonshine, I. (2020, January 8). OrCam Technologies to debut AI-based, wearable hearing and optical aides. *The Jerusalem Post.* https://www.jpost.com/jpost-tech/business-and-innovation/orcam-technologies-to-debut-ai-based-wearable-hearing-and-optical-aides-613480

Zuboff, S. (2019). *The Age of Surveillance Capital: The Fight for a Human Future at the New Frontier of Power*. Profile Book. https://www.bookdepository.com/Age-Surveillance-Capitalism-Shoshana-Zuboff/9781781256848

Zubrin, R. (2019). *The Case for Space: How the Revolution in Spaceflight Opens Up a Future of Limitless Possibility*. Prometheus.

Zuniga, J. M., Carson, A. M., Peck, J. M., Kalina, T., Srivastava, R. M., & Peck, K. (2016). The development of a low-cost three-dimensional printed shoulder, arm, and hand prostheses for children. *Prosthetics and Orthotics International*, 1–5.

Zur Hausen, H., Schulte-Holthausen, H., & Klein, G. (1970). Epstein–Barr virus in Burkitt's lymphoma and nasopharyngeal carcinoma: EBV DNA in biopsies of Burkitt tumours and anaplastic carcinomas of the nasopharynx. *Nature*, *228*(5276), 1056–1058. doi:10.1038/2281056a0 PMID:4320657

About the Contributors

Angus Hooke is Professor of Business at the Australian Institute of Higher Education in Sydney, Australia. He has extensive experience as a practicing economist, including 15 years in the IMF, two years in Seoul as Economic Advisor to the Minister of Finance of South Korea, two years in Paris as IMF Representative to the OECD, two years in Bali on a tourism project for the World Bank and six years in Beijing helping the Chinese government join and then implement the policies of the WTO. Professor Hooke also worked for 10 years at the University of Nottingham, Ningbo, China where he was Director of the Entrepreneurship and Tourism Department and Head of the Business School. He has been lead editor of one book for IGI and is currently working with IGI on the book "Technological Breakthroughs and Future Business Opportunities in Education, Health, and Outer Space".

* * *

Lauren Alati is a manager with Argon & Co, world leading operations consultants, headquartered in Paris with global offices including London, Mumbai, and Atlanta. Lauren brings operational and consulting experience in distribution logistics and strategic network design working with leading manufacturers and retailers across Australia, New Zealand, and Oceania. Lauren has global supply chain credentials; certified in Transport Logistics and Distribution by the Association for Supply Chain Management with further tertiary degrees earned in Australia, China, and the USA.

Nicholas Baldwin is currently a policy analyst in the New Zealand Government sector, having graduated from the University of Canterbury with a Master's in Policy and Governance (First-class Honours) in 2018 and a Bachelor of Arts in philosophy and history in 2017. Previously, Nicholas contributed as a research assistant and co-author on a project that forecasted scenarios for autonomous vehicles in the New Zealand transport sector (externally funded by the Ministry of Business, Innovation, and Employment as part of the National Science Challenge). He has also mentored at the 2018 Act in Space Hackathon in Christchurch. His current research interests are focused on how humans define themselves and their environment through the use of technology.

Enric Bas (PhD in Sociology) is Professor of Social Change and Communication, and Participatory Foresight and Social Innovation, at the School of Economics of the University of Alicante (Spain) and Director of FUTURLAB-Creative Futures at UA. Guest Professor/ Visiting Researcher at the UHH-University of Hamburg / Max Planck Institute Cluster of Excellence: CLICCS- Climate Change and Society for the academic year 2020/21. Editorial Board Member of multiple scientific journals / book collections (European Journal of Futures Research, Journal of Futures Studies, Springer-Nature, Plaza & Valdés, etc…). Former member of the WFSF-World Futures Studies Federation Executive Board (2000-2005) and the European Futurists Conference Lucerne-Switzerland (2005-2012). Member of Techcast-Global (GWU) since early 200s, and the SFRI-Strategic Foresight for Research and Innovation experts group of the European Commission. Recent outstanding publications would include "FLUX-3D, Participatory Foresight for Social Innovation" (Technological Forecasting and Social Change, 2013) or "The Future Internet; Alternative Visions" (Springer, 2015). Currently interested in Climate Change management, Social Change (Lifestyles) related with ICTs, and Circular Economy (the future of sharing and collaborative business).

Dennis Bushnell is Chief Scientist at the NASA Langley Research Center. He is responsible for Technical Oversight and Advanced Program Formulation with emphasis in the areas of Atmospheric Sciences and Structures, Materials, Acoustics, Flight Electronics/Control/Software, Instruments, Aerodynamics, Aerothermodynamics, Hypersonic Airbreathing Propulsion, Computational Sciences and Systems Optimization for Aeronautics, Spacecraft, Space Exploration and Space Access. His technical specialties include flow modeling and control across the speed range, advanced configuration aeronautics, aeronautical facilities, advanced power and energy, planetary exploration and hypersonic airbreathing propulsion. Dennis is a Member of the National Academy of Engineering, Honorary Fellow of the American Institute of Aeronautics and Astronautics, and a Fellow of the American Society of Mechanical Engineers and the Royal Aeronautical Society. He was responsible for the invention/ development of the "Riblet" approach to turbulent drag reduction, high speed "Quiet Tunnels" for flight-applicable boundary layer transition research, advanced computational approaches for laminar flow control, channel wing with circulation control for VSTOL, regenerative aerobraking for Martian entry, electron beam-free form fabrication and advanced hypervelocity airbreathing and aeronautical concepts with revolutionary performance potential including the truss braced wing long haul transport. His contributions to national programs include Sprint, supersonic transports, FASTSHIP, Gemini, Apollo, RAM, Viking, X15, F-18E/F [Dennis is patent holder for the "fix" to the wing-drop problem], shuttle propulsion and TPS, hypersonic national aerospace plane, submarine/torpedo technology, America's Cup racers, navy rail gun, MAGLEV trains, national security and planetary exploration.

José Luis Cordeiro, PhD, is an international fellow of the World Academy of Art and Science, vicechair of HumanityPlus, director of The Millennium Project, founding faculty at Singularity University in NASA Research Park, Silicon Valley, and former director of the Club of Rome (Venezuela Chapter), the World Transhumanist Association and the Extropy Institute. He has also been invited faculty at many institutions like the Institute of Developing Economies IDE – JETRO in Tokyo, Japan, the Moscow Institute of Physics and Technology (MIPT) and the Higher School of Economics (HSE) in Russia. José studied engineering at the Massachusetts Institute of Technology (MIT) in Cambridge, MA, economics at Georgetown University in Washington, DC, management at INSEAD in Fontainebleau, France, and science at Universidad Simon Bolivar in Caracas, Venezuela. He is a leading expert on technological change and future trends. He has published over a dozen books in different languages and appeared in programs with the BBC, CNN, Discovery Channel, History Channel and NatGeo, among many other international media interviews. He is a member of the Sigma Xi (ΣΞ) and Tau Beta Pi (ΤΒΠ) honor societies, has received many prizes, including the Spanish Health Award by Instituto Europeo for promoting research on life extension, and has been a Spanish candidate to the European Parliament.

Nina Dorata is Professor of Accounting at St. John's University Certified Public Accountant (New York) Chair, NYSSCPA, Committee on the Future of Accounting Education Director, BS Business Program, St. John's University Former Chairperson, Department of Accountancy, St. John's University Dr. Dorata earned her Ph.D. in Management with a concentration in Accounting and Finance from Rutgers, the State University of New Jersey following a career in public and private accounting.

Kevin Dow began his academic career two decades ago, receiving his PhD from the University of South Carolina and his MS in Analytics from Columbia University, and he has become a leading scholar and teacher in the field of Applied Accounting Analytics. He has held faculty posts on three continents, and has taught executives on four continents. Kevin's interests lie in the intersection of accounting and data analytics and focuses on the use of predictive analytics and machine learning to help companies become learning organizations -- specifically in the area of the design and use of accounting information for managing costs and evaluating business risk and value. His research has been published in journals that are consistently ranked internationally as high-quality with an international readership. His productivity and success in high-level journals shows that he has obtained an international reputation in Accounting Systems research.

Robert Finkelstein has been President of Robotic Technology Inc., a creative professional services firm that provides technical analyses, systems engineering, technology assessments, operations research, business development, and other services in the field of autonomous and unmanned vehicles, intelligent systems, artificial intelligence, and robotics for civil and military applications. He is also an adjunct professor teaching graduate courses in systems engineering at the University of Maryland Global Campus, and is a Director of the Intelligent Systems Laboratory in the University Of Maryland Clark School Of Engineering. He recently wrote two chapters for the SAGE Handbook of Evolutionary Psychology for publication in 2020: "Evolutionary Psychology and Robotics" and "Evolutionary Psychology and Cyberwarfare". He is currently developing an autonomous intelligent system of systems in a project for the Defense Advanced Research Projects Agency (DARPA) and the Office of Naval Research (ONR). Robert earned a Doctorate in Business Administration, with the primary field being Systems Theory and Cybernetics (and with the supporting field in the Management of Science, Technology, and Innovation) from the George Washington University (GWU); an Ap.Sci. (Applied Scientist degree) in Operations Research (GWU); an M.S. in Operations Research (GWU); an M.S. in Physics (University of Massachusetts); and a B.A. in Physics (Temple University). He also earned an LL.B. in American Law and Procedure (LaSalle Extension University); Diplomas from the U.S. Army Missile School and U.S. Army Ordnance School; Certificates from the University of Tennessee Space Institute (course in Combat Obscuration Modeling) and University of California, Los Angeles (course in Battlefield Robotics), and he completed post-graduate courses in Physics at MIT. Robert served as an Army ordnance officer with the Missile Intelligence Agency.

Amy Fletcher is a professional futurist and independent scholar. Her areas of interest and expertise include science and technology policy, with a current emphasis on the social and ethical implications of advanced artificial intelligence and the techno-politics of life extension. She also lectures and gives public talks on contemporary United States politics and is a co-investigator on a Marsden-funded research project that investigates the global debate on the use and regulation of lethal autonomous weapons. Prior to launching a freelance research consultancy (A Fletcher Strategies LLC), Amy was a faculty member in the Department of Political Science and International Relations at the University of Canterbury in Christchurch New Zealand (1999-2021). As of February 2021, Amy is a Board Member of the Lifeboat Foundation.

Parvind Gambhir holds postgraduate degrees in medicine from Patiala Medical College in India and from Sydney University in Australia. He has worked as clinician in hospitals in India and as manager at Sydney Childrens Hospital and Concord Hospital. Parvind also teaches medicine at Sydney University.

William E. Halal, PhD, is professor emeritus of management, technology, and innovation at George Washington University, with degrees from Purdue and UC Berkeley. He has published 7 books and hundreds of articles, consults to corporations and governments, and is a frequent speaker, once substituting for Peter Drucker. Halal is founder of The TechCast Project and also co-founded the Institute for Knowledge & Innovation. Previously, he served as a major in the Air Force, an aerospace engineer on the Apollo Program and a manager in Silicon Valley. He received the 1977 Mitchell Prize for his article "Beyond the Profit-Motive," and was cited by the Encyclopedia of the Future as one of the top 100 futurists in the world. His article, "Through the MegaCrisis," was awarded Outstanding Paper of 2013 by Emerald Publishing.

Lester Ingber has published over 100 papers and books in theoretical physics, neuroscience, finance, optimization, combat analysis, karate, and education. As CEO of Physical Studies Institute LLC (PSI) in Ashland OR he develops and consults on projects documented in the https://www.ingber.com/ archive. Prof. Ingber received: his diploma from Brooklyn Technical High School in 1958; his B.S. in physics from Caltech in 1962; his Ph.D. in theoretical nuclear physics from UC San Diego in 1967 while studying at the Niels Bohr Institute in Copenhagen and consulting at RAND in Santa Monica CA. Prof. Ingber has held positions in academia, government and industry: National Science Foundation Postdoctoral Fellow at UC Berkeley and UC Los Angeles; Assistant Professor in physics at SUNY at Stony Brook; Research Physicist in the Physics department and in the Institute for Pure and Applied Physical Sciences at UC San Diego; Research Associate at UC San Diego in the Music department; twice Senior Research Associate of the National Research Council of the National Academy of Sciences; Professor of Physics at the Naval Postgraduate School in Monterey CA and the US Army Concepts Analysis Agency in Bethesda MD; Research Professor of Mathematics at The George Washington University in DC; Director of Research and Development at trading firm DRW Trading in Chicago IL and hedge fund DUNN Capital Management in Stuart FL; Editor-in-Chief at Research Publisher for: Current Progress Journal (timely topics in science), Graduate Journal of Research, and Undergraduate Journal of Research, and associated e-conferences for these three journals; Partner in Pion Capital, a hedge-fund partnership of Caltech alumni; PI of a NSF.gov XSEDE.org physics project.

Jian Lei is a PhD candidate, and has achieved postgraduate degree in computer vision at Beihang University, China. He received his BS degree in Mathematics from Sino-French Engineering School of Beihang University. His current research interest focus on exchange rate pricing/financial accounting quantity (via ML) in financial field, and human pose estimation in CS filed.

Fei Luo, Professor of psychology, Department of psychology, University of the Chinese Academy of Sciences. Dr. Luo received his doctorate in the Department of Physiology, Beijing Medical University in 1995. Later he worked in this Department as a lecturer, associate professor and professor. In 2005, Dr. Luo moved his laboratory to the Institute of psychology, Chinese Academy of Sciences. His early research interests were pain, emotion and cognition. Later he began to study cognitive regulation on sensations and emotions, including pain and depression. His lab also focuses on empathy, mindfulness and attention bias. He has written more than 170 research papers and 16 books or chapters.

Qing-Ping Ma currently works at Nottingham University Business School China. He has a PhD in Physiology/Neuroscience from Peking University and a PhD in Economics from the University of London. After receiving his PhD in Physiology/Neuroscience, Qing-Ping worked at Peking University, University College London, and Merck Sharp & Dohme Research Laboratories, where his research interests were the mechanisms of pain, analgesia, anxiety and depression. After completing his PhD in Economics, he joined Nottingham University Business School where he conducts research in the areas of pension economics, China's economic growth, industrial revolutions and technological progress.

Don MacRae has held senior research and management positions in Australia's leading commercial and public institutions. His early work experience was with BHP, the company with the largest market capitalisation on the Australian Securities Exchange (ASX). Later, he moved to the Australian Government as an Under Secretary for environmental research to CSIRO, Australia's largest science and research organisation. As Director of Corporate Planning at CSIRO, Don coordinated the organisation's workshop-based R&D prioritisation process. He is now a private consultant with considerable experience in economic, energy, environmental, and urban & regional development projects in 28 countries in Africa, Asia, the Caribbean, North America, Europe, and Oceania. He has also directed major projects in global modelling, international drylands management, and R&D prioritisation for agencies in Australia and worldwide.

Alan Manly is a founding Director and Chief Executive Officer of the GCA group of companies. He is responsible for the corporate planning and governance of the group, which includes an Independent Higher Education Institute and Universal Business School Sydney (UBSS), which offers a Business degree, an Accounting degree and an MBA. Alan has published several books and written numerous articles on entrepreneurship and education. He has over thirty years of experience in the technology and education industries and twenty years of experience as a Company Director with private, public and NFP companies. Alan has been active in representing the education industry and is a former Board member of two education industry peak bodies.

Kathleen Quilty is an independent writer, consultant and researcher working at the intersection of health, education, business and technology to imagine and create better futures for a diverse range of communities. Kathleen brings unique strategic insights and innovative perspectives to every project, drawing on an educational background in psychology, philosophy, and business management as well as eclectic professional and personal interests and experiences.

Vincent J. Shea, II, PhD, is an Associate Professor of Accountancy at St John's University in Queens, NY. He received his Ph.D. from Kent State University and is a CPA in New York and Florida. His current research and teaching areas are Accounting Information Systems and accounting analytics.

Xinwei Shi is a Postdoctoral Researcher at Tsinghua University in Beijing. He has a Ph.D in Management from the University of Durham in the UK. His research interests are focused on emerging-country multinational enterprise FDI strategy, business ecosystem strategy, the ecosystem buildings of new emerging technologies (e.g., artificial intelligence, internet of things). He has published papers in known journals including Journal of Business Research.

W. Kathy Tannous is a senior lecturer in Economics and Finance in the School of Business at Western Sydney University. She has over sixty publications comprising of journal articles, book chapters and a textbook. She has an extensive research program on economic impact of critical incidents and has been recently examining effects of experiences of transition countries on refugees' mental health. She has previously undertaken work for the United Nations Economic and Social Commission for Asia on projects in the Middle East and North Africa (MENA) countries.

Greg Whateley has academic qualifications in music, psychology, and education (his doctorate was in Virtual Pedagogy). He commenced work in the tertiary sector as Lecturer in Education at the University of Tasmania. He then moved to Griffith University where he served as Head of School – Arts. Later he became Dean of the College at Western Sydney University. Professor Whateley has published more than 100 articles in refereed academic journals. Currently, he is Deputy Vice Chancellor, Group Colleges Australia (GCA). He has maintained a strong interest in online teaching and learning for a number of years, co-founding The Virtual Conservatorium in 2000 and overseeing the transition of Universal Business School Sydney (UBSS) to online learning in 2020.

Index

F

G

H

I

L

M

N

O

P

R

S

T

U

V

W

X

Z

www.ingramcontent.com/pod-product-compliance
Lightning Source LLC
LaVergne TN
LVHW060412300926
849162LV00006B/1408